P9-DVF-164

HOUSES

Illustrations:	Peter A. Farbach
	Joseph Jaqua
	Ronald Noe
	U.S. Savings and Loan League
Editing:	Helene Berlin
	Llani O'Connor
	Peg Keilholz
Design:	Sherman Mutchnick
	Libby Marschke

Houses

The Illustrated Guide to Construction, Design and Systems

Henry S. Harrison

NATIONAL ASSOCIATION OF REALTORS®
developed in cooperation with its affiliate, the
REALTORS NATIONAL MARKETING INSTITUTE®
of the NATIONAL ASSOCIATION OF REALTORS®
Chicago, Illinois

To: Eve and Julie

International Standard Book Number: 0-913652-05-9
Library of Congress Catalog Card Number: 73-84029
REALTORS NATIONAL MARKETING INSTITUTE® Catalog Number: BK 117

Printed in the United States of America
First printing, 1973, 10,123
Second printing, 1974, 20,324
First revision, 1976, 10,653
Second printing, 1977, 9,912
Third printing, 1978, 10,560
Fourth printing, 1979, 10,000

Foreword

About the Book

The real estate broker and salesperson enjoy a unique position in the business world. They are part of an extremely large, complex professional community, yet they deal directly with clients on a daily basis and assume most of the responsibility for helping people make what is probably the most important and expensive single purchase they will ever make: a home.

The responsibility of the real estate professional is to help each family select the house that will provide maximum benefits in the form of shelter and amenities for the price they can afford to pay.

The Marketing Institute has published *Houses—The Illustrated Guide to Construction, Design and Systems* to help real estate professionals discharge this responsibility through understanding the single family residence and how it fills a family's needs.

Houses begins with a look at communities, neighborhoods and sites, then delves into the detail of interior and exterior house design and styles. The book next extensively covers the construction of a house, the materials that make up a house and the mechanical systems that make a house "live."

Obviously, this book does not include all phases of real estate. It does not cover basic selling principles and practices or how to list, show and sell a house or make an appraisal. *Houses* restricts itself to the one-family dwelling, excluding multiple-family residences entirely. Although the book does not cover financing and mortgages, bankers and lenders will find it a useful way to increase their overall knowledge about the houses they help finance. Similarly, appraising and assessing are not directly covered, yet professionals in these fields will find this book an educational and useful resource.

The book is also valuable as a guide to Federal Housing Administration's *Minimum Property Standards for One and Two Living Units,* quoted liberally and generally referred to as *"MPS."* Because the MPS are constantly being revised there may be differences between standards quoted and current MPS.

The potential homebuyer can use this book as a self education tool to help himself buy a house with confidence, knowing that, within his budget, he is getting the home that best suits his family. In fact, any group of people interested in houses, their design, construction and mechanical features, will find this volume useful.

It is, however, primarily for the REALTOR® and real estate salespeople everywhere that the Marketing Institute publishes this book.

About the Author

To the reader of *Houses,* Henry S. Harrison brings over 20 years of experience in the sale, appraisal, development, management and rental of residential, commercial and industrial real estate. He is the owner of the Henry S. Harrison Appraisal Company in New Haven, Connecticut. Mr. Harrison's expertise in the real estate field is well-known and respected and he is often called in by companies, lending institutions and individuals to serve as a real estate consultant.

Mr. Harrison is a REALTOR®, holds the M.A.I. designation of the American Institute of Real Estate Appraisers and the S.R.P.A. designation of the Society of Real Estate Appraisers. He is past president of the Connecticut Chapter of the Society of Real Estate Appraisers and holds offices in the Connecticut Chapter of the American Institute of Real Estate Appraisers and the Greater New Haven Board of REALTORS®. He often serves as a Committee Chairman of the National Appraisal Organizations.

He writes a regular column in The Society of Real Estate Appraisers journal *The Real Estate Appraiser* and in *The Residential Appraiser.* His book *How To Appraise A House* will be released in 1976 by the American Institute of Real Estate Appraisers.

The author has taught real estate and real estate appraising courses at Quinnipiac College, the University of Connecticut, Purdue University, Indiana University and the Chicago Appraisal Laboratory. He is also a certified member of the National Faculty and the National Examiners of the American Institute of Real Estate Appraisers.

In addition Mr. Harrison has presented his Houses, Houses, Houses Seminar at two NAR Conventions and throughout the country for The American Institute of Real Estate Appraisers and various state REALTOR® associations.

He has a B.S. degree in Economics from the Wharton School of Finance and Commerce of the University of Pennsylvania and a M.A. degree in Adult Education-Real Estate from Goddard College. Mr. Harrison is a retired Major of the United States Air Force Reserve.

About the Publisher

The REALTORS® National Marketing Institute, the educational arm of the NATIONAL ASSOCIATION OF REALTORS®, exists solely to educate and serve the real estate professional. With a membership of more than 33,000 real estate brokers and salespeople, the Marketing Institute published *Houses* in its 50th year of furthering professionalism and serving the real estate community.

Preface

Reflecting upon the past three years spent in preparing this book I am struck by the number of people and organizations who offered their assistance and encouragement. They provided me with the adrenalin needed to finish this project which grew far beyond the original idea.

Beverly F. Dordick, the NATIONAL ASSOCIATION OF REALTORS® librarian, after patiently listening to my idea, gave me the final push that convinced me to go ahead. Thereafter she and her fine staff supplied me with a substantial amount of material and encouragement.

Lawrence Liebman acted as my legal advisor and also made the first publisher contact.

It was my good fortune to find three talented illustrators. Joseph Jaqua of Kentfield, California, drew all of the house style pictures. Peter A. Farbach, of New Haven, Connecticut, an architect, drew the construction and house parts illustrations. Ronald Noe, also an architect, of New Haven, Connecticut, drew all other illustrations except those supplied through the cooperation of Harold Olin of the United States Savings and Loan League. The book was expertly designed by Sherman Mutchnick and Libby Marshke of Chicago.

Jean Schildgen organized and edited the construction and materials sections. Margery B. Leonard researched and organized the historical material on the rooms of the house and prepared the initial draft of the layout and circulation section. Suzanne Darrach, my secretary, organized the material for the neighborhood section and did much of the typing and initial editing.

Jack K. Mann reviewed the book's outline and suggested I contact NIREB to publish it. Sol and Helen Bennett, who are always behind me, concurred.

It never even occurred to my mother and father that the project would be anything but successful. My daughters, Julie and Eve, to whom this book is dedicated, willingly shared me with the typewriter.

My sister, Diane Johnson, located and negotiated with Joseph Jaqua 3,000 miles away to do the house style illustrations.

Bruce Lehman, my real estate investment partner and friend, used his extensive knowledge of graphics to develop a format for the architectural style section.

Special thanks to Marilyn Margolis, Louis Durocher, Richard Cohen and Florence Milano who kept the businesses going while I was writing and to Emily Emerson and Emma Kirby who helped with the typing. Special thanks also to Kathy Fiddes, Helen Leonard and Anita Alteri who pitched in without pay to help me meet deadlines. Thanks also to Bobbsie Hertz who worked on the Index.

Review and encouragement were received from many professional association staff members, especially Margaret O'Brien of the American Institute of Real Estate Appraisers; Donald E. Snyder, Virginia Clendenin and Virginia Skowronski of the Society of Real Estate Appraisers; Raymond Gates of the Connecticut Association of Real Estate Boards, Inc. and Albert E. Poirier of the Greater New Haven Board of REALTORS®, Inc.

The valuable opportunity to test some of the material in front of a class was afforded me by my co-instructors E. Roger Everett, Arlen C. Mills, Lewis Garber, Dr. George F. Bloom and John Hoppe.

Dr. Margaret White of Goddard College, as my advisor in their master's program, greatly increased my social awareness. John Farling, my other advisor, helped me understand how to present material effectively to adults.

Frank L. Spangler reviewed the heating and cooling section for technical accuracy. David M. Lewis personally educated me about nonzoning in Houston.

So many members of the NIREB professional staff, editorial department and editorial committees helped and advised along the way, it would be impossible to single them all out here. Special thanks must go to Llani O'Connor, NIREB Editorial Director, who took the project to her heart and shepherded it from its initial presentation to its final publication. Helene Berlin, NIREB Book Editor, proved that the saying, "G.I.G.O." (garbage in — garbage out), is not true; at least it does not apply to authors who work with talented and dedicated editors.

Continual encouragement and support came from Art S. Leitch, 1972 President of NIREB; Stanley R. Dybvig, 1973 NIREB President; the NIREB Board of Governors; the NIREB Editorial Committee, headed by William Patterson and the Book Development Subcommittee, chaired by Albert Mayer III. Lydia Franz, of the Book Development Subcommittee, supported me all the way, providing useful criticism and substantial data on the Hillside Ranch.

Other NIREB staff members who worked on the book were Executive Vice-President Jack Kleeman; Margaret Givhan, production assistant; June Kanner, editorial secretary and Douglas McDonough, Promotion Director.

To all of the above, my sincere thanks. I hope you will be as proud of this book as I am.

Note on revision

Since Houses was published in August, 1973, I have had the opportunity to travel throughout the United States numerous times lecturing about Houses and presenting the Houses, Houses, Houses seminar for the A.I.R.E.A. and also at the 1974 and 1975 N.A.R. conventions.

What pleased me most was to see the change in emphasis taking place about what a salesperson needs to know to be a professional. There has been an increasing emphasis on the importance of being knowledgeable about the product being sold, namely the single family home.

The chapter "Common Problems of Houses" was written for the first edition but could not be used for lack of space. The Society of Real Estate Appraisers published it in *The Real Estate Appraiser* in 1975.

The Appendix "How to Read a House Plan" was made into an educational slide show for the American Institute of Real Estate Appraisers. The material from this show was incorporated into the new expanded version of this appendix.

After several years of testing, the CTS System, a uniform system for describing houses, was assembled into workable form. An article explaining the system appeared in the January, 1976, issue of *real estate today*®. This revision of Houses provides complete instructions on how to use the system.

Special thanks goes to Ruth Lambert who helped extensively with this revision. Peg Keilholz has succeeded Helene Berlin as Editor.

Henry S. Harrison
New Haven, Connecticut

Contents

Chapter 1 Location

1

Chapter 2 House Types

Chapter 3 Interior Design

Chapter 4 Architectural Styles 105

Chapter 5 Basic Construction

Chapter 7 Mechanical Systems   **271**

Chapter 8 Materials Manufacture 327

Appendix 343

Chapter 1
Location

Choosing the location of the family home can be the most important decision homeowners make and the most important decision that a real estate salesperson can help them make. It will affect the family happiness by playing a major role in shaping its life style. From a financial point of view, people profit from an investment in a house, in a community and in a neighborhood which are growing in value, community spirit and desirability.

Most families are attracted to a region because of available employment, proximity to relatives or the climate. It is beyond the scope of this book to explore relative advantages and disadvantages of the many regions of the United States. This chapter, however, will explore the factors that should be taken into consideration when selecting a community, neighborhood and site within a specific area.

Community

Economic Base of the Community

Much of the economic health of a community depends upon its economic base. This consists of those industries, businesses and institutions which export goods and services out of the community. It is the money received by them from outside the community that supports the non-basic employers who provide local goods and services.

Economists generally agree that the higher the ratio of basic employment to non-basic employment, the better the economy of the community. Another factor that must also be considered is the diversity and stability of the basic employers.

It is best for the community to have a variety of basic employers including several different industries, businesses, government agencies (regional, state or national). The base is further enhanced if the community has educational facilities that draw students from a wide area or if it is a regional trade center, a farm trade center, a tourist trade center or a transportation hub.

By having a diversity of basic employers, the economic health is not tied to one source which, if it should fail, would have a major depressing impact upon the community. If the homebuyer has a choice of locating in any one of several communities in an area, he should pick the one with the best economic base, all other things important to him being equal.

Government and Municipal Services

Municipal government affects all the neighborhoods in the community in a fundamental way. A stable administration contributes solidarity to the neighborhoods in the form of well-written and strictly enforced codes, a good inspection program (building, plumbing, etc.), effective zoning regulations, efficient fire and police protection and maintenance (street repairing and cleaning, garbage pickup).

Municipal services are also very important in contributing to the value of a neighborhood. Is municipal trash and garbage collection available? Is its cost included in property taxes or is it a separate expense for the homeowner? How often is the trash picked up and how far must it be carried to the pickup point? Street cleaning and snowplowing services should also be carefully scrutinized.

The quality of police and fire protection is not easy to evaluate, but the quantity is not hard to ascertain, and quantity can be important. Usually, the more vehicles the police and fire departments have, the larger they are and, theoretically, the better equipped to cover the city and its neighborhoods. Quality of police and fire protection can be checked by talking to local residents. A check of fire insurance rates in the area is one way to evaluate the quality of the fire department.

Utilities are discussed in more detail later in the book. Here, however, it is appropriate to point out that ready availability of utilities is a major plus factor in favor of any neighborhood.

Education

Public and private schools are of immediate interest to all families with children. Even homeowners who have no children would do well to take interest in the local schools since there is every possibility that the future buyer of their house will have children and will be concerned about schools. When the quality of education declines, it is often accompanied by a decline in house values. Quality of educational facilities is one of the principal reasons for the exodus of the affluent from the core cities to the suburbs.

Finding out the name and location of the school that serves the neighborhood is the way to begin assessing the educational system. It should be determined whether a school bus is necessary and, if so, available. A visit to the school will indicate the condition of the facilities. The buildings should be in good repair. There should be an indoor gym, a spacious playground outside and hot lunches served at noon. Whether or not the school, for lack of funds or facilities, is on double session or is scheduled for them in the foreseeable future is also significant.

A call to the superintendent of schools will answer one very important question: how many dollars per student are allocated? The more money spent per pupil, the better the teachers and facilities are likely to be. Another general rule is that high property taxes are a sign of good schools. If they are notably low, the local schools may not be in especially good shape unless industry in the area is helping to defray school costs by paying a large share of local taxes.

Convenience and Recreation

It is generally agreed that a variety of conveniences such as food stores, laundries, and repair shops should be readily accessible in the

3

community, yet they must not inflict a commercial atmosphere on the residential neighborhoods. Most people do not want to travel too far for shopping. On the other hand, they don't want to live right next door to a bustling shopping center. In addition, a direct route to a shopping area passing through a neighborhood would not be attractive to many people.

People differ widely in their specific requirements for recreational facilities and services that are within walking distance and those that are located a little further from their homes in the surrounding community. A community can be planned well, yet still not contain every kind of convenience, service or recreation. The prospective home-owner has to decide which of those services are most important to him and his family and then determine if the community provides easy access to them.

The list of such services is practically endless but it might include some of the following: airport, gas station, movie theater, concert hall, art museum, skating rink, sports stadium, swimming pool, golf course, doctor's office, dentist's office, hospital, park, playground, river or lake, shopping center and employment centers. The important factor to be aware of is not necessarily how far away any of these may be but rather how long it takes to get to them, either by car or by public transportation.

Taxes

Local taxes, too, may be important. If they are similar to those of the surrounding communities, they play little or no role in the area selection process. If they are substantially lower than those in the general area, they may attract people, and if they are substantially higher, they may repel potential new residents.

Neighborhoods

Definition of Neighborhood

Traditionally, a residential neighborhood has been defined as an area within a community that contains houses of about the same age, quality, size and value range. Often the houses are also of similar architectural styles or of styles that complement one another. In the traditional definition, residents of a neighborhood have about the same earning power and the same interests and concerns as the majority of their neighbors. In other words, a neighborhood gives a noticeable impression of unity, exemplified by the appearance of the structures and by the people who live in them.

Houses are not necessarily the only structures in the residential neighborhood. There may be neighborhood services such as retail stores and professional offices and community facilities such as parks, recreational areas and schools. All these are considered part of the residential neighborhood when they are located within the neighborhood boundaries and when they serve its residents (although they may serve residents from other neighborhoods as well).

The new neighborhoods created by redevelopment in the cities and found in planned urban communities in the suburbs do not fit this traditional definition, however. These neighborhoods consist of a more heterogeneous blend of houses and people. Old is mixed with new, big with small, rich with poor. Here, it is possible to mix residential, commercial and industrial uses together and still maintain a harmonious neighborhood.

Boundaries and Streets

Neighborhood boundaries are usually quite definite. Expressways, railroad tracks, rivers, hills, the city's edge—all can serve as neighborhood limits. A boundary can be created when land use changes as when a commercial or industrial zone is built up next to a residential neighborhood. Sometimes neighborhoods blend into one another and have no particular boundaries at all. Explicit boundaries, however, are usually a favorable sign. They help preserve the neighborhood's unity, and unity, in turn, helps preserve real estate values.

Boundaries also make it easier to determine the size of a neighborhood, although size has no special significance unless a neighborhood covers an unusually large area. Very large neighborhoods easily lose their cohesiveness and character.

Streets are the entrances and exits of the neighborhood. They cut into it, divide it up, provide a way to travel within it and give access to the neighborhood's houses. If the streets are wide, curving and tree-lined, they can be a true source of beauty.

Streets, however, are especially significant because they have an inherent quality of danger. In some older neighborhoods, the effects of poor planning can be seen where streets run in grids or plain, square blocks. When streets are laid out in grids, they are all about equally convenient and city traffic tends to spill through the neighborhood. This is especially true where there are traffic lights along an adjoining boulevard. In such a neighborhood, traffic can be a hazard that impedes communication and diminishes the feeling of unity.

Contemporary planned neighborhoods, on the other hand, make use of curving streets, cul-de-sacs with generous turnaround space and circular drives as deterrents to through-traffic. Ideally, expressways and boulevards outside the immediate neighborhood (but easy to reach from local streets) should handle the heavy traffic. Traffic within the neighborhood should be primarily local and should move easily and slowly.

Physical Characteristics

Certain physical characteristics of neighborhoods can be considered desirable by some people and undesirable by others. While one's individual taste may not be in agreement with the opinions expressed **5** here, it must be emphasized that property values are governed by the tastes of the majority.

Paved streets, shade trees, an atrractive view and, in urban areas, sidewalks and curbs are usually desirable characteristics. The preferred topography is a rolling terrain so that the whole neighborhood is on higher ground than the surrounding area. Values tend to go up with the height of the land in most areas. Curved and dead-end streets that are lit and have street signs are what people like.

On the other hand, people generally do not like very flat land or excessively rugged terrain without reasonable access. They worry when they live on straight streets that carry through-traffic and appear to present traffic hazards. Neighborhoods with flooding danger, fire hazards, stagnant ponds, marshes, poor surface drainage and excess dampness are understandably unpopular.

Urban Neighborhoods

Urban neighborhoods have certain advantages over suburban neighborhoods. They are situated near metropolitan centers and offer easy access to theaters, museums, special schools, libraries and large medical centers. These are the kinds of cultural and intellectual institutions that usually only a city can support. People whose cultural or social needs are more specialized or whose natures respond to the excitement and complexity offered by the city may well be happiest in an urban neighborhood. Necessary shopping and services can usually be found close by in urban neighborhoods. In many city neighborhoods, the yards and houses have had a chance to become natural in feeling and to overcome the "just planted" look that a development yard and house may have.

While life in the city offers these advantages and personal satisfactions to many people, city neighborhoods also have their drawbacks. Although a few urban neighborhoods were planned, most were not. Parking and traffic congestion can be big problems. Houses are likely to be large, but old, and may contain outdated fixtures and mechanical systems. In addition, most cities have sociological problems that are not nearly as prevalent in the suburbs.

Many older neighborhoods (and indeed, cities) in the United States have grown up in a haphazard, uncontrolled manner. If some of these neighborhoods look well or work well, it is probably more because of good luck than good planning. Some cities, aided by the federal Model Cities program and various other redevelopment programs, have taken steps to revitalize their neighborhoods through renovation and new construction, with the guidance of city planners and redevelopment agencies. There is strong evidence that such efforts bring good results.

Neighborhood Life Cycles

Every neighborhood, whether urban or suburban, passes through several phases in its life cycle. The first is a period in which it grows most rapidly. Homebuyers are attracted to it and prices are on the rise. When prices reach a more or less steady state, the neighborhood enters its second stage. This stable period may last up to 40 to 50 years.

Finally and inevitably, the neighborhood enters a period of decline as newer areas beginning their own patterns of growth become more attractive to residents. Now the neighborhood's houses may become accessible to a lower-income group less able to spend money on maintenance and improvements, with a resulting deterioration of the neighborhood. Houses may be converted to other uses such as apartments, boarding houses or offices. They may be completely razed to make way for large apartment or office buildings. In any case, the character of the neighborhood changes drastically.

Since neighborhoods are dynamic by nature, homebuyers must be able to tell what stage a given neighborhood is in. The first factor to check is population. The growth in population for each year in the last five or ten years is a good indication of where in the cycle the neighborhood is. If population is rising, the neighborhood is probably on the way up. The population should also be studied in terms of the occupations represented in the neighborhood. A healthy percentage of junior executives, foremen, skilled craftsmen and professional people is usually a good sign.

Land prices in the neighborhood should also be checked. If they are consistently rising, the neighborhood can usually be assumed to be in a stage of growth. Another sign of neighborhood upsurge is the scarcity of vacant houses. If the turnover of houses is slow, residents are obviously satisfied.

There are several early warning signs that foretell a neighborhood's period of decline. The population may consist of high percentages of unskilled workers and workers representing only one industry. Another sign may be a high percentage of older couples who are long-time residents and whose children are grown and have left home. An area where development of available vacant land has stopped and where large houses are being converted into apartments and rooming houses is also probably a bad risk. In addition, factors such as absentee owners, houses in bad repair, houses in foreclosure, an overabundance of "for sale" signs, a decrease in rental rates and a breakdown in the enforcement of zoning regulations and deed restrictions are often warnings of an impending period of decline.

When a new transportation system, such as a throughway or a subway, or a new industry is introduced into an area, it is certain to have a major impact on any nearby residential neigborhood. Often it is difficult to tell in advance whether the impact will be to stimulate neighborhood growth, slow it down or reverse it. For instance, a throughway that provides new access to a metropolitan area may make new job markets available to residents, but if the new highway cuts the neighborhood off from its own city center, there will be negative ramifications. A neighborhood, in a situation like this, will probably benefit and

7

suffer at the same time. The question is whether the advantages will outweigh the disadvantages; the final answer is sometimes not immediately evident.

Quality of Neighborhood Life

How does the neighborhood look? Are community areas well-kept and clean? Do yards and houses reflect ongoing interest and care by their owners? The way a neighborhood looks tells a lot about the people who live there. Talking to those same people is another good way to learn about their neighborhood. They can tell what they feel are the neighborhood's special attractions and defects. People are usually willing to discuss neighborhood pro's and con's with an interested listener.

People actually can do more to make or break a neighborhood than anything else can. The people who live in a good neighborhood do much to hold it together by their shared interest in events and conditions that affect it and sometimes by their attempts to change these conditions. Underlying this practical relationship is a deeper basis for neighborhood unity: people's acceptance, consideration and acknowledgment of each other as members of a living community.

Taking walks around a neighborhood at different hours of the day can help form a feeling for the quality of neighborhood life and the presence of any adverse influences or nuisances. For instance, is there a lot of noise? If so, what kind of noise is it? If the noise is made by children playing outside, it could be music to the ears of a buyer with four children of his own. But if the noise is from a commercial airliner whose flight pattern lies directly over or near the neighborhood, a person should know about that before he buys.

What about smoke and odors? Some of the typical offenders are factories, stockyards, rendering plants, garbage disposal areas, dumps, stables, kennels and chemical plants. The pollutant may be many blocks away, but winds can carry smoke and odors quite a distance. It is wise to check for pollution under various weather conditions or at various times of the day or week.

Some additional factors that tend to decrease house values when they are very close by are heavy industry, mills, vacant houses, airports, railroad tracks, cemeteries, funeral homes, hotels and motels, utility wires and pipelines, outdoor advertising billboards, hospitals, fire stations, apartment houses, commercial buildings and taverns. Of course, having these services and businesses a relatively short distance away from a residential neighborhood can be a convenient plus-factor.

Fair Housing

Neighborhoods that exclude members of minority races and religions may soon be so scarce that a discussion about them will be of academic interest only.

The two final blows that ended any legal segregation were struck by Congress and the Supreme Court in 1968. Prior to that, many local and state governments had passed fair housing and other similar legislation that had made many forms of segregation in many areas illegal. Various federal laws and agency regulations also prohibited discrimination. The NATIONAL ASSOCIATION OF REALTORS® has taken many positive steps to encourage equal opportunities in housing, through its Code of Ethics and other regulations. Through its publications, committees and speakers, the National Association seeks to provide leadership in fair housing both to its own membership and to the public.

In 1968 the U. S . Congress passed the Civil Rights Act. The stated policy was ". . . to provide, within constitutional limitations, for fair housing throughout the United States." The Supreme Court, in a case known as "Jones vs. Mayer," ruled that a federal statute passed in 1866 was still valid and enforceable. This statute stated in very clear language that it was illegal to discriminate against anyone and that all citizens had the same rights to inherit, purchase, lease, sell, hold and convey any property.

In the 1950's several articles appeared in national magazines reporting alleged high profits made by real estate agents and others through the tactics of blockbusting, scare selling and panic selling. The thrust of the articles was that it was possible to convince the residents of a non-integrated neighborhood that it was about to become integrated, that values would decline and therefore, that residents should sell before it was too late. If the panic selling developed, these articles said, the operators would buy the properties and sell them to members of minority groups at an excessive profit.

While there have been some unfortunate incidents of scare-tactic selling by unethical people both within and outside the real estate field, the NATIONAL ASSOCIATION OF REALTORS® does not condone any of these actions. The National Association's Code of Ethics condemns such practices and the National Association acts against any members who participate in questionable selling practices.

The consensus of numerous studies conducted and reported on in appraisal, banking and economic journals and other publications, is that through harassment, scare tactics and excessive pressure, owners may be convinced to sell their properties below the market value. If enough of these sales are forced to take place, values in the neighborhood will become depressed. However, what soon happens in most cases is that the properties are resold at higher prices, the depressed selling stops and values return to normal.

As previously discussed, when a neighborhood reaches the third stage of its life cycle, values will decline, although the rate of decline may be fast or slow. Neighborhoods in this cycle are characterized by the influx of residents of a lower income, education and social status than those who formerly lived there. It is overly simplistic to blame this natural decline of value on integration.

9

Land Use Planning

Nuisance Laws and Litigation

Hundreds of years ago when English Common Law was being formed, the idea developed that one landowner could not conduct obnoxious activities on his land that would adversely affect his neighbor. Five basic principles of land use control developed which still hold true today. While these principles are fairly consistent throughout the country, enforcement procedures vary from state to state.

The first principle is that there is a right of action against somebody who causes any physical nuisance to cross from his property onto his neighbor's. Typical physical nuisances include the danger of fire or explosion, all types of air and water pollution and excess noises and vibrations. With the current emphasis on ecology, there has been an increased awareness of the damage caused by these nuisances.

The second legal principle is that these nuisances are an invasion of property rights and, to be aggrieved, one has to have the property right to start with. It would seem that, since tenants, like landowners, have property rights, they might be able to seek remedy. However, there is not much legal history to look to in order to judge their success to date.

The third principle is that there is no automatic right to stop a nuisance. One must look to the courts to weigh the landowner's rights to enjoy the use of his property against the adjoining property owner's right not to be annoyed. Since many acceptable land uses create some nuisances, it becomes a matter of convincing the courts that the nuisance is severe enough to warrant interference with the landowner's right to use his property. Every time something burns, smoke is created, noise is all around and most restaurants would have to close if odors had to be confined within property lines.

The fourth principle is that the courts consider a judgment against a nuisance to be a drastic one. These court orders, when they are issued, require the landowner to cease creating the nuisance or at least to modify his action. The courts offer him no compensation for the decreased use of his land. Therefore, these judgments are almost never awarded before the activity that creates the nuisance actually takes place and unless it can be proved that it really is a substantial nuisance.

Fifth and finally, since the courts consider the remedy against nuisances to be drastic, they have been insistent upon maintaining full control and have resisted actions by legislatures to make laws defining what constitutes a nuisance.

In spite of the difficulty of proving a nuisance case in court, there have been some good examples of courts stopping nuisances under these common law principles. Some examples are requiring sport parks and commercial users to stop shining their lights on adjoining property, producing excess noise with their public address systems and creating excess dust by the use of unpaved areas for parking.

There is a series of actions in which airports have been required to alter their activities, flight paths and schedules where nuisances have been created. In one interesting case, a funeral parlor was held to be a nuisance when constructed in the middle of a beautiful single family historical neighborhood in Litchfield, Connecticut.

It is quite possible that there will be an increase in the number of such actions to stop nuisances as concerned citizens become more aware of the available remedies.

Private Covenants

Private covenants are agreements made between property owners to restrict the use of their properties, resulting in mutual advantage. When a restrictive covenant is entered into, each property owner is required to comply with the covenant's restrictions. One owner has the right to enforce compliance with the restriction by the other parties to the agreement. Usually, these restrictions are established by a developer of a residential subdivision, a redevelopment agency or by members of a neighborhood association.

Normally it is the intention of those who enter into such agreements to make them run with the land, meaning the agreement is binding not only on the parties who sign it but also on successive owners.

There are many differences from state to state in covenant laws. However, some principles have evolved which have become fairly standard. The most important standard requirement is that notice must be made of the restrictions. Often this notice is accomplished by making it part of the deed.

The second principle is that normally the only one who can bring an action to enforce a restrictive covenant is a property owner in the affected area. Another more obscure standard requirement is that the restrictive covenant must concern the real estate itself, although most of them meet this test.

Finally, the restrictive covenants must be negative in nature rather than affirmative: they can stop the property owners from doing certain things but cannot place a burden upon them to take any kind of positive action. There has been a tendency of the courts to overlook some or all of these technical principles at times.

Not all restrictive covenants are enforceable. If, for instance, the court can be convinced that there has been a substantial change in the neighborhood it will consider this to be a valid reason to void the covenants. Racial and religious covenants are also not enforceable.

In spite of the problems of enforcement and other shortcomings, restrictive covenants continue to produce a great deal of litigation and are an important method used to control the use of the land.

Zoning

One of the greatest visual experiences one can have is to fly over a major city on a clear night or to view it from the top of its highest building. The strings of multicolored lights from horizon to horizon form patterns of spectacular beauty.

When viewing the vastness of the city, one feels small and insignificant. The city seems to be of superhuman size; control of its growth seems beyond man's capabilities. It appears that no legislation or other act of man outside of an atomic bomb could make a change.

Fortunately, this is not the case. We are learning how to control our cities and to plan for the future. Today, zoning is the widely accepted method by which society, in general, controls the use of the land. It is one of the means by which it controls the development of the physical environment.

This was not always so. Zoning was unknown in the 19th century, the period in which many cities developed. Up to World War I, the only protections a property owner and the public had were the nuisance laws, private restrictions and the building codes.

The results of this lack of control and planning are cities that developed with congested streets, overcrowded buildings, poor light and air and a mixture of uses, each hurting the other. From this disorganization came the deteriorated commercial areas, slums and urban blight of today.

Zoning History

In the 1700's and early 1800's, growth in this country took place primarily in the uncrowded towns and villages and in the farm areas. Many of the houses built were two-story detached houses.

By the middle of the 1800's the growth pattern had shifted to the cities. Streets were lined with row houses and tenements. Multiple story single family houses were lined up side by side on crowded lots. Streets were laid out in checkerboard fashion. There was little separation between industrial and commercial activities and residential neighborhoods.

In the early 1900's the cities were already feeling the pains of overcrowding, traffic congestion and the infiltration of non-residential uses into the residential neighborhoods. Uncontrolled social pressures increased still further the population density which resulted in the construction of large apartment houses that occupied almost all of their lots. At the same time, those people with enough money to do so began to flee the cities into the newly developing suburbs. This pattern of development still exists today.

In the early 1900's, Los Angeles passed some laws that controlled the use of the land and Boston enacted some similar regulations. In 1916, New York City passed the first true, comprehensive zoning law. Included in it were use districts, control of building heights and area coverage regulations. Shortly thereafter other cities passed similar laws. These stood up against many court challenges and by the middle 1920's their constitutionality, based on the government's right to use its police power to regulate private property, was well established.

The city zoning regulations of the 1920's placed emphasis on height regulations and front, side and rear yard requirements, heavily weighting considerations of light and air in overcrowded city areas.

Then, in the late 1920's, the emphasis of zoning laws gradually changed from fresh air and light considerations to the separation of the actual uses of property from one another. Separate industrial, commercial and residential zones were created. High density (apartment) residential areas were also segregated from low density (single-family) residential areas. Many regulations were passed in suburban and semi-rural areas, in attempts to preserve the residential characteristics of these areas.

Since World War II, there has been a new thrust in city zoning. These newer zoning regulations place emphasis on direct control of development and design in an effort to prevent any further spread of urban congestion and poor building design.

Zoning Goals Today

The goals of planning and zoning today for residential areas have become well recognized and are strongly advocated by those who believe in zoning as the best method of land use control. Unfortunately, there are also other, undeclared goals that some try to accomplish.

Nobody will argue the merit of protecting a residential neighborhood against the hazards of fire and explosion that accompany many industrial and commercial land uses. This can be accomplished through zoning. Protection against the common law nuisances of excessive noises, lighting, vibrations and pollution can also be included as part of an enforceable zoning regulation.

A residential neighborhood can also be protected, through zoning, against traffic that comes from the outside and does not directly serve the neighborhood. Protection against ugly visual elements such as garbage dumps and flashing neon billboards, however, is often difficult to achieve through zoning.

All residents in a neighborhood should have adequate light, air and reasonable privacy. This can be accomplished through both specific and general zoning controls which provide protection against congestion and make provisions for open spaces.

People sometimes try to use zoning to protect themselves against change in the form of development and community expansion. They try to accomplish this by increasing minimum lot requirements to unrealistic sizes and by imposing other requirements that make it impossible to economically develop the land. There are, however, growing social pressures to eliminate the use of zoning laws for these purposes. Ideally, zoning should not be used to stop development but should be used to control it so that school facilities and other necessary municipal services can be expanded in an orderly fashion.

Increasing density and economic and social pressures have been strong influences in directing communities to find new ways to use

13

land efficiently. Communities are discovering that they must make their own provisions for orderly expansion. In some areas, courts have been setting aside overly restrictive zoning regulations and state legislatures have been investigating regional planning and zoning. It is possible that these acts would diminish the local community's right to plan and zone itself. Many communities will retain their right to plan by exploring new methods of land use and integrating them effectively into community life.

Land Usage

One of the first reactions to increasing population and the cry for more housing was to build several million new houses each year in subdivisions at the edges of urban centers. Later, zoning restrictions were somewhat relaxed and the apartment boom began. The boom reached its peak in the 1960's when almost half of all housing built was in the form of apartments.

By this time, counter pressures were building up. Environmentalists, backed by concerned citizens, were expressing shock at the way in which some subdividers were developing the land. Long-time residents of suburbia also resisted what they considered to be intrusions into their communities in the form of apartment buildings and new subdivisions. Opposing these forces were social and economic pressures to decongest the urban centers and to get people out of the cities and into the "fresh air" of the suburbs.

Temporary solutions and compromises to land use problems were reached, such as improving and controlling subdivisions, requiring better site and building specifications for apartments and better land planning in general. In the late 1950's and early 1960's several things happened which provided the next ten years with better solutions to the problem.

The condominium form of ownership was introduced and the public liked it. With FHA mortgages on condominiums allowed by federal law, the concept spread across the country and condominium projects sprouted like mushrooms. Some condominium statutes provide considerably more protection for the public than do others. However, since this book concerns single family houses, a more detailed discussion of the complexities, advantages and disadvantages of condominium ownership is omitted.

It is important to note here the condominium's profound effect on land use and the general housing problem. Up to the 1960's, fee simple ownership forced achitects and land planners to restrict themselves to putting single or multiple unit dwellings on individual lots. The cooperative form of ownership did exist, but it worked best for urban apartments and never became very popular outside the big cities as the condominium has. The condominium also offered a solution to the problem the subdivider often encountered when he tried to use cluster

planning in layouts for apartment communities. The clusters often resulted only in open land that the communities refused to accept and maintain. With condominium ownership, however, the property owners accept the responsibility of maintaining property that, although used by the community, is in fact owned by them.

By 1965, cluster-planned subdivisions, condominium ownership and generally well-sited and well-built apartment communities were all firmly established. Now the architects and land planners were free to use their new, advanced concepts of land use: open spaces, cluster planning and common recreation and park facilities.

One of the earlier developments that helped bring about the acceptance of condominium ownership and the planned unit development (PUD) was Heritage Village in Southbury, Connecticut. From a 1000-acre estate, the developer created a beautifully planned urban community. The project was the first PUD to receive major national publicity. Through national news coverage, people learned what PUD's could offer. They saw visible proof that it is possible for a developer to take a large tract of land and successfully develop it with high density housing.

In Heritage Village, more than 2,500 families live in clustered apartments and townhouses in an attractive rural-like setting. The fact that Heritage Village was a financial success also helped to spur on other developers and lenders to produce more large PUD's and induced more communities to accept them.

The PUD, as it has evolved today, consists of high density housing, mainly apartments and townhouses (often sold as condominiums) and a few single family dwellings. The residences are grouped into clusters with a substantial amount of remaining land left as open space or developed for recreational purposes.

In its most advanced stage, the PUD can actually be an entire city by itself as it is in Columbia, Maryland, and Reston, Virginia. In these larger developments (as well as in some smaller PUD's) some of the nonresidential land is used for nonrecreational purposes such as shopping centers, office buildings and industrial plants.

Variances

No sooner will a zoning ordinance be passed than some property owners will, for any number of reasons, want to obtain permission to construct or alter a building contrary to the requirements of the ordinance. They will contend that the zoning ordinance imposes a hardship upon them and in their particular case they should be permitted to do what they want to do in spite of the ordinance.

The mechanism for obtaining permission to build something other than that which is permitted varies from area to area. Generally, such an appeal must be taken to a government body, usually the zoning board or the zoning board of appeals.

Hardship is the only basis, in most areas, on which a variance can be granted; economic hardship is not, by law, considered to be a valid **15** reason. Most people who request a variance will claim the zoning imposes upon them a severe hardship and that the granting of the variance is almost a matter of life or death. A more objective examination of the facts would probably lead to the conclusion that the real question is one of money and that by granting the variance the zoning board will allow the owner to increase his profits.

Zoning boards have a long history of granting a substantial number of the variances requested. They do this in spite of the fact that they are letting the property owner do something his neighbors are prohibited from doing. If the zoning ordinance was carefully thought out from its inception, the prohibition, theoretically, would have been for the community's general good and the purpose of the zoning ordinance would be defeated each time a variance is granted.

Special Exceptions

The typical zoning code takes into consideration the fact that it is in the community interest to permit certain uses in a zone that do not conform with the predominant permitted use.

For example, a school, religious facility or certain types of local commercial services, such as a drug store or small food outlet might be desirable special exceptions for a one-family zone. It is important that these exceptions be carefully controlled, which is usually left to the local zoning board.

The main difference between a variance and a special exception is that to obtain a variance it is necessary to prove hardship while this is not necessary to obtain a special exception. A special exception is considered to be a right of a property owner and is usually spelled out in the zoning regulation. The zoning authority must grant the exception unless it can be shown that the proposed use is detrimental to the neighborhood.

Special Use Permits

The most significant difference between special use permits and special exceptions is that the special use permit is not considered to be a right of the property owner. In order to get a special use permit, a person must prove that a need exists for the use of property that he is proposing and that it will be to the community's advantage to allow the proposed use.

An example of this kind of special use would be at an intersection of a major highway where development of shopping centers and automobile service stations would be desirable even though they may not be permitted under the zoning for that specific piece of land. The property owner would go before the local zoning board and attempt to prove that the proposed use would benefit the community. He would have to provide detailed plans to the zoning board, including

feasibility and marketing studies. In addition, the property owner usually must submit a technical report on the effect of the proposed land use on traffic circulation and the surrounding land uses.

Nonzoning

Inherent in zoning is that the use of the land is restricted. This tends to create a shortage of land available for some types of uses and a surplus of land available for other uses. Zoning definitely does not abide by the principle of supply and demand.

Critics of zoning contend that it restricts innovation, produces stereotype developments and reduces the number of imaginative projects that might otherwise have been created. They further claim that because zoning is a legislative function and administrated by politically appointed boards, the legislation and enforcement are highly susceptible to political influence. Critics also point out that extensive red tape in zoning slows down or even stops many building projects.

These are persuasive arguments when it is realized that many communities around the country, in spite of zoning ordinances, have had a poor development pattern by most any standards used to judge them.

One of the best anti-zoning arguments is Houston, Texas. Without zoning, Houston has developed into one of the most modern, exciting cities in the country. Several times the citizens of Houston have been asked to vote on whether they should continue without zoning or change to a traditionally zoned city. In every voting test, they have voted to remain an unzoned community.

The key to Houston's orderly development has been the use of deed restrictions and the enforcement of a rigid building code which includes control of density. By special act of the Texas legislature, the enforcement of deed restrictions is the responsibility of the city attorney rather than the individual property owner. The deed restrictions are also enforced by the local lending institutions who refuse to loan money for purposes that violate deed restrictions. Between 150 and 200 civic clubs in Houston police and help enforce deed restrictions.

There are problems, of course. Federal funds for urban renewal were withheld for a time. In many areas, the mixture of uses has caused severe loss of value to adjoining properties and older residential areas tend to be hastened through the last phase of their life cycle.

Advocates of nonzoning claim that land values and land uses depend upon the land's natural highest and best use and should not be controlled by what they claim are the sometimes misguided whims of local political appointees.

Site and Site Improvements

Zoning relates a piece of property to the larger community around it. This is important to the homeowner and real estate professional. Of equal importance are the location of a lot within a neighborhood, its physical characteristics, its relationship to neighboring sites, available utilities and services, location of the house on the lot and the site improvements.

If the boundaries of a lot are not readily apparent by visual inspection, steps should be taken to ascertain exactly where they are. The best and most accurate way to get this information is to have a professional survey made and have the surveyor set boundary markers into the ground. The surveyor will work from a legal description found in the deed.

In many situations the owner or real estate broker, with this same information, may establish the lot boundaries accurately enough for immediate practical purposes, especially when no construction is contemplated and all the abutting property owners seem to be in general agreement as to where the boundaries are or if the boundaries are naturally marked by fences, sidewalks, streams, walls and other permanent monuments shown on the survey. More and more lending institutions, however, are wisely requiring a survey for security whenever they lend money for a house mortgage.

A title examination will reveal any private deed restrictions that may have been imposed upon the property by former owners concerning the use and development of the land. There are also public restrictions on the use of the land imposed by laws, the most common of which are zoning laws, building codes, fire ordinances and health codes.

Whichever is more stringent, the private deed restriction or the public restriction, takes precedence. Generally it is complicated and expensive to obtain a change in these restrictions. If they appear to impose a hardship, a lawyer familiar with the area and real estate laws could be the logical person to turn to for advice.

Size and Shape of the Site

Lot size is a characteristic on which there seems to be no universal agreement. Many feel that small house lots are substandard and that anything under an acre is inadequate. On the other hand, a great many people feel that large lots require more care than they wish to provide and that very satisfactory living conditions can be obtained from a well-planned small lot.

However, it is generally agreed that, with the exception of townhouses which require a lot only the width of the house, a minimum satisfactory lot is 50 feet wide in the North and 60 to 70 feet in the South and the West. It is possible to make an acceptable layout on a lot as shallow as 90 feet deep if the front yard setback requirement is only 15 feet and the living area is oriented to the rear rather than to the front of the lot.

The MPS no longer has rigid requirements for minimum lot sizes, front, rear and side yard dimensions and percentage of lot coverage. Now it recognizes that the size, needed parking spaces, recreation and other open space requirements should be determined by the

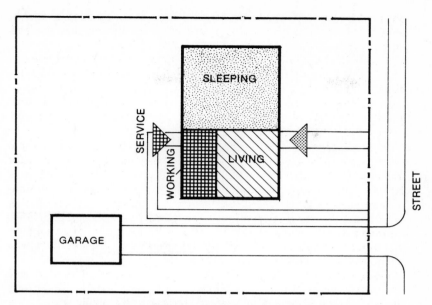

OLD WAY TO PLACE HOUSE ON INSIDE LOT

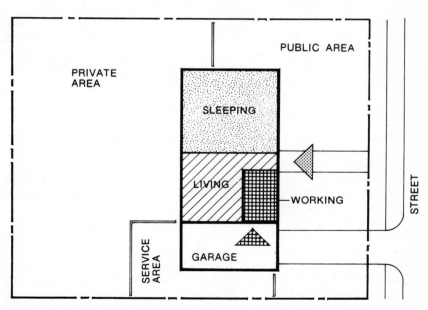

GOOD HOUSE PLACEMENT AND LAYOUT FOR INSIDE LOT

19

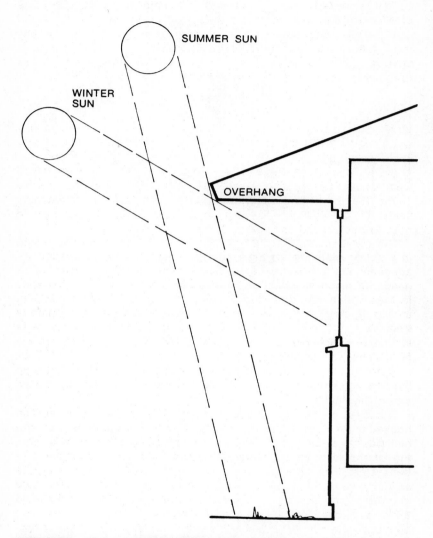

SUMMER AND WINTER SUN ANGLES ON SOUTH SIDE OF
HOUSE WITH ROOF OVERHANG

characteristics of the site, its location, land cost and acceptability by the community.

Generally, a minimum front yard setback of 25 feet is necessary and 50 feet is needed when the house is on a major highway. Side yards of at least 10 feet each improve privacy.

In the past, more lots were rectangular, except where topography made this shape impossible. However, due to new thinking in sub-division layout and street design, many lots now being created are irregular in shape.

Utilities

Water may be supplied by a well, usually on the site, or a community water supply. Wells and water supply systems are discussed in Chapter 7, but it should be emphasized here that if a well is needed along with a waste disposal system, the lot must be large enough to permit at least a 50-foot separation between the well and a septic tank and a 100 to 150-foot separation between a well and cesspool or leaching fields. It is obvious that the minimum lot sizes discussed above would not be sufficient under these conditions.

The requirements for water storage, main capacity and pressure in a public system are usually set according to what is needed for adequate fire protection rather than according to domestic water needs. Minimum fire insurance rates are obtained when a house is located within 500 feet of a hydrant. Therefore, no public system should have hydrants more than 1,000 feet apart. A better standard is from 300 to 500 feet apart. A domestic water supply will operate on as low as 20 pounds per square inch of pressure, but hydrants require 50 to 60 per square inch.

Water mains may be located under the street, under the grass strip between the street and the sidewalk or under the grass on the house side of the sidewalk.

The best method of sewerage disposal is a community system hooked up to a public treatment plant. (Septic tanks and other on-site methods are discussed in Chapter 7.) When an on-site system is used, the lot should be large enough to accommodate all the elements of the system. How large an on-site system is needed depends on the soil conditions. If the soil is extremely poor, the lot must be very large in order to provide an alternate location for the leaching field should the original one clog up and need replacement. Provisions must also be made to handle the increased amount of water runoff that is produced after a lot is built up and some of it covered with pavement.

The best method is a separate storm water drainage system tied into separate storm sewers in the street. The next most desirable method is to channel the water into a natural water course. If neither of these alternatives is possible, the water should drain off the land at the same point it was naturally doing so before the house was built.

Electricity should be brought to an area via power lines strung across the rear of the lots and connected to houses by underground service lines. This is a substantial visual improvement over the old method of bringing the power lines overhead along the street and connecting them to the house with three overhead wires. Compromises being offered by some utility companies are a combined wire that wraps the three wires together so it looks like just one wire or an underground connection from overhead wires in front of the house.

Gas is usually piped through mains under the street from a public utility company to the house site. The gas either may be manufactured locally by the utility company or piped in from the gas field by long distance gas transmission lines. The gas main is tapped into with a pipe that runs from it to the house.

House Location and Orientation

Unfortunately, the vast majority of houses are poorly located and oriented on their lots. Awareness by planners and developers of the best way to locate houses for maximum livability has been very slow in coming. The public's awareness has been even slower. As long as the public will buy houses lined up in a row facing the street in spite of the proven advantages of alternate methods of site planning, that is what developers will continue to build.

Just as a house should be divided into zones for good planning, so should a lot. There are three zones of a house lot — the public zone, the private zone and the service zone.

The public zone is the area visible from the street — usually the land in front of the house. A vast expanse of well-trimmed front lawn and a long driveway are important to many people. However, a large front lawn makes for high maintenance costs, lack of recreational use of this area and high snow removal costs. One argument supporting a large front yard is that setting the house back from the street reduces noise. This, however, is not true. For all practical purposes, noise 100 feet from the street is the same as 25 feet away.

If avoiding highway noise is particularly important, however, the house should be at least a mile from a busy highway. If this is not possible, street noise can be substantially reduced by orienting the house·away from the street and eliminating or reducing the number and size of the windows on the street side.

An important rule of good house location on the lot is to keep the front yard — public area small. This is done by bringing the house as far forward on the lot as possible. If this is not possible, then some sort of screening, such as fencing or shrubs, should be considered.

The service zone consists of the sidewalks and driveways plus trash storage and clothes drying areas. Like the public area, the service zone should be kept as small as possible.

The private zone is where the children play, where the patio is located and where the vegetable gardens grow. This zone, of course, should be large.

When a house is very well designed it takes advantage of the fact that in the hot summer months the sun rises in the northeast, travels in a high arc across the sky and sets toward the northwest. In the cold winter months it rises in the southeast, travels across the sky in a low arc and sets in the southwest. As a result, the south side of the house, when protected by a large roof overhang, will receive much more sun in the winter than in the summer. The opposite is true of the east and west sides of the house. All other factors, such as street location, topography and view being equal, the best direction in which to face a house is with the broad side containing the large windows towards the south.

Topography

The ideal lot is one with gently rolling topography which is attractive and supplies natural drainage of surface water. A steep, sloping lot looks spectacular at first and children will love rolling down the hill. Unfortunately, the novelty will soon wear off and the practical problems of sloping driveways, playing on a hill and placement of lawn furniture will offset the visual advantages.

A flat lot is easy to build on. However, drainage on a flat lot presents a problem, and results in a wet basement, flooding and seas of water on the lawns.

Grading

Lots must be graded to divert water away from the house, prevent standing water and soil saturation, provide for disposal of water, preserve desirable site features and provide grades for safe and convenient access to and around house and lot for their use and maintenance, MPS specifies. A lot should also have one or more areas conveniently located and of sufficient size and shape to adequately provide for outdoor living, children's play areas and such service functions as laundry drying and refuse storage. At least 400 square feet is needed for this.

Drainage is accomplished by sloping the lot toward the point where the surface water is to run off the lot or into the drain or by the construction of drainage swales designed to lead the water to the runoff point or drain.

Landscaping

The beauty of a well-landscaped lot is priceless. Some of the major advantages of an older house in many areas are the large trees, well-established lawns, shrubs and attractive flower beds. Fortunately, many builders are realizing the value of trees and taking the necessary steps during construction to preserve them, even to the extent of relocating the house or garage. MPS requires that every house have at least one shade tree, preferably on the southwest side of the lot. This

23

is, of course, a minimum requirement and additional trees always enhance the beauty and value of a lot and house.

Lawns, in addition to their attractive appearance, also prevent erosion in areas where the land has been disturbed and no other suitable vegetation is present.

Driveways and Parking Spaces

A driveway may serve as an access from the street to a garage or carport or as an on-site parking area. It should be at least 8 feet wide or, if the ribbon type, should consist of two strips 5 feet on center, each at least 2 feet wide. Most driveways are made of bituminous pavement or concrete, although there are other acceptable paving materials indigenous to various local areas.

Portions used for parking spaces should be at least 10 feet wide and 22 feet long. Concrete driveways should be at least 4 inches thick. Expansion joints are required every 10 feet. Where the driveway joins the garage, a carport slab or sidewalk or curb, additional joints are required.

Bituminous driveways should also be at least 4 inches thick with a wearing surface of at least 1½ inches. It is dangerous and sometimes illegal to back into a public street, so there must be room to turn a car around in the driveway. The driveway should be designed so that cars parked in it will not obstruct the walkway to the house.

Walks

MPS requires, as a minimum, "A main walk extending from the front entrance of the dwelling to the street pavement, public sidewalk or driveway connected to the street." In addition, except when the main entrance also serves as a service entrance, there should be a walk from the service entrance to the main walk, driveway, public sidewalk or street pavement. The minimum required width is 3 feet for the main walk and 2 feet for the service walk. In climates subject to freezing, the grade should not exceed 5 percent (⅝ inch per foot) and in warm climates, 14 percent (1¾ inches per foot).

Concrete walks should be 4 inches thick and have expansion joints where they connect to entrance platforms, driveways and sidewalks. They should have construction joints at 4-foot intervals. Other materials besides concrete are often used for walks, but generally they are cost-saving substitutes. Brick, tile and other similar materials are sometimes used because of their decorative appearance.

Exterior Stairs

When the topography of the lot is so steep that walks cannot be constructed within the maximum grades permitted, stairs should be used.

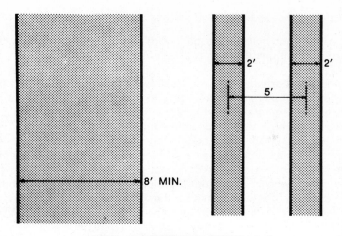

MPS DRIVEWAY SIZES

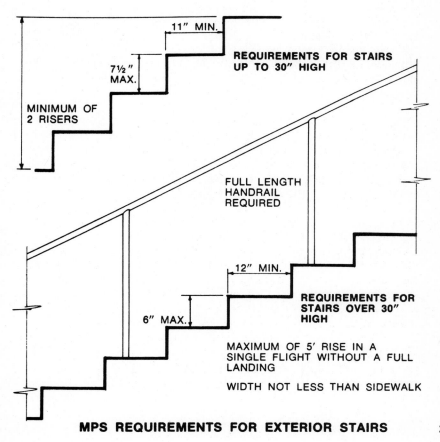

11" MIN.

7½" MAX.

REQUIREMENTS FOR STAIRS UP TO 30" HIGH

MINIMUM OF 2 RISERS

FULL LENGTH HANDRAIL REQUIRED

12" MIN.

REQUIREMENTS FOR STAIRS OVER 30" HIGH

6" MAX.

MAXIMUM OF 5' RISE IN A SINGLE FLIGHT WITHOUT A FULL LANDING

WIDTH NOT LESS THAN SIDEWALK

MPS REQUIREMENTS FOR EXTERIOR STAIRS

25

Security and Privacy

Millions of tract houses are built with large picture windows facing the street which, unless covered with closed drapes, result in a major loss of privacy.

A check for privacy may be made by walking slowly by the house and then around the lot line, trying to look inside the various areas of the house. Sometimes a substantial improvement in privacy can be made by planting a few shrubs or building a fence.

Electric, gas and water meters should be outside the house to eliminate the need for strangers to enter the house to read them.

The danger of opening the door to a stranger is obvious. In order to safely check who is at the door without opening it, there should be some clear or one-way glass in the door or sidelight, a peephole or an intercom system. An additional safety feature is a door chain that permits the door to be opened a crack to talk with a stranger before deciding to let him in.

Most professional thieves will get into a house no matter what kind of locks there are and what other precautions have been taken. However, many burglars are not professionals and good hardware with locks that are difficult to pick will often act as a satisfactory deterrent.

An automatic burglar alarm system connected to the police station provides maximum burglary protection. A system that sounds an alarm on the premises may scare away a prowler before he has a chance to take anything.

Chapter 2
House Types

Chapter 2 together with Chapters 3 and 4 explain the basic principles of good house design. This chapter shows how houses can be classified and described by their structural configuration. This is known as the house "type."

The following chapters discuss circulation within the house by dividing it into three zones and showing their relationship to each other. Each room will be discussed in terms of how to design it to best serve its special purpose.

A well-designed house takes careful planning by an architect whose services should be utilized whenever possible. The ideal method is to have each house designed for the individual owner by an architect who is familiar with both the site and the owner's needs and tastes. When this is not possible, a set of architect-designed stock plans, prepared for the kind of lot available, should be used. These plans must be studied carefully to select one that will produce a home that will fill the family's needs and appeal to its esthetic tastes. Of course, it is possible for an owner, builder or broker to design a house. However, these people usually are not trained to consider all the elements that make up good design and the result is usually mediocre.

There are nine basic types of houses that make up the vast majority of new homes and many older homes.

The discussion of good design begins with an illustration of each of the basic types of houses and comments on their inherent advantages and disadvantages.

The most common type being built today in most areas is the one-story ranch style house. Prior to the 1960s the one and one-half story Cape Cod style was the national favorite back to the 1920s. From then back to pre-Revolutionary days the two-story Colonial style was the most popular house. Much newer on the scene is the split level which started as a solution to a sloping lot, caught on in popularity after World War II and was built in vast numbers on level lots. Its popularity is fading, however, and it is being replaced by the split-entry or bi-level type that started in the Northeast and is working its way west and south.

There is a great deal of confusion in the real estate industry on how to describe and classify houses. The words "type" and "style" are used interchangeably and with little uniformity. The CTS System which is described in detail in the Appendix is a proposed solution to this problem. On the following pages are pictures of each of the basic house types, their advantages, disadvantages and common problems. The CTS System identification information is shown on each page, the narrative form first, followed by the abbreviation and the computer code.

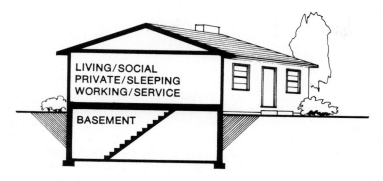

One-Story, Ranch, Rambler (1 Story - 1)

Advantages

Glorious lack of stair-climbing, especially appealing to invalids, elderly people and mothers with small children.

Adapts itself to indoor-outdoor style of living.

Porches, patios, terraces and planters can be designed for any room.

Roof can be low-pitched as no headroom is required above ceiling.

Entire exterior easy to maintain as whole house can be reached from ground or short ladder (for screen and storm window jobs, painting, fixing roof, cleaning gutters, etc.)

Low height simplifies construction.

High acceptance of design results in good resale value.

Disadvantages

Difficult to entertain without waking up the children as noise tends to be transmitted from living and service areas to bedroom areas.

Children tend to play at mother's heels.

Requires larger lot.

Of all types of houses, requires the largest ratio of walls, slab, basement or crawl area, foundation and roof to total living area and therefore costly to build for equivalent square footage, compared to other types of houses.

In smaller models, difficult to maintain privacy in bedroom areas.

More costly to heat and cool than two-story house with same living area.

Common Problems

If basement is built several feet off the ground, the house looks poor, discourages indoor-outdoor living and the stairs create a safety hazard.

Poor interior design may negate many potential advantages.

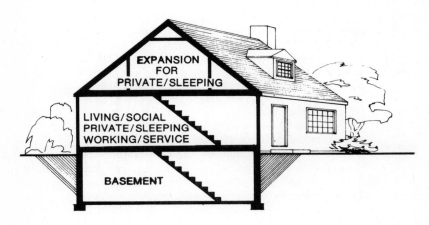

One and One-Half Story (1½ Story - 2)

Advantages

Second floor rooms can be completed as needed.
Heating cost low because of small perimeter compared to enclosed
   living area.

Disadvantages

Second floor rooms often small and cramped.
Second floor window area often limited without expensive dormers.
Ceilings on second floor low.
Second floor often hot in summer and cold in winter.
Stair-climbing.

Common Problems

Insufficient insulation under roof.
Insufficient ventilation for second floor.
No provision to heat second floor or heating plant too small to heat
   second floor when finished.
No water pipes or waste lines roughed into second floor for future
   bathroom.

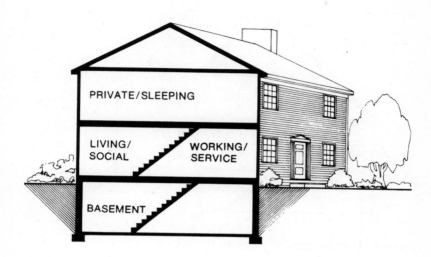

Two-Story (2 Story - 3)

Advantages

First floor rooms and sleeping areas are separated.
When built on small lot, permits maximum living space.
Economical to build because of small ratio of exterior wall,
    roof and foundation to total living area.
Many people feel there is more privacy and safety when bedrooms
    are on second floor.
Because of its roots in American history and its long association
    with gracious living, it still has strong appeal to many people.

Disadvantages

Continuous need for climbing stairs.
Space used for stairs is wasted.
Upstairs bedrooms and playrooms do not have direct access to
    outside.
Most designs are difficult to expand.

Common Problems

Poor stairway location.

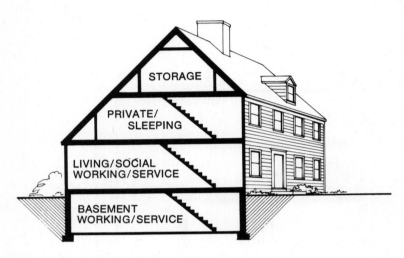

Two and One-Half Story (2½ Story - 4)

Advantages

First floor rooms and sleeping areas are separated.

When built on small lot, permits maximum living space.

Economical to build because of small ratio of exterior wall, roof and foundation to total living area.

Many people feel there is more privacy and safety when bedrooms are on second floor.

Because of its roots in American history and its long association with gracious living, it still has strong appeal to many people.

Attic space usable for sleeping room, recreation and storage.

Disadvantages

Continuous need for climbing stairs.

Space used for stairs is wasted.

Upstairs bedrooms and playrooms do not have direct access to outside.

Most designs are difficult to expand.

Expensive to heat.

Common Problems

Poor stairway location.

Poor light, heat and ventilation in attic rooms.

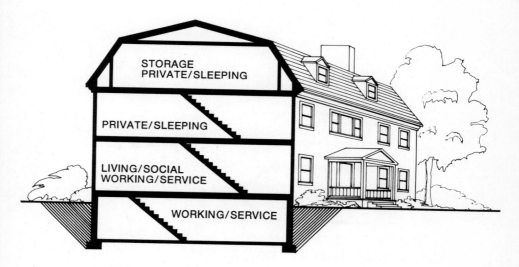

Three or More Stories (3 Story - 5)

Advantages

Often allows more than one family to occupy house.

When built on small lot, permits maximum space.

If occupied by a single family provides many extra sleeping, recreational and storage rooms.

If occupied as a multiple-family provides separate floor or floors for each family.

Disadvantages

Continuous need to climb stairs to reach upper stories.

Space used for stairs is wasted.

Upstairs rooms do not have direct access to outside.

Common Problems

Living units too small or poorly laid out.

Noise transmission between floors.

Difficulty heating evenly.

33

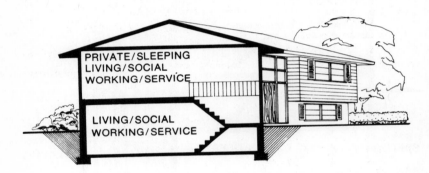

PRIVATE/SLEEPING
LIVING/SOCIAL
WORKING/SERVICE

LIVING/SOCIAL
WORKING/SERVICE

**Bi-Level, Raised Ranch, Split Entry, Split Foyer
(Bi Lev or R Ranch or Splt Ent or Splt Foy-6)**

Advantages

By raising basement out of ground several feet, greater window
depth results which gives better light to basement and makes it
usable as living space.
Layout usually keeps traffic out of living room.

Disadvantages

Half of living space is in basement which may be cold and damp.
Exposed exterior of basement wall often unattractive.
Rear porch often at awkward level.
May be a fad design that will lose popularity.
Upon entering the house some stair-climbing required to get any-
where (poor for invalids).

Common Problems

Difficult to keep basement rooms at even temperatures unless insulation and waterproofing are of top quality.

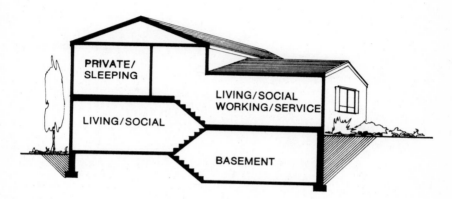

Split Levels — Side To Side, Back To Front, Front To Back (Splt Lev - 7)

Advantages

Good design for a sloping lot.

Cost per square foot of finished living space is lower than a ranch.

Each functional area of the house is separated on a different level and the design lends itself to good interior circulation.

The distance from one level to another is less than a two-story house.

Disadvantages

When built on a level lot, a poor indoor-outdoor relationship develops and many stairs are needed.

It is necessary to go up and down stairs in going from one interior zone to another.

Bedrooms on an upper level are often hot from heat rising through the house.

Looks poor on a level lot.

Common Problems

It is difficult to design a good small split level house; they work out much better when they are at least 1,500 square feet. It is more difficult to construct the frame of this type of house than for a one or two-story house.

35

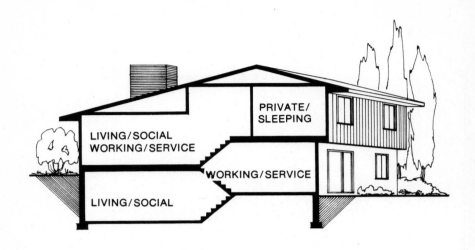

SPLIT LEVEL — BACK TO FRONT

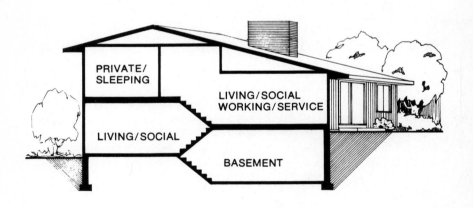

SPLIT LEVEL — FRONT TO BACK

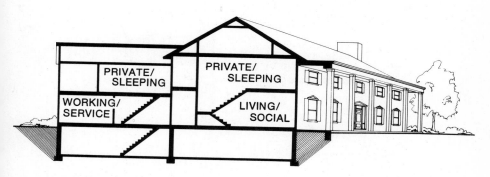

Mansion (Mansion - 8)

Advantages

Many extra rooms for recreation, entertaining, guests and servant quarters.
May enhance social prestige of owner or occupant.
A vehicle for gracious living for those who can afford it.

Disadvantages

Expensive to physically maintain.
High taxes.
High utility costs.
Difficult to keep clean.
May have more rooms than needed.

Common Problems

If old, tends to need constant repairs.
Harder to sell.

It is hard to define precisely what a mansion is but the term is commonly used to describe a house of considerable size or pretension.

Other (Other - 9)

Illustrated

Hillside Ranch
Cellar or Foundation House
Cliff Side House
Town House with English Basement

Not Illustrated

Barn
Tree House
Grass and Sod House

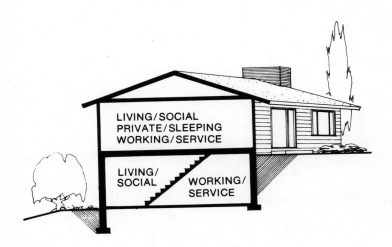

The Hillside Ranch

Advantages

Good design for a sloping lot.
Basement can be finished into living area at a lower cost than adding
    additional first floor space.
Convenient direct access from basement to outside.
House is attractive from the front.

Disadvantages

Finished living space in basement may be cold or damp.
Tends to be unattractive when viewed from the rear.
When basement is finished there is a loss of storage space.
House is often designed so that the only service entrance is through
    the basement and up the stairs in the kitchen.

Common Problems

Difficult to keep basement warm and dry unless foundation has been
    carefully constructed with adequate drains, no cracks and ample
    insulation.

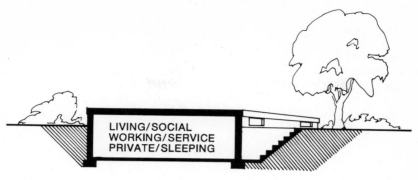

Cellar

LIVING/SOCIAL
WORKING/SERVICE
PRIVATE/SLEEPING

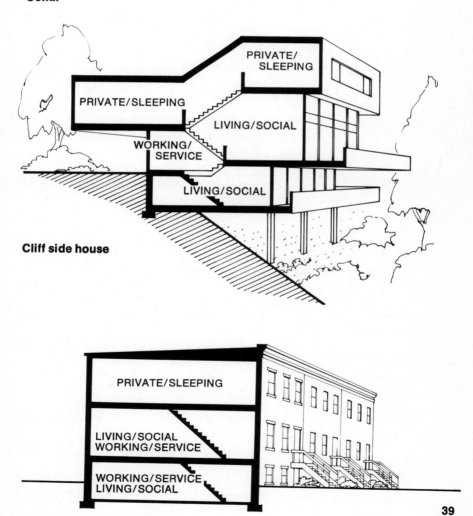

PRIVATE/SLEEPING

PRIVATE/SLEEPING

LIVING/SOCIAL

WORKING/SERVICE

LIVING/SOCIAL

Cliff side house

PRIVATE/SLEEPING

LIVING/SOCIAL
WORKING/SERVICE

WORKING/SERVICE
LIVING/SOCIAL

39

Townhouse w/Eng. bsmt.

Chapter 3
Interior
Design

House architecture means different things to different people. In one sense it can be described as the method by which we enclose living space in such a way as to provide maximum utility to satisfy the needs of the occupants, together with maximum amenities to please the senses, on the selected site within the budget available. Utility requirements and budget allowances of occupants vary as does what aesthetically appeals to them. Therefore, house designs will also vary to meet these needs, tastes and budgets.

People generally have a certain natural appreciation of beauty. However, this sense of appreciation can be substantially increased by exposure to and study of beautiful things. Without study one can enjoy music and art, but the enjoyment increases when one knows more about them. The same principle applies to architecture.

This country has come through an era of horrible community planning and massive construction of ugly, poorly designed and poorly constructed houses. The fad house of today and yesterday will be the eyesore of tomorrow if it is not already. When the shine of newness wears off, the blight of ugliness shows through.

Fortunately people are becoming more aware of the need for good community planning and the merits of good design. Hopefully, the future will be brighter than the past in this regard. If this book makes any contribution towards that end, then the efforts of the author will be well rewarded.

Zones Within a House

A good way to think about the interior layout of a house is to divide it into zones. The private/sleeping zone contains the bedrooms, bathrooms and dressing rooms. The living/social zone consists of the living room, dining room, recreation room, den or enclosed porch. The working/service zone consists of the kitchen, laundry, pantry and other work areas. In addition to the three zones there are circulation areas consisting of halls and stairs plus guest and family entrances. The three zones should be separated from each other so that activities in one zone do not interfere with activities in another.

The private/sleeping zone should be located so that it is insulated from the noises of the other two zones. It should be possible to move from the bedrooms to the bathroom in this zone without being seen from the other zones.

In a two-story or split-level house this can be accomplished by putting this zone on two separate levels or floors. In a two-story house it is easier to isolate the zone visually than it is to keep out the noise. Noise transmission from the working and living zones to the private/sleeping zone through the floor and ceiling is one of the major disadvantages of a two-story house. Another common layout mistake is to place the bathroom at the head of the stairs, clearly visible from the downstairs.

The working/service zone is the nerve center of the house. From here the housewife directs the household activities. The cliche, "a

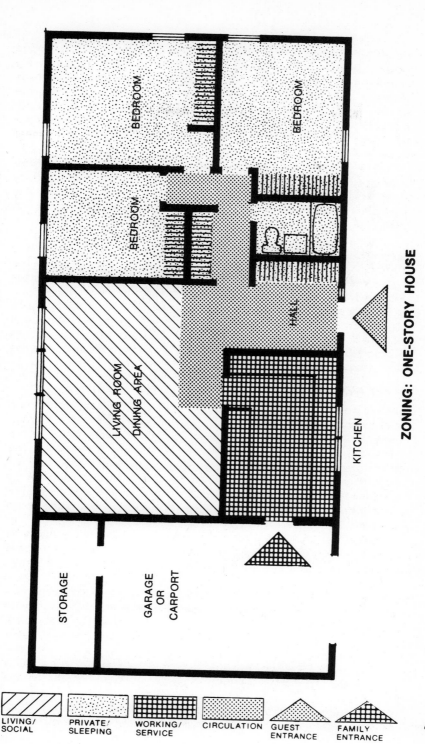

ZONING: ONE-STORY HOUSE

BEDROOM

BEDROOM

BEDROOM

HALL

LIVING ROOM
DINING AREA

KITCHEN

STORAGE

GARAGE
OR
CARPORT

LIVING/
SOCIAL

PRIVATE/
SLEEPING

WORKING/
SERVICE

CIRCULATION

GUEST
ENTRANCE

FAMILY
ENTRANCE

43

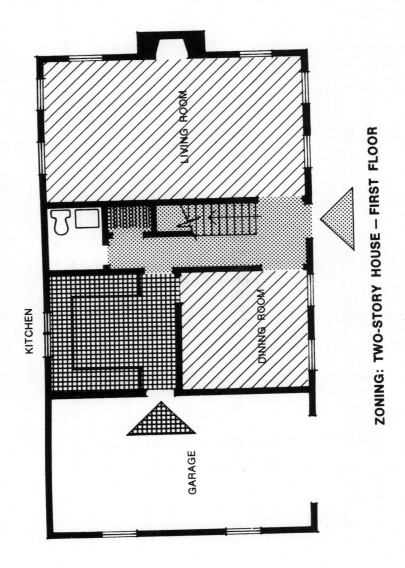

LIVING ROOM

KITCHEN

DINING ROOM

GARAGE

ZONING: TWO-STORY HOUSE — FIRST FLOOR

LIVING/
SOCIAL

PRIVATE/
SLEEPING

WORKING/
SERVICE

CIRCULATION

GUEST
ENTRANCE

FAMILY
ENTRANCE

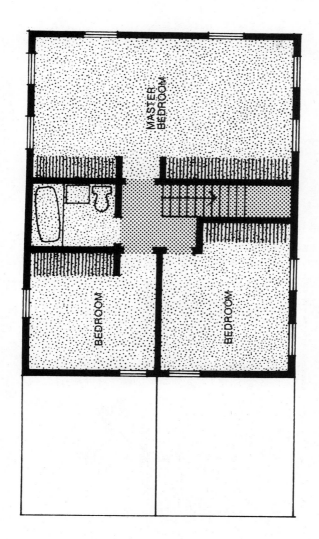

MASTER BEDROOM

BEDROOM

BEDROOM

ZONING: TWO-STORY HOUSE — SECOND FLOOR

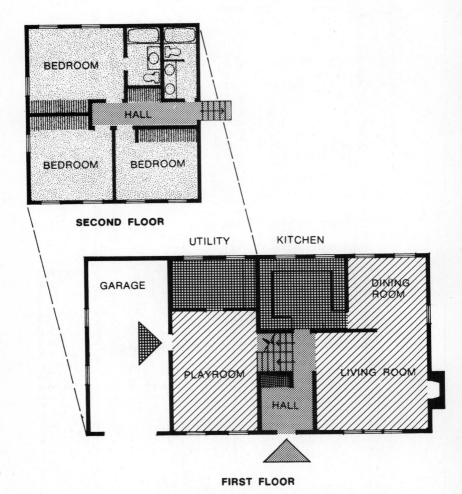

SECOND FLOOR

BEDROOM

BEDROOM

HALL

BEDROOM

UTILITY   KITCHEN

GARAGE

DINING
ROOM

PLAYROOM

LIVING ROOM

HALL

FIRST FLOOR

ZONING: SPLIT LEVEL HOUSE

 LIVING/
SOCIAL

 PRIVATE/
SLEEPING

 WORKING/
SERVICE

 CIRCULATION

 GUEST
ENTRANCE

 FAMILY
ENTRANCE

46

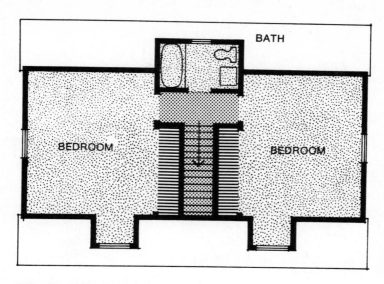

SECOND FLOOR PLAN

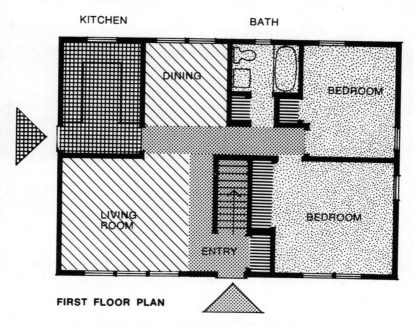

FIRST FLOOR PLAN

ZONING: ONE AND ONE HALF-STORY HOUSE

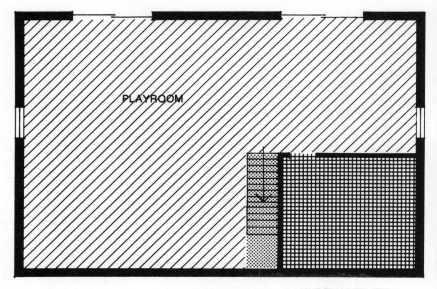

PLAYROOM

BASEMENT   LAUNDRY/ UTILITY

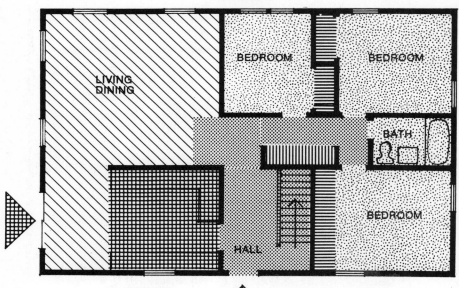

LIVING
DINING

BEDROOM   BEDROOM

BATH

BEDROOM

HALL

KITCHEN   **FIRST FLOOR PLAN**

ZONING: HILLSIDE RANCH

 LIVING/
SOCIAL

 PRIVATE/
SLEEPING

WORKING/
SERVICE

CIRCULATION

GUEST
ENTRANCE

 FAMILY
ENTRANCE

48

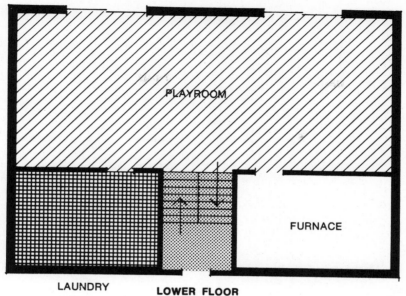

LAUNDRY   **LOWER FLOOR**

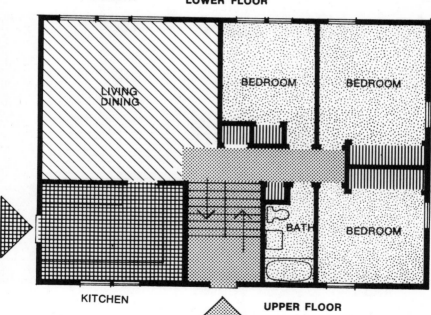

KITCHEN   **UPPER FLOOR**

ZONING: SPLIT ENTRY

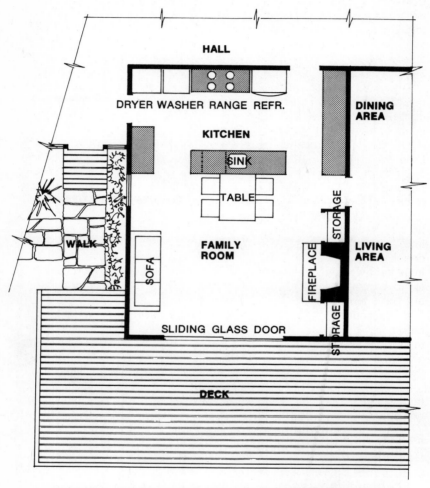

MODERN HOUSE WITH GOOD WORKING/SERVICE AND LIVING/SOCIAL ZONING

mother's place is in the kitchen" is quite true when she is at home and not at some other job or activity.

Making the kitchen efficient and pleasant is discussed later. However, it should be noted that the relationship between the kitchen and the rest of the house is the most important key to good interior layout. From the kitchen it should be possible to control both the guest and family entrances, activities in the sleeping/private and living/social zone plus activities in the porch, patio and back yard areas. As difficult a requirement as this may seem, it is not impossible. Illustrated are a series of basic layout plans which accomplish these goals quite well.

The guest entrance should lead into the center of the house. From here there should be direct access to the living areas, guest closet and guest lavatory. A noise and visibility barrier should exist between the guest entrance and the sleeping/private area.

The family entrance ideally should be from the garage, carport or breezeway into the kitchen or a circulation area directly connecting to the kitchen. However, traffic from this entrance should not have to penetrate the work triangle of the kitchen. A hallway or small room (called a mud room in some areas) at the family entrance prevents a lot of dirt from being tracked into the house. An alternate location, but not as satisfactory, is a family entrance from the back yard or driveway into the kitchen. The main problem in this situation is carrying groceries from the automobile into the house without getting wet in inclement weather. The circulation should be such that it is possible to move from the work/service zone to the private/sleeping zone without going through the living/social zone.

If the house has a basement it should have a separate outside entrance. The inside entrance should lead into a circulation area that in turn has access to the private/sleeping zone and living/social zone and both the guest and family entrances without going through the living room or the kitchen work triangle.

There is really no such thing as a perfect layout. Individual sites, family needs, individual preferences and budget must all be taken into consideration, priorities established and compromises made.

Poor Floor Plan

Here is a list of some of the most common floor plan deficiencies:
1. Front door entering directly into living room.
2. No front hall closet.
3. No direct access from front door to kitchen, bathroom and bedrooms without passing through other rooms.
4. Rear door not convenient to kitchen and difficult to reach from street, driveway and garage.
5. No comfortable space for family to eat in or near kitchen.
6. A separate dining area or dining room not easily reached from kitchen.
7. Stairway between levels off a room rather than a hallway or foyer.

8. Bedrooms and bathrooms so located as to be visible from living room or foyer.
9. Walls between bedrooms not soundproof (best way to accomplish this is to have them separated by a bathroom or closets).
10. Recreation room or family room poorly located.
11. No access to the basement from outside the house.
12. Outdoor living areas not accessible from kitchen.
13. Walls cut up by doors and windows making it difficult to place furniture around room.

Living Room

In the Golden Age of Pericles, home was only of secondary importance. The affluent family might have eight or ten rooms grouped around the courtyard, one side of which contained the living rooms and were used primarily by the women. Meanwhile, the men would be on the way to the forum for public debates, sporting events and other entertainment which could consume most of the day.

Contrast this to the tower houses in Afghanistan centuries later (and similarly in other houses of central Asia) where the "living room" was at the top of the building. In all probability these rooms were converted from watch towers to living areas after the Mongolian invasion in the 14th century. Remodeling and good use of existing space was as much a family concern then as it is in America today.

Still another example of a different living room was in the 14th and 15th century Italian house, which actually was a house and business combined. The shop was on the first floor and bordered directly on the street. Upstairs, one room was set aside for living and dining and the rest for bedrooms for the family and help. Often these homes would be multi-storied and its owner a man of affluence and respect. During this era in Poland, middle class houses were very similar but slightly smaller in area, with the living rooms always on the upper floors.

In the 17th century, the New England living room was the main room of the house—and living room it was! It was the center of all family life and activity. It served as the kitchen, dining room, bedroom, sitting room and storeroom all at once.

A document of the period describing a typical Colonial home in Ipswich, Massachusetts in the mid-1600's shows this inventory of furnishings in the living room:

"One table, three benches, and one stool; one bed with a small bolster; one set of curtains and bedhangings; one mattress and a feather pillow; one straw sack and a woolen pillow; one white blanket; one pair of sheets; one green carpet; one other bed and mattress; one complete set of curtains and covers; one hemp mattress; one feather cushion; one pair of old sheets; two old blankets; one red carpet."

There also must have been some fireplace equipment and cooking utensils as well as dishes, pewter pieces and candlesticks.

By the 18th century, the finer homes in the Colonies were modified to the point that all those functions were moved to a room separate for each. There were nearly always four rooms on each floor and an exact center entrance. The living room could be either to the right or left but was firmly established by now as a special place for reading, needlework, family get-togethers or company. With the exception of a few refinements, such as adding the fireplace, the living room remained very much the same for almost two centuries.

In the past several decades, however, the function and status of the living room has undergone an almost imperceptible change. The realization that the living room no longer retains its exalted place as the family living center is just beginning to dawn on our society. The old "front room" is no longer in the front, figuratively and often literally as well.

Today, the family room, the patio and the kitchen are more lived-in and much more likely to be the locations for relaxing and entertaining. As these areas grew and developed, the size and importance of the living room diminished proportionately.

But the living room has survived our sociological change and still maintains its importance. Millions of Americans find the "living" room indispensable. They are the ones who lay great stress on the furnishings and functions of this inner-sanctum. There is no doubt that graceful living is still very much a part of cherished traditions of many families.

Architect Frank Lloyd Wright, whose philosophy was that the whole house was for living in, was well ahead of the times in his thinking. In one of his houses, the ground level was no longer the main floor but instead was designed for bedrooms and children's playroom. The living area of the house was on the upper floor where the rooms all centered on a large living room. This in 1905!

As is the case with the other rooms in the well-planned house, the size and use of the living room should be determined by the size and make-up of the family and, hence, by its needs. If there is also a family room, this too must be considered in relationship to the living room.

Location

The old rules governing where the living room should be located within the house have all but been discarded since the family room has become so overwhelmingly popular. This is almost certainly attributable to the population explosion and the trend to having larger families directly after World War II. The children of these families are now of parental age and, because of new population philosophies, not quite so inclined toward reproducing on the grand scale. For this reason, it is remotely possible that the family room will diminish in importance in another decade or two, but most probably it will be modified rather than disappear entirely. In either case, the living room will live on and several basic rules about its location should be applied and adhered to.

The location should be appropriate to the house and family. It should not be a passageway for traffic, but normal circulation should be easy. The room should be in a position to supplement the dining and outdoor entertaining areas. Very often, one end of the living room is the dining area, so it must have good juxtaposition with the kitchen-service area.

In the "old" days it was simple: the living room was the front room and that was that. It should not necessarily follow that the living room will lose its status by being located to the side or back of the house. Perhaps the view is better or it is quieter there. However, if there is a large family room (and large family) the current tendency is to give that room the location of best view and accessibility to the most attractive outdoor area where the patio is usually located.

It is not enough to just "have" a fireplace in the living room. It should be well located, out of the way of traffic and with allowable space for chair groupings. Other personal requirements such as ease in bringing in wood and/or other types of fireplace fuel need consideration, as does disposal of the ashes.

The sophisticated refinements which can add to the enjoyment of the living room are a matter of personal choice once these basic requirements are met.

Size and Layout

It is more difficult to recommend specific room dimensions in a living room than in a bedroom. The bedroom serves a very clear function; the living room has become more vague and varied. Consideration of proportion has replaced size in fulfilling the purpose of the living room for each particular family. Shape and layout with special interest considerations are also every bit as important as square footage.

There are some good general guidelines which can help develop the coordination of the living room and its usage with the rest of the house. For example, in a three-bedroom house the living room should have minimum dimensions of 11 by 16 feet or at least 170 square feet. The recommended dimensions, however, are for a room 12 by 18 feet. In any case, the width should never be less than 11 feet.

Square rooms make good furniture arrangement difficult and tend to be uninteresting as a result. If there is a dining area at one end of the room, dimensions may go up to 16 by 26 feet or more. A maximum width of 14 feet is recommended for furniture arrangements at the proper distance across the room. Where there is necessary traffic through the room, a width of 15 or 16 feet could conceivably be utilized to advantage by routing the traffic outside a conversation circle made with the furniture.

Ideally, there should be no traffic through the room; it should be a dead-end area. Good planning puts the exit at only one end of the room. This is now modified somewhat by the widespread use of glass doors which open onto a porch, patio or lanai. Even so, their placement should not set up a cross-traffic pattern.

In general, there will be one wall for the fireplace and perhaps book **54** shelves or built-in cabinets, one long wall for the large divan, one wall

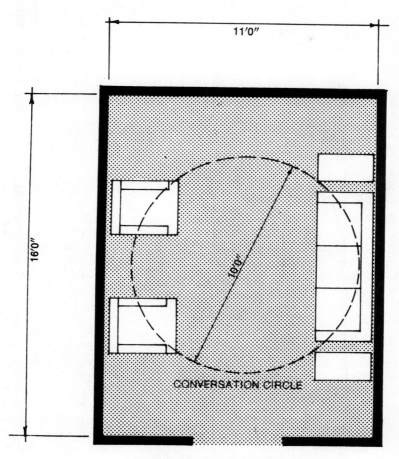

11'0"

16'0"

10'0"

CONVERSATION CIRCLE

MINIMUM SIZE LIVING ROOM

55

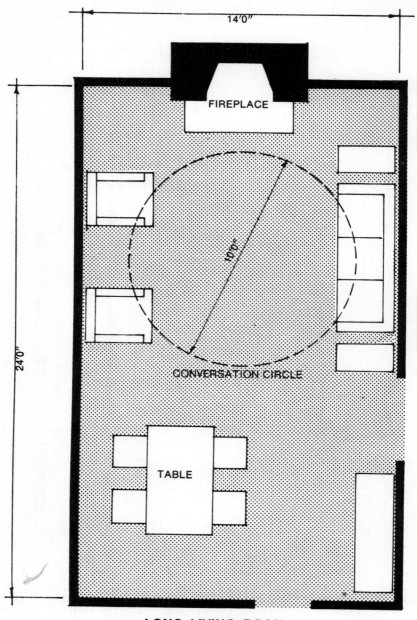

14'0"

24'0"

FIREPLACE

10'0"

CONVERSATION CIRCLE

TABLE

LONG LIVING ROOM

56

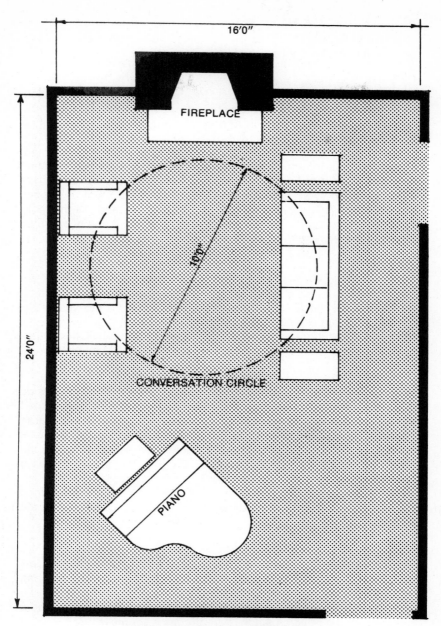

16'0"

24'0"

FIREPLACE

10'0"

CONVERSATION CIRCLE

PIANO

LONG AND WIDE LIVING ROOM

57

for the entrance (archway or door) and the fourth for windows or sliding doors. There may be various combinations of these to some degree, depending on individual taste or necessity.

Storage

Storage for this room or area should include a closet outside the main entrance for coats and other street clothing, umbrellas and raincoats. It should be spacious enough to accommodate guests' clothing, particularly outerwear.

Many living rooms have or need storage space for books, radio, stereo, records, newspapers and magazines, ash trays and coasters and stationery and writing equipment. The television set no longer holds as much sway over the environs of the living room as it once did. An interesting survey has revealed that in two-thirds of the more affluent homes, there is no TV set in the living room. It had been transferred to the recreation room, family room or bedroom.

Ventilation

Cross-ventilation in the living room is desirable but, where there are large glass doors which slide open, not always necessary. Ventilation areas should exceed 10 percent of the floor area. Proper ventilation is very important where there is a fireplace, for both proper draw and fresh air circulation. A door may be considered part of the cross-ventilation system since it too creates desired air circulation.

Heating

In newer houses baseboard and other similar types of heating units do not interfere with furniture arrangements. In older homes the forced hot air unit or the cold air return duct should not be blocked by large furniture pieces like chairs, divans or chests. New and attractive metal coverings for the old cast iron steam or hot water radiator enhance the room and do not interfere with the heat. But they should not otherwise be blocked if they are to perform well in their old age.

The thermostat should never be placed close to the fireplace since the heat causes it to fluctuate and send improper impulses back to the furnace.

Lighting

Natural lighting can be achieved best by correct and attractive window treatment. Unfortunately this cannot be a 24-hour-a-day arrangement.

The newest innovations in artificial light add tremendously to the aesthetics of the living room. If they can be incorporated into the room while it is in the building stage, even the most sophisticated arrangements and fixtures are not extremely costly.

Cove lighting is usually mounted on a wall and directs light upward to the ceiling where it is reflected back into the room. Soffit lighting is used in the underside of any architectural feature such as the area over a mirror, a corner niche or a beam. This type of structural lighting can be very flattering, particularly when combined with amber, pink or blue-toned lighting. These mood effects are restful and low key but do not meet the needs of the family scholar who wishes an easy chair by the fire. A lamp should be strategically situated on a table or stand to provide quality lighting.

In general, living rooms have a low intensity-to-brightness ratio (i.e., the difference between surfaces) because of furnishings like draperies, rugs and upholstered furniture, and tend to look dull. Specific lighting such as at chairside or desk not only provides the proper intensity of light but also enhances the mood and atmosphere by creating centers of brightness. An attractive lamp will also add to the decor.

Adequate electric outlets should be readily available in the living room so that there is flexibility in arranging furniture and lamps to serve the seating plans. Nothing detracts more from good decor than lengthy wires running along the floor.

Conversation Areas

The ideal conversation area is a circle approximately 10 feet in diameter. Furniture groupings should be arranged with this in mind. The family and/or guests are more comfortable when seated within this range since most people like to congregate in small groups whether in a large or small room. Within the 10-foot-diameter circle they can see each other easily and communicate comfortably without shouting. A large room may require several conversation centers of this size if this principle is followed. At a party, where guests tend to stand around to converse, even less space is desirable. The ideal diameter of a standing conversational circle need only be about 6 feet. A living room more than 22 feet long by the correct 11 to 14-foot width mentioned previously provides for several seating and standing circles of conversation and activity. When the living room has a fireplace it is desirable to make it the focal point of furniture in one of the conversation circles.

Munching on hors d'oeuvres and fondue, finding elusive ash trays or circulating from one conversation group to another takes maneuverability. The living room should provide this.

Kitchen and Dining Room

History

Some time around 40 B.C., one Gaius Sallustius Crispus wrote that the war with Catiline was "on behalf of their country, their children, their altars and their hearths." Note that he did not say on behalf of their bed chambers, grape arbors, the atrium or the arena. His priority after God, country and the family was the hearth.

After 20 centuries men are still fighting for exactly the same reasons, the only difference being that the hearth is now called the kitchen. It remains, still, the "heart" of the home.

History gives us very little documentation on the evolution of the kitchen. We do know, however, that its metamorphosis must have taken place very slowly, until the comparatively recent revolution in kitchen equipment. Oddly enough most of what we do know has been revealed through the art of particular periods. Paintings, sketches and frescoes reveal that Egypt had its kitchens on the side of the courtyard away from the rest of the house, complete with drains for the removal of water. The water closet was installed next to it in order to make use of the same amenities. The early Romans had an oven of sorts which has become one of the most important of all archeological finds—it has been uncovered from Scotland to Persia.

In the Middle Ages, a fire in the living room served to heat the house and poor families did their cooking there. It was not a separate room but an area important chiefly because of the fire. Those who could afford it had a brazier, a large metal pot which held hot coals for cooking. Only the very wealthy had a kitchen, which was a separate building slightly removed from the castle. Even so, that kitchen and the ones we know today bear little resemblance to each other.

As the kitchen moved into the house it was still necessary to have servants to carry in the firewood and water and to prepare the food. Cleanliness was not a requirement. The floor was hard dirt, packed down and often slippery with fish scales and stray entrails which the dogs or chickens walking around underfoot may have overlooked. The dogs and chickens also contributed somewhat to the slippery conditions. Lack of ventilation didn't help any. In order for the considerable amount of smoke to escape, holes were cut in the roof. An enormous roast would be slipped onto a metal "spit" which was supported by hooks and this would be constantly turned for hours by a servant. After awhile, both the meat and the servant would be covered with smoke and roasted.

In the wealthier homes, the chickens were parted from their feathers in a separate room called a scullery and much of the food preparation was done there. It is interesting to note that we "regress" in our cooking habits by cooking on spits in sophisticated electric ovens and gas grills or broiling meats over a hibachi, which by any name is a brazier.

The Renaissance period found the kitchen demoted to the cellar or lower region of the house. These areas in a palace or mansion were cold, damp, dark and enormous. The food was brought up by back stairways or, later, sent up by small lifts. During this period the kitchen could just barely be considered a part of the house.

This is quite different from the home of today. It is impossible to imagine a house without a kitchen. The preparation of food has become only one of its many functions. It is a gathering place for the family and most likely the place where the family eats. The kitchen is often a laundry center. It can also serve as an "office" or business center for family financial planning. The kitchen can be a marvel of versatility, constituting the major part of the work area in the average house. As such, it deserves the greatest attention in planning.

Location

The location of the kitchen in the house should be planned to fit into modern family life. Whether it should be in the front or back of the house should be determined by style, site, view, exposure or other variables. There should be direct access to the dining area and to the front or guest entrance. Access to an outdoor eating area, patio or porch should be from the kitchen and/or dining room. If there are children in the family the kitchen should be located adjacent to or have a good view of the play area. It is also desirable that the kitchen entrance be close to the garage, breezeway or carport.

If there is a family room, the kitchen should be open to it wherever possible. Like the play area the family room should be within the view of the mother for safety's sake. Furthermore, recent psychological studies of children's play habits have proven that small children want and need to have the mother within their view. Space for extras such as storage, utility cabinets, space for eating, laundry facilities, a serving counter or a planning area with a desk, should be incorporated whenever possible.

What a kitchen should not be is a main thoroughfare for the rest of the house. Any traffic through the house should bypass this important work area.

Size and Layout

The size of the kitchen depends on factors such as the space available, the number of people in the family, the kind of equipment desired and what activities other than those directly associated with food preparation will be carried on there.

FHA-MPS standards for houses with a floor area of less than 1,000 square feet are a small kitchen of 8 by 10 feet to 10 by 10 feet. The standards for houses with 1,000 to 1,400 square feet are 10 by 10 feet to 10 by 12 feet. Liberal standards for houses over 1,400 square feet in area are 10 by 12 feet and over.

Ten percent or more of the cost of a new home is spent on the kitchen. After years of careful study and research (by an astronomical number of independent groups with a variety of motives), some basic requirements emerged: adequate storage, appliance space, counters and activity space all carefully arranged with maximum efficiency in mind.

When kitchen layout finally caught up to the progress made in kitchen equipment, the term "triangle" emerged. It has yet to be improved upon. The triangle is applicable to each of the five basic kitchen shapes: the L-shape, the U-shape, the corridor (gallery) shape, the broken U-shape and the straight line shape. Other good kitchen shapes are possible and, as long as there is sufficient wall space for appliances and the necessary cabinet and counter space, the all-important work center relationships can then be accomplished.

There are three essential work areas of use and activity. These "centers" are the refrigerator area, sink-wash-preparation area and

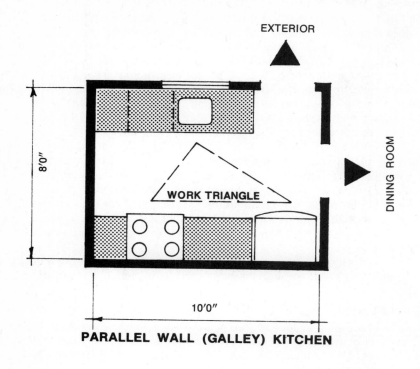

PARALLEL WALL (GALLEY) KITCHEN

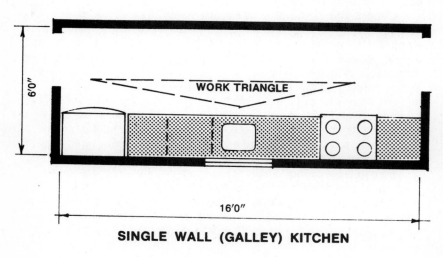

SINGLE WALL (GALLEY) KITCHEN

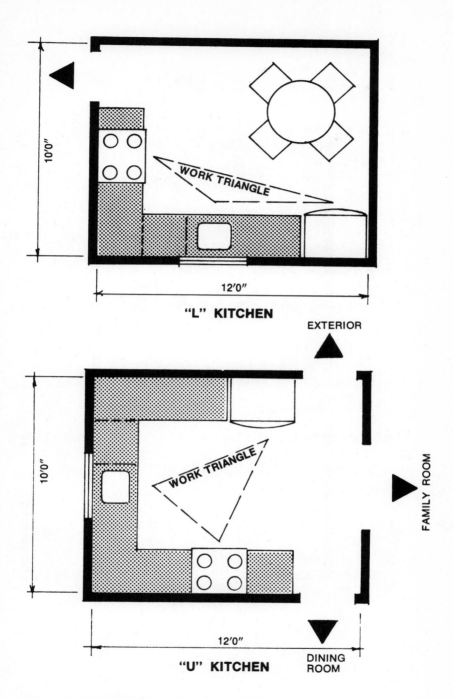

"L" KITCHEN

"U" KITCHEN

The total of the three sides of the Work Triangle
should not exceed 22 feet.

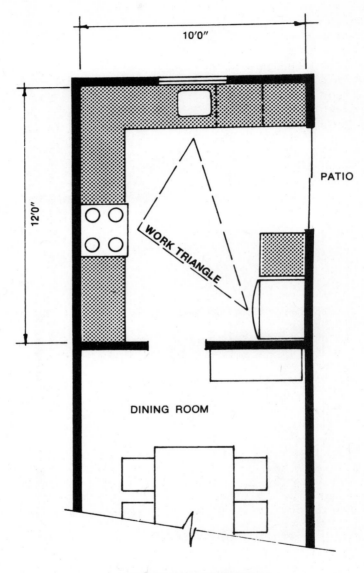

10'0"

12'0"

PATIO

WORK TRIANGLE

DINING ROOM

BROKEN "U" KITCHEN

The total of the three sides of the Work Triangle
should not exceed 22 feet.

range-serve area. The centers can be arranged in any logical way, determined by the space available and the personal preference for one particular center over another. A fourth separate center is often added, for "planning." It usually includes a desk for menus, cookbooks, stationery, a family activities calendar and the telephone. Generally this area is located outside the triangle.

However the kitchen is arranged, work should flow in a normal sequence from one center to another. Ideally, no traffic should move through the work center. Properly establishing the location of the windows and doors and the traffic pattern ensures efficient use of the centers. The counters of two or more centers can be combined, but dual-use counters should be wider wherever possible.

A kitchen comes alive when it is more than just four walls and a model of efficiency. It should be bright and well ventilated, warm and inviting in winter, cool and beckoning in summer.

The amount of wall space available for cabinets in the kitchen is affected by the placement of the windows. Most building standards require that the glass area should equal at least 10 percent of the floor area of the room; 15 to 20 percent is preferable. At least one section of a work counter should have a window over it with provision for controlling direct sunlight. Many women prefer to have a window located over the sink. (For reasons of both safety and good housekeeping the range should never be located under a window.) Good design brings the window down as low as possible to the sink to form a neat panel of light.

If the kitchen is part of a dining unit, the windows need not be in the kitchen work area but should equal at least 10 percent of the total area. For the best ventilation, windows should be placed away from exterior corners. The window-over-the-sink concept is still the best plan, aesthetically and otherwise. The average kitchen ventilation is achieved with one or possibly two windows and the "back" door.

Artificial as well as natural ventilation should be provided for the kitchen. In cold or inclement weather, windows cannot remain open. FHA-MPS requires the installation of a kitchen ventilation system when a kitchen does not have the required natural ventilation: "Air shall be exhausted from kitchens directly to outdoor air by a range hood or by a ceiling or wall fan."

There is more flexibility for location of fan installation when the kitchen has the required 10 to 15 percent window area ventilation, but the fan above the range must not be more than 6 feet off the side of the center line of the range itself.

Range hoods are desirable and are now being installed in at least seven out of ten homes (possibly a conservative estimate). A fan which exhausts 300 cubic feet of air a minute changes the air in the room completely every four minutes. In one hour, the air is changed 15 times, the rate considered necessary for good ventilation.

If a kitchen has an air-conditioner, it should be remembered that the primary function of an air-conditioner is to cool. Many people confuse this with ventilation. Air-conditioning helps ventilation by removing moisture and by circulating the cooled air, but it does not actually ventilate.

Counters and Cabinets

Cabinet space should be planned with care and thought. It takes little extra effort to make the difference between good utilization of space or expensive waste. Some common and glaring errors found in a sampling of kitchens (mostly smaller homes) are insufficient base and wall cabinet storage, too little cabinet space, wasted wall space, no counter beside the refrigerator and insufficient space in front of the cabinets and appliances. All these can be avoided with good sense and effective utilization of space. Cabinet space needs depend on what and how much is to be stored. Food, utensils and dishes should be stored where they are first used.

The most common base cabinet is 36 inches high over a drawer and two fixed shelves. This is not, however, storage space at its best since it is difficult to reach far back. The ideal arrangement is to have drawers of the proper size for the items to be stored and with adjustable dividers. Corner cabinets are difficult to utilize unless they are equipped with lazy Susans (turn-around shelves). A recent refinement in cabinet height is to adapt it to suit the owner, who may be taller or shorter than average. This can be a resale hazard, however, and should be considered carefully.

Kitchen storage is adequate or not when judged by its "base cabinet frontage," the length of accessible cabinet space in total. Space under the sink, although useful, is not included in base-cabinet frontage since it does not have multiple shelves. Turn-around or lazy-Susan corner cabinets add extra storage space equal to an additional 6 inches in base-cabinet length. General recommended standards for base cabinet frontage area are: minimum, 6 feet; medium, 8 feet and liberal, 10 feet. The standard counter width is 25 inches.

Wall cabinets should be installed 15 inches above the counter top of the base cabinet. This provides sufficient room for blenders, toasters, mixers and all the other luxury items-turned-necessities which the modern homemaker enjoys.

Wall cabinets over the refrigerator or oven range surface (stove or built-in) are not easily accessible and should not be added when frontage measurements are taken. Naturally, one would not store everyday dishes in these. It bears repetition that cabinets are not desirable over the sink. Cabinets over the range should be separated from the surface by a hood (and fan) or some type of insulation. The FHA-MPS requirements are for installation at least 24 inches above the surface. Anything closer than that may be permitted, except over ranges, if it does not interfere with the use of the counter top. Wall cabinets should be located above the counter for each "center," as discussed above.

Quite possibly there will never be a family which has "enough" cabinet or counter space. Fortunately, the new concepts have taken kitchen planning a giant step toward narrowing down the number of people who feel that their kitchens are merely adequate.

Various research studies have indicated that (in small houses, particularly) two-thirds of the kitchens have too little counter space, which also indicates too little base cabinet space. More than half of them had insufficient counter space beside the refrigerator and range and little or no counter space on either side of the sink.

Frequently, counters are used for more than one function and, in combination, save space. Whenever two (or more) counters are combined, this counter area should be equal in length to the longest counter plus one foot. Whatever the combination, it should not reduce the recommended base-cabinet frontage. Lest the homemaker has grand illusions of infinite cabinet and counter space, it is wise to remember that any exceedingly liberal dimensions are self-defeating and result in excessive walking and excessive work.

There should be 15 to 18 inches of counter top on the handle side of the refrigerator (this is the counter that is most often missing). There should be 2 to 3 feet of counter top on each side of the sink. On each side of the built-in range and oven there should be 1 to 2 feet of counter top. In addition, there should be another 3 to 4 feet for general food preparation, mixer, toaster and other small appliances.

A wide selection of kitchen counter top material is available (some of them are discussed in more detail later in the book). The most common ones in order of cost are:

Stainless steel
Ceramic tile set in mortar
Ceramic tile attached with adhesive to wood
Hardwood-laminated
Melamine-laminated (Formica) with molded edges
Melamine-laminated (Formica) with stainless steel or cut
   Melamine edge
Vinyl
Laminated polyester
Linoleum
Tempered hardboard

Stainless steel has excellent durability and is impervious to moisture (although it does stain, most stains can be removed with cleanser or alcohol). High heat from a pan will mark it but not damage it. Sharp knives will scratch it but not cut into it (many scratches can be removed with fine steel wool). It is not affected by sunlight. However, stainless steel makes a poor cutting surface as it dulls knives, will dent, is noisy, conducts electricity and may be a shock hazard.

Ceramic tile is hard and durable. It is not affected by heat or fire. Sharp knives will not cut it. It is very stain resistant and the color is not affected by sunlight, but it will crack when a heavy pan is dropped on it. Because ceramic tile is so hard, it will dull knives and break dishes and glasses. The grout between the tiles will stain.

Hardwood is usually used for only a portion of the counter top. It makes an excellent cutting surface and is difficult to damage. When it becomes dented, stained or burned it can be refinished. Hardwood must be cared for and kept dry or it will roughen and become difficult to clean. In addition, sunlight will change its color.

Laminated melamine, of which Formica is a popular brand, is now the most popular counter surface. Its surface is smooth, hard and durable, it is easy to clean and it resists stains and wears very well. However, it will scorch from cigarettes and hot pans, sharp knives will cut it and it will dent or crack when struck by a heavy object. Laminated polyester is less durable than melamine but has similar properties.

Vinyl makes a smooth, quiet surface. However, some foods stain it, sharp knives cut it and it is highly susceptible to heat damage. These characteristics limit its use as a kitchen counter top material.

Linoleum was, for many years prior to World War II, the most popular kitchen counter top material. It provides a smooth, quiet, resilient surface, but it will stain from some foods, detergents and bleach. Alkalies damage it and it will mildew if left constantly wet (as from a slow leak in the kitchen faucet or water under the dish drain board). Sunlight tends to fade it. Today its use is usually a sign of cost cutting.

Tempered hardboard is inferior to laminated hardboard and is also used as a cost cutting substitute.

Inadequate Kitchens

Many kitchens suffer from one or more of the following inadequacies (listed in order of most common occurrence):
1. Insufficient base cabinet storage space.
2. Insufficient wall cabinet storage.
3. Insufficient counter space.
4. No counter beside the refrigerator.
5. Not enough window area (at least 10 percent of floor area).
6. Poorly placed doors that waste wall space.
7. Traffic through work area.
8. Too little counter on either right or left side of sink.
9. No counter beside range.
10. Insufficient space in front of cabinets.
11. Distance between sink, range and refrigerator too great.
12. Range under window (unsafe).

Dining Room versus Dining Area

In the second half of the 18th century, the concept of the dining room as we know it evolved, particularly in Italy, Germany and France. It was not always properly allied with the kitchen; instead it became a separate room for dining.

In 1877 the *Domestic Encyclopedia,* published in New York, printed a list of no fewer than 200 items which it considered indispensable to a

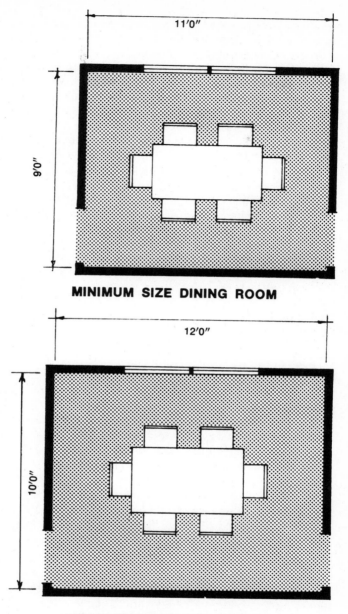

MINIMUM SIZE DINING ROOM

RECOMMENDED SIZE DINING ROOM

69

"modest" household. In the dining room it said, "the table should extend to 12 places: a coffee pot of Brittania metal, a steel carving knife and a brass bell are recommended."

The question in the 1970's is not what to put into the dining room but whether to have one at all. If space permits, it is a room desired by many homemakers. However, the trend is toward "space in which to dine," since the separate dining room is now considered an amenity and not really necessary. Now Grandma sets a table for 12 or more comfortably on Thanksgiving in space which the rest of the year serves as a living area for her sewing, bridge and social activities.

Acceptable alternatives to the dining room are dining areas, extra-large kitchens which provide space for family meals, sometimes the family room or living room and the patio in season. A common compromise is a dining "L" between the living room and kitchen. This is usually furnished with a drop leaf or other expandable type of table, several chairs and, if possible, one other piece of furniture such as a serving cart, cabinet, hutch or lowboy to provide storage space for linens, dishes and silverware.

The priority of requirements for the dining room or dining area are identical. Access from the kitchen service area to the dining table should be short and direct. Ten to 15 percent of the floor area should be given to windows in order to provide adequate light and ventilation.

Window placement is usually standardized. Often a window is placed on each of two walls which form one corner of the room. Another wall has the door or opening. One wall should be "free" to ensure that not all of the wall area is cut up, thereby precluding good furniture arrangement.

The size of the separate dining room or area obviously depends upon what circulation exists through the room, as well as its function for dining. The FHA-MPS minimum room sizes are set forth in two categories: separate rooms and combined rooms.

According to FHA, for separate rooms in a house with two bedrooms, the dining room must be 100 square feet; 110 square feet for three-bedroom houses and 120 square feet for four-bedroom houses.

When the living room and dining area are combined the required size in a two-bedroom house is 210 square feet, 230 square feet for a three-bedroom house and 250 square feet for four-bedroom houses.

When the living room, dining area and kitchen are combined, the required size in a two-bedroom house is 270 square feet, 300 square feet in a three-bedroom house and 330 square feet for four-bedroom houses.

Since ideal seating room is a clear 3½ feet all around the table and there is invariably other furniture in the dining room, a 10 by 12-foot room scarcely suffices. Assuming that there also is some traffic, a better minimum would be 12 by 12 feet. There are no minimum requirements for view. If there is a good one, so much the better.

Bedroom

One of the basic human needs of man everywhere and in every time is sleep — that blessed time which "knits up the ravel'd sleeve of care" while refreshing the physical machinery of the anatomy.

This need has been satisfied in many ways and styles throughout history, fluctuating from the comfortably simple to the uncomfortably stylish, from the practical to the profane. It is interesting that we still retain a little of each of these qualities today, but basically the 20th century bedroom has gradually evolved into a haven for rest and relaxation.

The caveman improvised with piles of leaves and straw or the more fortunate threw down some fur hides in a spot sheltered from the elements but probably near a fire. He simply made bed-room. The evolution of the separate room in which to sleep came slowly and with it, the idea that beds should be elevated from the floor. Although some Oriental cultures still rest on comfortable mats on the floor, they are in a bedroom. Other cultures, such as ours, have gone from 'caves to waves' with the latest innovation, the water bed.

There is considerable evidence that very early civilizations had ornate beds of choice woods, precious metals and even ivory. After 1000 A.D., man began to see the bed as something more than utilitarian and began to devise means of warmth and sometimes privacy, which in itself did not seem to be of prime consideration until later. When curtains were first placed around beds (alone or in an alcove) it was to provide warmth. The idea of privacy seemed to follow and became a fringe benefit.

The Middle Ages shows the greatest advance in the separate bed-room concept, with one or two exceptions. Cairo, with its Moslem "citadel arrangement," had separate "sleeping" quarters on upper floors for the harem but no bedrooms as such. Contrast to this the bedroom of a wealthy French home in 1550 which contained among other things: "chairs upholstered in morocco and petit point; two wooden bedsteads with any number of pillows, mattresses, quilts and blankets; a Turkish carpet; a framed oil painting; a walnut chessboard with its pieces; tapestries and, in the dressing room next door, a wardrobe and four chests" (from document of period).

In Germany at this time, there was the canopied bed. The hangings served to ward off drafts, since, it is thought, night clothing of that period was not warm. The hangings also provided some privacy from the personal servants who slept in the same room, as was the custom. From this point on, the history of the bedroom up to today is one of familiar documentation of social and economic growth — in a word, refinement.

Still, there remain vestiges of centuries past, when literature tells us of the great Bed of Ware in which:

"Four couples might cosily lie side by side,
And thus without touching each other abide."

71

Bob and Carol and Ted and Alice only serves to prove that there is really not much new under the sun.

Number

From the area with fur hides, to the alcove with draperies and finally to the comfort and privacy of today's bedroom, one thing constantly remains true: people sleep about one-third of their lives. We still scatter fur rugs on top of the wall-to-wall carpeting and we drape the windows and occasionally have bed alcoves or canopy beds. However, today we are also concerned with how the bedroom function is applicable to that two-thirds of our lives when we are awake.

A bedroom is no longer just a place in which to sleep or dress. It is a place where one enjoys reading or lazy contemplation, a place to unwind and relax during the waking hours. For children it harbors games, books and other treasures.

The number of people in the family usually determines the number of bedrooms in the house they buy. Other variables are the number of adults and children, what their particular requirements in a bedroom are and the inevitable budget. As construction costs have escalated, the size of the bedroom has diminished. Therefore, the most important consideration is still that of its main function: to accommodate beds for sleeping and to allow for their use and care.

The maximum number of persons to a bedroom should be two. Three bedrooms for a family are considered minimum, with two-bedroom houses becoming obsolete. Usually, the better the house the more bedrooms there are. There may also be a guest room but such a room, per se, is rare today and it usually doubles as a study or den.

Location

Location of bedrooms shares equal importance with size in good planning. The relationship of the bedrooms to the other areas of the house and to each other varies with the style of the house.

Whatever the plan, however, bedrooms should be reached directly from a hall, be removed from the living/social zone and working/service zone and be directly accessible to the bathroom without observation in ordinary circumstances.

When these basics in good planning are achieved and finances allow, more sophisticated considerations can be made. Such considerations would be of a more extraordinary nature such as locating the bedrooms in the "quietest" part of the house, completely withdrawn from the street noises. Some homeowners desire that the adult bedroom-bath area be segregated from children's noises. This is usually provided for when the proximity of parents to very young children is no longer a necessity.

Even something as clinical as the FHA-MPS has, as one of its objectives, privacy and "the interior arrangement of rooms, particularly
with reference to access to bathrooms from bedrooms."

Sunshine is desirable, psychologically and aesthetically, but there are so many variables as to how to make the best use of the sun that it is almost impossible to set guidelines for its best utilization. A good generalization is that the best position for the house is a location on the site so that as many rooms as possible face south. A site which has the street on the north side is ideal. Because the south is the only elevation which has sun at all times of the day and year, it allows some degree of control. Of course, the homeowner should not despair if he cannot orient the house and bedrooms to this. There are ingenious architectural methods of including and excluding the sun regardless of ideal orientation.

Size and Layout

In the last analysis, bedroom size is determined by the make-up of the family. The square footage will vary in each room, but the need for space for adequate movement and circulation is a minimum requirement no matter what the size. This is also true of clearances, general outlines of closet, window locations and entry. A bath in the master bedroom suite requires a wall interruption of its own, for example.

The minimum bedroom size allowable under FHA-MPS is 80 square feet or 8 by 10 feet. This applies to two, three or four-bedroom homes. Usually only one of the three bedrooms in the average house will be that size. Childrens' room dimensions of 13 by 14 feet or 11 feet, 4 inches by 17 feet have proven to be suitable for both boys and girls with a large number of playthings and possessions.

Some guidelines as to room size beyond the minimum 8 by 10 feet are: small, 10 by 11 feet (either size room can accommodate one bed); medium small, 10 by 12, providing minimal space for two beds (medium room dimensions are considered to be 10 feet, 2 inches by 14 feet or 11 feet, 8 inches by 12 feet which in either case is 140 square feet, providing adequate space for two beds); above average, anything over 170 square feet, such as a room 12 by 14 feet, 2 inches; large, 190 square feet, such as a room 12 by 15; very large, anything over a room area of 220 square feet, with dimensions such as 12 by 18 feet, 4 inches or 15 by 14 feet, 8 inches.

As the size of the room increases, problems of layout decrease. There is more uninterrupted wall space in which to place furniture and allow for windows, doors and in some cases, radiators. The wall area or usable room perimeter length has to accommodate the bed or beds and usually a chest of drawers, a dresser and a chair. Other furniture often includes one or two night tables, a dressing table and perhaps a desk and a chaise.

Doors should be placed so as to open back against the wall and should never interfere with other doors. They also should be placed so as not to interfere with beds and other furniture.

Window placement should take advantage of good, natural light and afford cross-ventilation, leaving enough wall perimeter for large furniture. It is desirable not to have electrical outlets located directly

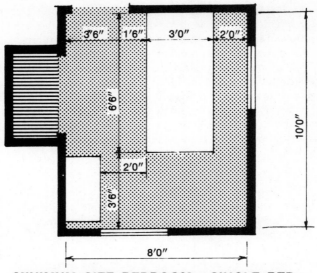

MINIMUM SIZE BEDROOM — SINGLE BED

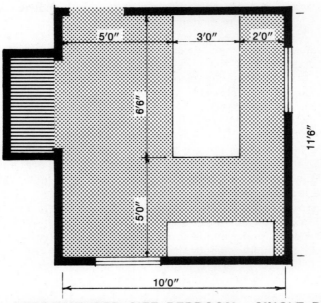

RECOMMENDED SIZE BEDROOM — SINGLE BED

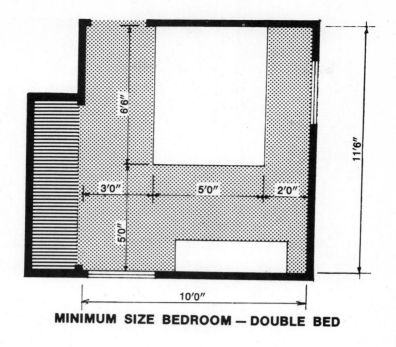

MINIMUM SIZE BEDROOM — DOUBLE BED

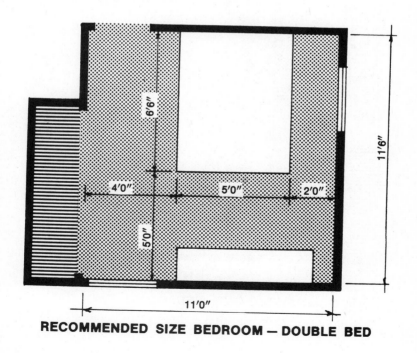

RECOMMENDED SIZE BEDROOM — DOUBLE BED

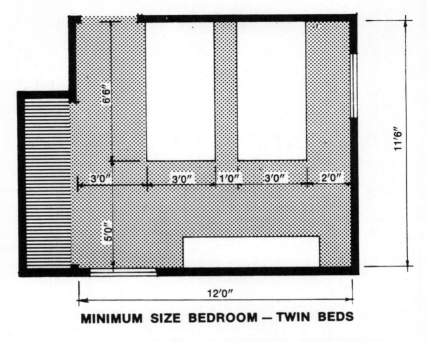

MINIMUM SIZE BEDROOM — TWIN BEDS

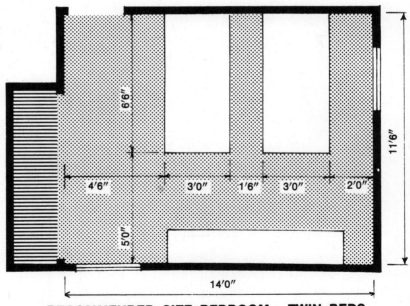

RECOMMENDED SIZE BEDROOM — TWIN BEDS

behind furniture except in the case of a night or dressing table for illumination there.

A single bed may be placed with one side along a wall since normally only one person occupies it and arises from it, but the double bed requires space on both sides in order to rise from it and make it. Space between twin beds is necessary for the same reasons. Twin beds pushed together to utilize one headboard are popular and in layout arrangements should be treated as a double bed or queen or king-size bed.

Additional space allowance must be added for making the bed and for cleaning under it. Therefore, specific clearance distances for use and circulation are as important in planning the layout as are furniture dimensions. Clearance space around the bed should be a minimum of 14 to 19 inches on the side and 2 feet at the foot. Space between twin beds should be from 18 inches to 2 feet. Twin beds require 9 feet, 8 inches to 10 feet wall space. If there is a night table on either side of a bed, the space between the bed and the wall should be from 18 inches to 2 feet.

At least 2½ feet of clearance should be provided in front of the dresser or chest for movement in pulling out drawers. Space for dressing between the bed or other furniture and a wall should be 3 to 4 feet. The space and floor area required for dressing should be convenient to the closets, dresser or chest and dressing table/chair and as near to the door of the room as possible. This allows for the most efficient use of the traffic area. Traffic routes ordinarily take up about 3 feet of space. It has been determined that the average adult needs an area of 42 inches in diameter for dressing. Some larger new homes accommodate this with the luxury of a dressing room in combination with the bedroom and bath.

Minimum code requirement of MPS for window area in relation to the bedroom area is 10 percent. In a 10 by 12-foot room this would be 120 square feet, with glass area of 12 square feet. Thus, a window 3 by 4 feet would satisfy the requirements but would be scarcely enough to provide adequate light under the best conditions. More windows are preferred, since they are often used as much in daylight hours as at night and thus should provide an atmosphere of comfort, quiet and illumination. This is best achieved with glass areas in excess of 20 percent of the floor area, if possible. At least one window should be placed not more than 3 feet from the floor in case of fire.

Windows should be placed to provide for the best use of light, sun and ventilation. They should be located to allow free wall space for the furniture and doors. The increase in popularity of bedroom air-conditioners might be a factor where window planning would be affected by installation of a cooling unit.

Ventilation must be planned in the treatment of the windows just as carefully as the other broad considerations of view, uniformity of light across the room and appearance. The ventilation area should exceed 10 percent of the bedroom area. MPS requires that windows should provide natural ventilation with a net area of not less than 4 percent of the floor area of the room or space. This barely provides what health authorities believe is adequate ventilation for sleeping purposes. Fortunately, this minimum is almost always exceeded in newer houses. Some of the more expensive homes today feature entire bedroom walls of glass which are designed to slide open with screen protection, providing excellent ventilation.

For best ventilation, windows should be placed away from corners on as long a diagonal as possible with one another when in adjoining walls. Cross-ventilation wherever possible is most desirable. This does not mean that the windows must be on two sides of the room. A door or correctly situated window in another part of the house can provide it.

The trend toward use of room air-conditioners and central air-conditioning is having a profound effect on the whole concept of ventilation. While air-conditioning does not provide fresh air, it cools and circulates air in order to produce the proper effect. Where cross-ventilation is achieved by mechanical means, often extra windows can be eliminated, thus producing more usable wall space. It is extremely doubtful, however, that the hum of the air-conditioner is the death knell for the window. Fresh air and light continue to be as much a part of the house as the foundation. They actually become substance when a house is built around them.

Once the type of heating system has been selected, the only other consideration as far as the bedroom goes is the location of the radiators or grilles. Care must be taken that beds or chests not be placed in front of any heating unit for proper convection and maximum efficiency.

Lighting

The decor of the bedroom is important to its function as a place of rest, be it a refuge for one hour or replenishment for eight. Proper lighting enhances the desired climate of relaxation. At the same time it must illuminate different areas more than others, the criterion being how much light is needed. There are three general sources of artificial light in a bedroom: the ceiling light, which should be of low intensity, brighter intensity lights for the dressing table or, if there is none, for the dresser-mirror and a light for the night table or tables (about 60 watts, if two are used simultaneously or 100 watts for one bedside lamp). A switch control at the room entrance should turn on one of these sources. Different intensities of light make the room more interesting and attractive. Glare and reflection are eliminated and the walls and draperies have a softer glow.

General Electric research recommends about 40 lumens per square foot for the bedroom. (A lumen is the measurement of the amount of light emitted by a source; i.e., various types of bulbs or fluorescent rods.) For comparison's sake, the kitchen should have 80 lumens, twice that of a bedroom.

In the more sophisticated and expensive houses, use of structural lighting, such as soffit lighting, luminous ceiling panels or lighting under valances is both attractively decorative and functional. Lighting can be controlled by dimmers to achieve the proper intensity desired at any given time. Diffused lighting, which is flattering to the room and its occupants, directs light to the ceiling or uses one of the many new types of diffusing materials with structural lighting.

Closets

If there is any single indication of how civilized we have become, it is evidenced in our total dedication to the shrine, the closet. If we were to judge its importance by the interest and concern shown in closet space by the average housewife, rooms would be closets and the closets would be rooms! It is the one thing that there is "never enough of," if universal complaints are any evidence. Certainly no group of women set FHA-MPS minimum closet requirements, which follow:

Each bedroom shall have at least one closet having a minimum:
1. Depth, 2 feet clear for required area.
2. Width, 3 feet clear;
3. Height, (a) minimum: adequate to permit 5 feet clear hanging space for at least the required width; (b) maximum: lower shelf shall not be over 74 inches above the floor of room.
4. One shelf and rod with at least 8 inches clear space over shelf.

Bedroom closets serve as clothing storage primarily but also store hobby or special articles, books, radios and built-in TV's. The children's bedrooms, ideally, should have clothes closets with movable hooks and shelves, a rod which can be elevated as children grow and clothing becomes larger plus storage for hobbies, games, books and equipment.

General bedroom storage includes space for bed linens, blankets, extra pillows and towels. More often than not there is a linen closet located in the hall near the bedroom which proves satisfactory. Occasionally there is room within the confines of the bedroom for built-in storage. If this is possible, the storage room volume of the drawers may be substituted for a maximum of 50 percent of shelf volume. The wardrobe closet and the walk-in closet are most commonly found in the average-size house. The newer and larger homes are incorporating both features into a dressing room area, obviously a luxury.

Good closet planning utilizes every square inch of space while allowing room for movement. The wardrobe closet is shallower than the walk-in and takes up more wall space, but it is not always possible to have the depth necessary for the walk-in. It must be calculated that the deeper the closet is, the more space is wasted because of the "traffic" movement.

The following statistics are a composite of studies on what makes for adequate, well-organized closet space for a medium-size bedroom:

Depth — for a closet having a rod running parallel with the opening, the inside dimension should be 24 inches.

Width — an arbitrary recommendation is 48 inches rod space per person for hanging clothes. This is calculated from the average rod space per garment of 2 inches for a woman's garment, 2½ inches for a man's and 4 inches for heavy clothing.

Rod Height — 64 inches for adult clothing; for robes and gowns, 72 inches. (In the children's bedroom closet, for those under 12 years of age, 48 inches.)

Shelves — at least one shelf 12 inches deep, for hats, purses, folded garments and other bulk-type apparel. Sometimes shoes are stored on this shelf. In other cases a shelf the width of the closet is installed 6 to 8 inches off the floor to hold shoes, as does the space underneath.

The distance between the top shelf and the clothes rod should be at least 2½ inches to allow for ease in raising and removing the hanger. If any kind of hooks are used against a wall (for hanging jackets, robes, etc.) they should be 6 to 12 inches apart. Built-in drawers or pull-out shelves add more usable closet space and are an asset to achieving an uncluttered look in the room.

If sliding doors are used, half of the closet can be exposed at once. Swinging or hinged doors (either singly or in a pair) expose the entire closet at once, but they open into the room and take up space. Care must be taken that they do not hit the furniture or another door.

Bedroom closets should have an electric light installed — preferably one that automatically turns off and on as the door is closed and opened. To avoid a fire hazard the light should be placed so that flammable material (i.e., plastic cleaner's bags) cannot come in contact with the bulb.

Bathrooms and Lavatories

History

The bathroom for washing is recorded in history as far back as 5,000 years ago in Eshnunna (a Sumerian city), as well as in early Egyptian history and in early Greece and Rome.

Washing, except for the hands and face, was unpopular in the Middle Ages and in the 1600's and 1700's many thought it was unhealthy. In the early 1800's the bath and bathing came back into fashion and most substantial houses of this period had bathrooms.

The toilet (non-plumbing varieties) had almost as long a history as the bath in some cities, but in general the necessary function was done outdoors, away from the house. Chamber pots and outdoor latrines were the American answer to the problem until the late 1800's when a version of the water closet that had first appeared in England almost 300 years before was invented by an English engineer, Thomas Crapper. Called "Crapper's Valveless Water Waste Preventer," it became popular and the bathroom as we know it today came into existence.

The American bathroom continues to grow in size and splendor. It has become a major selling point in new homes and next to the kitchen is the most important interior influence in selling houses. Features that were considered major luxuries and found only in expensive custom homes are becoming commonplace. They include partitioned toilets, twin wash basins and sunken or square tubs.

Number and Location

There is no universally accepted formula to establish the number of bathrooms and lavatories to be considered adequate for any given house. Such things as regional customs, social status, ethnic background, size, shape and layout of the house as well as the individual family size and habits must all be taken into consideration.

In most areas the minimum acceptable standard for a three-bedroom house is one full bath with a tub and shower combination plus a lavatory. New houses are still being built with one bath, but it is likely that these will soon be obsolete. Many three-bedroom houses have a bathroom off the master bedroom, another full bath off the hallway in the bedroom area of the house and a lavoratory near either the front or rear entrance. Another minimum standard is a bathroom or lavatory on each floor. If there is a maid's room or guest room there should be a separate bathroom in that area too. Custom-built luxury houses often have a full bathroom for each bedroom.

Interior bathrooms are growing in popularity as more people become accustomed to them from using them in hotels where they are the rule rather than the exception. Interior bathrooms have many advantages. Ventilation must be mechanical and therefore does not depend upon an open window which is drafty and during many types of weather must be kept closed. Cold and dirt will not come in through windows and no curtains, shades or blinds are needed. Additional wall space is available for cabinets, mirrors and towel bars.

A bathroom located between two bedrooms makes for nothing but trouble when there is a door directly into the bathroom from each bedroom. Cracks around the doors let light and noise into the bedroom when the bathroom is being used by the adjoining room's occupant. In addition, it is easy to forget to unlock the adjoining door when leaving, thereby locking the other user out.

Entrances to bathrooms should be private. It should be possible to get from each bedroom to a bathroom without being seen from another area of the house, especially living areas where guests are entertained. The soundproofing of the bathroom can be accomplished in several ways. If possible, walls between bathrooms and bedrooms should contain closets which make excellent sound barriers. If this is not possible, the wall may be soundproofed with an extra layer of sheet rock, staggered studs or insulation bats nailed between the studs — or ideally, all three of these methods.

Size and Layout

The minimum size for a bathroom containing a 5-foot tub and shower combination, basin and toilet is 5 by 7 feet or 35 square feet. This size

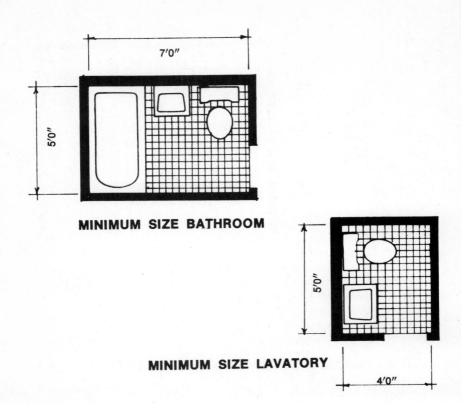

MINIMUM SIZE BATHROOM

7'0"

5'0"

MINIMUM SIZE LAVATORY

5'0"

4'0"

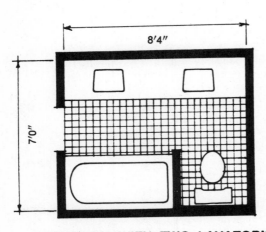

BATHROOM WITH TWO LAVATORIES

8'4"

7'0"

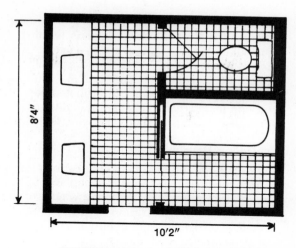

BATHROOM — COMPARTMENTED

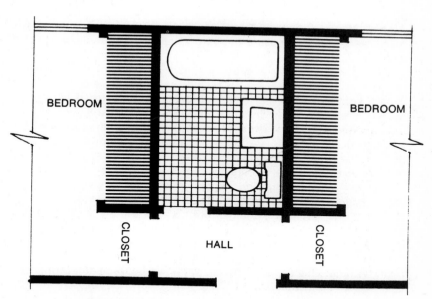

BATHROOM WITH CLOSETS AS SOUND BARRIERS

83

will allow the toilet to be on the wall opposite the tub rather than between the tub and basin, which is unsatisfactory. It will also allow the door to swing in without hitting a fixture. A 6 by 8-foot or 48 square-foot bathroom is much better and even a few more square feet improves the room substantially.

By increasing the size still further a separate toilet compartment can be created, two sinks can be built into the vanity counter and a separate dressing room or a dressing table included.

Ventilation, Heating and Windows

The bathroom requires the most heat and the best ventilation of any room in the house. It is equally acceptable to build a bathroom or lavatory in a central location without a window or on an outside wall with one or more windows.

Ventilation of an interior bathroom or lavatory must be accomplished with a mechanical ventilation fan ducted to the outside. This fan should be wired to the light switch so it automatically goes on when the room is in use and turns off automatically when the lights are turned out. The fan should be of sufficient size to change the air at least 12 times an hour or better, 15 to 20 times. Ideally there should be an intake vent too, but a satisfactory intake can be accomplished by cutting an inch off the bottom of the door. Since this is so much cheaper, it is the method commonly used. There are combination electrical fixtures being marketed that include the ventilation fan, a radiant heater and a light.

Natural ventilation and light are supplied from windows when the lavatory or bathroom is on an outside wall. Because windows are usually kept closed during cold and inclement weather, it is good to have a ventilation fan in an outside wall location as well.

The window location is very important. When over the tub, windows produce drafts and radiate cold which feel uncomfortable to a wet body in the process of taking a shower or bath. These windows are difficult to cover since the steam from the bath or shower wets the blinds, shades and curtains. Windows directly behind toilets also produce uncomfortable cold drafts and interfere with privacy. The best light and ventilation comes from a single window away from the tub and toilet that is air tight, opens from both the top and bottom to provide ample ventilation and has a good screen.

Keeping the bathroom or lavatory sufficiently warm may require auxiliary heat. This can be provided with an electric heater or infrared light bulb in the ceiling.

Accessories and Storage

Besides a toilet, bathtub and shower and basin, every bathroom also must have a medicine cabinet with a mirror, towel racks, toilet paper holder, soap holder, glass and toothbrush holder, linen storage space, dirty laundry storage and hooks for clothes.

Ceramic tile fixtures on a ceramic wall are excellent. They should be installed so as to withstand 300 pounds of pressure. The more towel bars the better. At least one for each person using the bathroom plus one extra is preferable. There should be a soap holder at the tub and sink and in the shower. A shelf on brackets is also a handy extra. The toilet paper holder should be on the wall at the side of the toilet and not behind it.

The door should have hardware with an emergency device on the outside to permit entry if a child locks himself in or an adult has an accident. An important safety feature is a firmly anchored grab bar in the shower stall and tub enclosure to help prevent falls. Another safety feature is a lockable medicine cabinet to keep children out. Medicine cabinets should be at least 20 by 30 inches.

A linen storage closet may be in the bathroom or in the hall directly outside. A hamper for dirty laundry can either be built-in or freestanding, if there is room. There must also be ample storage for medicine and beauty aids. This can be on shelves in the vanity or medicine cabinet or on bracketed shelves. If space is lacking, these items will end up on the window sill, toilet tank top, bathtub edge and vanity counter top.

A nice luxury is a full length mirror. Two installed on opposite walls permit visual inspection of the entire body.

A discussion of the various types of counter surfaces that are available is found in the section of this chapter about kitchens.

Family or Recreation Room

History

The family room as we have come to know it in 20th Century America is almost totally peculiar to our culture. Furthermore, its conception took place little more than 30 years ago. Its progress and popularity are well documented since World War II. Before that there are only bits and pieces of information which hint at how this addition to our life-style evolved.

However farfetched this idea may appear, some students of architecture feel that the concept is traceable to medieval days when a feudal castle had a primitive tower for its fortification, referred to as a "keep." The ground floor was used as a store for food and supplies and usually had no windows. Near the walls of the castle were the crude homes of the dependents of the feudal lord. Whenever there was a threat to these peasants they would take refuge inside the castle walls and would receive food and supplies from the keep.

During the Renaissance period, the tower disappeared, but the courtyard was retained and took on a great deal of splendor in the wealthier houses. The life of the home centered here and again it was a place for gathering and being supplied with food and drink. The French courtyard was also the heart of what might be either an elaborate or a simple house. In the more affluent ones there was a second courtyard within the servants' quarters which also accommodated the stables and storerooms and connected to the other yard for servicing.

Although the tower was gone, the term "keeping room" remained and spread. Its final distillation was the large room with an enormous fireplace where the "stores" were kept and the family gathered for food and warmth — "keeping" warm may also have contributed to this designation. The Pilgrims had their keeping rooms, not so much as a nice idea but as a necessity.

The heart of the home has changed little throughout the centuries except for refinement. The old country kitchen was still the center of activity while the food was being prepared. While Mother cooked, other family members might be reading, doing school work, playing games or working on needlepoint. It was not such a giant step to the family room of today.

Many things converged at once to stimulate the mushrooming popularity of a special room where games, recreation or hobbies could be pursued. The returning war veterans were marrying and quickly establishing families, larger families than had been the average for many years. They needed room. Then, too, there was the trend toward houses all on one floor (the "ranch" and houses without basements). Many hobbies which had been carried on in basements had to be pursued in a room on the main level.

Homes with basements were partitioned off and refinished into attractive family centers for the ever-increasing popularity of games, new hobbies and the trend toward more home entertaining. It was then only a step up to move this center to a room of its own on the same level as the living quarters and, at the same time, expand its functions.

The family room is alternately a den, a study, a sometime guest room, perhaps a library or a hobby center. Whatever, it is an asset.

Location

Wherever is a question of space and finances. Many homeowners in older houses have stayed with the cellar area where there is the most available space. Others have converted attics and garages or added a room to the house. It is difficult to say what is the ideal location for the family room in circumstances like those because, obviously, the best place for it is where space is available.

It should be noted that the FHA-MPS has some minimum standards here. "Finished rooms in basements or below grade . . . are considered habitable rooms and shall comply with building planning standards in the same manner as rooms above grade. When acceptable to the FHA field office, 'recreation rooms' in basements, auxiliary to the living unit, may be permitted with reduced ceiling height, light and ventilation and outside grade standards."

For the new homeowner the options are quite different and there are now criteria for the best location as the family room relates to the rest of the house. Because the family room carries a minimum of through-traffic, that does not have to be a prime consideration. Most planners and builders feel that the family room should be in the kitchen area, away from the sleeping area and separate from the living-dining area. This is especially desirable if there are teen-aged family members.

With small or pre-school children it might better be situated near the children's bedroom yet linked to the kitchen for better supervision by the mother and better vision of mother by the children. On the other hand, small children don't stay small for too many years, something which should affect plans for the permanent location of the family room.

How much entertaining the adult members of the family will do in this room is a factor in location. If it is frequent, the family room would best be near the kitchen. It is interesting to note that usually the larger the family room, the smaller the living room. One is informal, the other formal in use and furnishings.

Under ordinary circumstances it is more practical to have the family room located in the rear of the house. This allows for more privacy and places it in the patio or outdoor entertainment area while at the same time situating it near the kitchen. This is also a good arrangement if a sink or wet bar is to be installed in the family room.

Size and Layout

A good size for a small house is approximately 12 by 16 feet to 13 by 18 feet but, whatever the room area, the smallest wall dimension should be no shorter than 10½ linear feet. There is no maximum size and a larger home could conceivably have a family room of 16 by 26 feet, 18 by 30 feet or more.

Recreation or family room layout depends solely upon the desires of the family. Except the usual windows and doors, there is nothing used as standard procedure. Many homeowners plan for a fireplace first and fit the needs of the family's pursuits around it. If there is to be much reading or studying done there, bookshelves may be built alongside it or in another place. If the room is to serve as an occasional guest room, provision should be made for a studio or convertible type divan, figuring in plenty of space for when it is open as well as closed. There should also be some type of chest for clothing, either freestanding or built-in. It need not be large and can be utilized daily, guests or not.

For the music buff, an entire area might be devoted to a stereo set, its components, records and other equipment. Built-in shelves and cabinets are helpful because they allow for better furniture arrangement. A TV set is very nearly always found here. Other things which might be planned into the layout are a hobby center or a sewing area. A poker or pool table might be still another consideration. In short, the layout takes on the "character" of the family.

A clothes closet should be included in the overall plan for the family room. The dimensions of the average size bedroom closet are adequate as a guide. If a larger closet is possible, some of it should be for clothes and some for shelves or built-ins. Many people prefer to have the TV set encased in the closet. Or, in lieu of shelves, the space could hold the card table and folding chairs for the bridge-playing family. Small power tools and equipment can be stored here also, leaving the work center with an uncluttered look and free from danger to would-be handymen.

No attempt will be made to recommend cabinet size or counter space. The cabinet or unit is best built to suit its use. Measurements and guides are available at any reputable hardware or lumber store to aid in the tailor-made family room. Stamp albums, music albums, photograph albums, cassettes, paint brushes and canvasses, clip-boards, file cabinets and paper — each has a particular dimension and can be neatly housed with a little forethought.

Lighting

Good lighting is always a minimum requirement for health, safety and efficient use. Each area should have its own lamp — one or two for close intensity work. A cove light over work or game areas produces good quality lighting. Thoughtful lighting can be the "frosting on the cake" in the warm, friendly, truly recreation-oriented room.

Patios

History

The word "patio" is derived from the Latin "patere" — to lie open. It is defined by Webster as: (1) a courtyard or inner area open to the sky, common in Spanish and Spanish-American architecture; (2) a terrace.

In America today, where we have become geared to leisure time-long weekend-holiday entertaining, the patio has become an integral part of the house although it is actually outside the walls. A majority of our houses have one, be it an elaborate terrace, a side porch, a slab of concrete in the back yard or 100 square feet of grass with a grill. It is affectionately referred to as the patio and we cherish and care for it as though we had given birth to the idea.

Not so. Although the patio of today bears little resemblance to the original, with the exception of its outdoor location, it is not new. History records one patio-type area as beginning in Iran during the 11th century. In upper class houses, there was a courtyard (or some-times several), with rooms opening off it. The Spanish adopted this concept and improved upon it by giving it more importance, usefulness and attractiveness, the idea probably having been carried to Spain by the Moslems. By the 15th century the house and its galleries and porticos surrounding the square central patio had become both very ornate and very popular. The Alhambra in Granada is a classic ex-ample of this structural architecture.

With the migration of the Spaniards to Central and South America came the patio. Here, as early as the 16th century, houses were built with two arcaded patios. In the poorer colonies, of course, the house had none. In the West Indies, the patio was the heart of the house, with galleries, arcades and screens. The tropical sun is bright and hot and although air-conditioning has long since elimated the use of screens, they are often still kept for decorative purposes.

Gradually the idea of the patio crept into our architecture and culture and perhaps it does bear more than a little resemblance to the first by being the warm weather center for family-oriented activities of a festive nature. The 20th Century American patio comes in all shapes and sizes and all but gift wrapped, so much do we revere it. It is an area which the homeowner usually "designs" for himself to fit his needs, whims and pleasure. Anything goes.

Location

There are certain basics which are recommended for patio location. The south side of the house usually makes an excellent location. The patio should also be removed from the noise of the street by being in the back of the house. This ensures the necessary privacy. The most important consideration in locating the patio is that there be direct passage from the service area to the patio for serving. The fewest steps from the kitchen for the transporting of food, drink, dishes and other cookout supplies should be the top priority. If this area can be "tied" to the house by extending the flooring material of the terrace to the interior of the house for an indoor-outdoor effect, so much the better. Slate flooring, for example, is often used. Glass doors are ordinarily installed to achieve the proper effect and give good access.

It is also desirable that there be a walk or some kind of passage from the front of the house to the patio so that the guests can carry their offerings of potato salad and watermelon around without going in the front door and through the house. The food may then be transferred directly into the kitchen or service area until served. Serving food outdoors is pleasurable and cookouts one of America's delights, but it takes work. When things are carried out they must also be carried back in, so patio placement should be considered carefully.

Size and Layout

Some patios have clearly defined areas for different functions: the cooking area, the eating area and the relaxing or sunbathing area. There is no hard and fast rule for arrangement. Sometimes it is dictated by trees or the landscape, the view or simply by the number of people in the family. An adjacent swimming pool can influence the set-up. The arrangement is purely individual and the possibilities for making the area a haven are endless.

Since it is an outdoor area, the elements will affect the use of the patio. The rising and setting of the sun cannot be controlled, but we can use the sun's rays where we wish to and control them where we do not. Sunshine is welcome in the sunbathing area and undesirable at the picnic table. Usually shade trees are sufficient for control, but manufactured means such as trellises, ornamental "roofs" or awnings are both attractive and functional. The awnings can be permanent, such as the new plastic ones, or the familiar temporary canvas ones which also are practical and colorful. Aside from being assets, the primary function of any of these is to ensure that anyone using the patio can always be able to move in and out of the sun as he pleases.

Occasionally wind control is necessary in certain areas of the country. The erection of screens such as dense plantings, certain types of fencing or a combination of both usually eradicates the problem adequately; where practical, glass may also be used.

The kind of material selected for the "floor" of the patio most frequently combines durability with economy for the average home-owner. Concrete fits these specifications best. Patterns can be created with it; it can be colored or not, yet it provides an even, inexpensive, long-lasting surface for the least amount of money. There are variations on the use of concrete such as adding pebbles or other decorative materials to the surface but these add to the cost slightly.

Brick makes an attractive patio flooring but takes more time and money. Used brick is slightly cheaper and very effective. Brick flooring can also be carried into the house area as discussed above. Slate and flagstones are among the more popular surfaces used. They come in a variety of shapes, sizes and colors and can be installed by the homeowner.

One of the newer types of surface is the round wooden block. The blocks are literally sliced from the tree trunk just as one would slice up a carrot. They are treated to resist rot, moisture and insects and are very attractive when set into the ground. Some kind of "fill" is necessary to go with them because of their circular shape.

Where the architecture of the house or the slope of the land is a factor, a popular patio is the wooden deck. Quite frequently this is an extension of the living room although it may also be an integral part of the family room or kitchen, since it goes across the entire side of the house. Again we frequently find glass doors a companion feature.

On occasion there is a need or desire for a raised wooden patio deck at ground level. Interesting patterns can be formed, either in sections or as a whole. This deck must be set on top of some type of footing and must have some space between the boards for drainage. Otherwise it would be impractical.

The more expensive types of patio flooring are ceramic tiles (usually near the pool), terrazzo, slate, unglazed quarry tile and a number of new types of composition materials. All these require expert if not professional installation but, where finances allow, repay the owner in beauty and "mileage."

Ingenuity has provided more surfaces for patios than one could list. They run the gamut from old railroad ties to the large discarded spools from the telephone company.

Occasionally one will find a homeowner who wishes to have his patio entirely screened. There, however, are other methods of insect

control besides this one, such as special electric light bulbs which repel insects and a particular blue type of light which attracts them "to
death." Careful placement of either of these two types of lighting can
achieve the desired effect of an insect-free area, for evening entertaining in particular. It is advisable to have them since much of our outdoor
living takes place in warm weather and in the evening, both of which
beckon the uninvited bug.

The patio area should be properly wired for both pleasure and the
aforementioned insect control via electric light blubs. Electric outlets
help provide far wider and more versatile use and ultimately, successful entertaining. Local building codes should be carefully followed
regarding the installation of electric outlets for the patio and pool area.
Many cooking appliances, such as the electric barbecue spit, are
popular. A popcorn popper, deep fryer or stereo all enhance a party
when the electric outlet is situated near the cooking area. "Mood"
lighting becomes possible with the imaginative use of different colored
bulbs to achieve softening effects. Skill in their arrangement can set
a charming scene.

A more practical aspect is that of easing the care and maintenance
of the patio area. Electrical outlets can be used for the hedge trimmer,
garden edger and other electric tools. Good lighting is necessary for
walks and garden steps and filtered floodlights can achieve a beautiful
environment for safety and many hours of enjoyment.

Porches

The porch has been on the architectural scene for centuries. In the
equatorial regions it appeared on the larger buildings above and
around the entrance and was supported by ornamental brackets. In
ancient Greece, even a simple hut would have a porch added to the
front. In the Orient, the porch was a natural because of the projection
of the roof so commonly found in the architecture of the houses and
buildings. The porch was then under the overhang rather than being
an addition.

The Renaissance period brought the wide use of porches, particularly by the French and Italians. These porches did not, however,
always enhance the architecture. It remained for the English to refine
the porch into a structure for the protection of the front door. From
there it was but a short step to its obvious place at the back door.
The porch really came into its own in the United States during the
19th century. Then, porches were grand, long, wide and popular. There
had been porches in colonial architecture, of course, but the emphasis
was elsewhere.

Millions of older Americans living today would have been appalled
and unbelieving if, when they were children, someone had told them
that porches would all but disappear from houses by mid-century. But
it happened, for several reasons, and the front porch has approached
obsolescence.

The great housing boom, the population explosion and the mush-rooming development of the ranch-split level community conspired to eradicate the porch from the drawing board. As if that weren't enough, the trend toward outdoor living and entertaining was growing. The disappearance of the porch was in direct ratio to the appearance of the patio. The small, token porch, usually formed from the deep overhang above the front door and the one or two-step concrete slab just below it, is about all that remains.

There are exceptions, of course. In the larger or more expensive homes the "side porch" (covered or not) off a living, dining or family room is not unusual, nor is the sun porch. Often these have to double as patios where lot size and other considerations make it necessary or just desirable. In Hawaii, the lanai, a covered patio (or an open porch), continues to be popular.

Many young people who are seeking the most house space for the least amount of money are buying older houses and remodeling them. When finances permit, the first thing to go usually is the front porch. Again, there are exceptions and occasionally a young family will dis-cover the joys of sitting, rocking and watching the neighbors or cars go by. Another exception is the beach or summer house, particularly in the northern areas of the United States. Here, the screened-in porch is common and popular, frequently doubling as sleeping space.

The FHA has revised its minimum qualifications to include among others: "Unless otherwise acceptable to FHA field office, floor of . . . porch shall be at least 4 inches below floor of dwelling; least dimension of porch, 3½ feet; to be considered a protected entrance, least dimen-sion of porch, entrance hood or extended roof overhang, 3 feet."

It is unlikely that these dimensions will be extravagantly exceeded for some time. However, because styles in architecture, as in other art media, have a way of coming full circle every now and then, it is also unlikely that the porch will ever completely disappear.

Laundry

History

Running water has always been a prerequisite for washing clothes. To the American housewife, "running water" conjures up an image of the stainless steel sink or a sparkling porcelain washer and dryer standing side by side in all their glory. The source of water is at her fingertips.

Running water to the primitive woman and up to colonial days was just that — water running in a nearby stream or brook. It could be said facetiously that the source was at her fingertips too, but any resem-blance to today ends there. The garments of the era were brought to the stream, immersed in the cold water, pummeled, pounded and rubbed with coarse sand as a washing agent. Great improvement was made over this procedure when water was heated in large pots over the open fire and the mother discovered that this was a more efficient and pleasant way to keep the family fresh-smelling.

The next refinement was to bring the tub inside a "wash house" structure and place it on a low table near a small stove which heated the water. First the scrub board was used, followed by a crude hand-operated wringer. It is difficult for the modern woman (or man) to comprehend what the invention of lye soap must have meant unless, perhaps, it is compared to the corrugated scrub board's relation to the automatic washer.

From the "wash house" to the wash or laundry room was a relatively short step. The washing machine was often in the cellar with a floor drain improvised for the waste water. Indoor plumbing "elevated" the washer to the big kitchen where water could be heated and not have to be carried so far. It could also be disposed of at the sink. Today, because it is filled and emptied of water at precisely the right temperature and time, the automatic washer can be installed anywhere in a house where the plumbing is accessible.

Location

The average house has four general areas in which the laundry facilities may be installed: the kitchen or service area; the "first level" or utility room; the bedroom-bath area; an attached garage, enclosed porch, basement or other type of utility space.

The location is determined by space, practicality and preference, not always in that order. In new houses the kitchen is a commonly used location. Because the automatic washer and dryer are compact and attractive they fit in well with the decor of any kitchen. The plumbing is already there, as is cabinet space to house the necessary laundry supplies. Occasionally they have to be installed under a counter, but the front loading machines eliminate this difficulty. The counter top serves as a useful holding and folding area. Women with young children seem to prefer the laundry to be in or near the kitchen and convenient to the service entrance to save steps when supervising the children. Careful placement of the machinery should eliminate the necessity of traffic through the kitchen "triangle" work center. Machines should also be out of the way of the kitchen to dining and living room traffic. In the newest houses these appliances are also frequently found tucked into a closet type space behind folding, sliding or more commonly, shutter doors.

The attached garage can provide an excellent utility room situated between the garage and the kitchen and easily accessible to both. Many women prefer to have the laundry clutter completely removed from the kitchen, but they like the other advantages of the first level utility area. The attached garage arrangement gives easy access to the back yard play area and if there is no dryer, to the clothes line. The plumbing is already installed and usually a toilet and sink can be found here for added convenience. When the children come in from play they can shed their muddy clothes within inches of the washing machine. The mud room-laundry room is very advantageous to busy mothers.

Larger, more expensive houses often permit the luxury of a separate utility room off the kitchen, with its own lavatory, counters, cabinets, appliances, ironer, ironing board and perhaps a sewing machine.

The subject of placing the laundry facilities near the bedroom-bathroom area is controversial. Some feel that this is the ideal location since that is where the soiled clothing and linen emanate from and have to be returned to. A combination washer-dryer is particularly adaptable to the smaller area. The availability of the plumbing makes this a suitable location. Others feel the noise, dirt and odors emitted from the laundry require location away from this area.

Where space is limited, there are other possibilities for installation which can be considered. In general these would be in the attached garage, an enclosed porch, the basement or any other usable space which is available in any particular home. Even the space under the stairway has been utilized where plumbing makes it feasible. The attached garage is close to the kitchen area and provides an adequate solution except in very cold weather. An enclosed porch is slightly preferable as a general rule because it frequently receives some heat.

The basement is a good solution for laundry placement; in fact, many women choose it to be there. Space here is not in short supply as it is upstairs and soiled linen and laundry supplies do not interfere with the aesthetics of the home decor. Many new houses with basements have a walk-out entrance. In a multi-level house, a clothes chute from the bedroom-bath area to the basement is desirable.

Size

The size of the laundry is not as fundamentally important as are the other room sizes; efficiency and location are. The neat, compact washer and dryer can be housed within a very small space. Nevertheless, it is advantageous to have some counter space and shelves. Warmth and attractiveness cannot be measured but are very important.

Each homeowner will probably tailor his own area to individual specifications. Since washers and dryers come in standard heights, this and the attendant counter top height do not need to be considered. In general, the shelving should not exceed 24 inches deep nor be too high, so that there is easy access to supplies. Boxes of soap powder should always be stored off the floor to eliminate any absorption of dampness, particularly in a basement area.

It should be noted that there are some non-automatic type washers in use and these require a large tub or two. This necessitates a larger area in which to work. A space roughly 7½ by 8½ feet is adequate.

Proper ventilation in the laundry room or area is important both within the room and for the dryer. The dryer vent, be it for an electric or gas type dryer, removes moisture and humidity to the outside and is necessary and highly recommended. Counter surfaces and flooring should be of finishes that withstand high humidities and moisture from liquid supplies, which occasionally get dropped or spilled. Where the walls are covered or decorated these too should be moisture-resistant.

Most new houses come with the necessary water hook-ups for the automatic washer and provide for the elimination of waste water. The same applies to wiring. In most new houses the wiring provides for the necessary outlets and electric lights. Good artificial light is also a requirement. A well-lighted laundry room is attractive and makes the usual laundry routine not only more pleasant but also much safer.

Storage Areas

The days of storing everything in the basement or attic have long since gone. In some cases even the basement and the attic are gone. However, there are literally dozens of storage areas to be found in today's house. The basement and attic are still popular but do not seem to evoke a vision of mysterious trunks, old letters, dust and cobwebs any more.

Storage has become a complex problem and each family's needs are individual and should be carefully considered. Convenience, accessibility and organization are the objectives. "Everything in its place" is fine, but the place must be provided and planned for in order that the family live a well-ordered existence.

Indoor Storage

Still, the basement is one of the best and most popular storage areas because it provides a lot of leeway in arranging for storage of large articles. This is possible only if the basement is dry and well ventilated. Good lighting on the stairs and near the door to the outside is necessary. MPS requires that basement storage space have: artificial light provided; natural light and ventilation provided; when basement area is divided into separate spaces, required area for natural light and ventilation provided for each space; if heating equipment is located in enclosed space, combustion air provided.

Basement storage in general often houses skis, skates, archery, fishing, camping and other types of sporting equipment. Seasonal items like suitcases and trunks, Christmas and other holiday decorations, screens, storm doors and summer furniture are also found here. Occasionally there is a pool or ping pong table and several bikes. It is also a favorite spot for unused furniture, children's toys, frames, old clothes, family treasures and just junk.

Regardless of the bulk or variety, if things are properly stored to utilize every available space, there is that much more living space within the home. Wall shelving, units combined to form a wall, racks and cabinets are ideal.

Utilizing existing space in a house is always more economical and easier than building an addition. The attic makes an excellent storage area and where existing conditions permit, it can be attractively remodeled into a bedroom, a study or a hobby center in combination with storage. Lockers and cabinets fit ideally under the lower portions of the roof. Closets for items other than clothing need not be at regulation height in order to hold storage shelves.

It is cheaper to provide electricity and plumbing to the existing attic than to put it into an addition. Heating problems are minimal because of the warmth of the house. MPS states, "if mechanical equipment such as an attic furnace is installed in attic space, access opening shall be of sufficient size to permit removal and.replacement of equipment."

The FHA lists minimum guidelines for "Attics and Spaces Between Roofs and Top Floor Ceilings":

1. Provide cross ventilation for each separate space by ventilating openings protected against the entrance of rain or snow.
2. Ratio of total net free ventilating area to area of ceiling shall be not less than 1/150, except that ratio may be 1/300, provided:
    a. a vapor barrier having the proper transmission rate not exceeding one perm is installed on the warm side of the ceiling; or
    b. at least 50 percent of the required ventilating area is provided by ventilators located in the upper portion of the space to be ventilated, with the balance of the required ventilation provided by eave or cornice vents.
3. Attic space which is accessible and suitable for future habitable rooms or walled-off storage space shall have at least 50 percent of the required ventilating area located in the upper part of the ventilated space as near the high point of the roof as practicable and above the probable level of any future ceiling.

Stairways for accessibility and safety, says MPS, shall be provided by means of conveniently located scuttles and disappearing or permanently installed stairways with a minimum opening of 14 by 24 inches. Obviously much more space than this is desirable if the attic is to be utilized as a room for daily use rather than just storage.

Outdoor Storage

Outdoor storage sheds have been increasing in popularity at the same rate as we acquire sophisticated equipment for the care of the home and its environs. Houses without basements are another cause contributing to their growth. Proper care of expensive equipment such as yard tractors, power mowers, snow blowers, leaf compacters and garden tools ensures their long and efficient use. When other space is nonexistent it is often necessary to house valuable bikes and other sporting equipment in an outdoor shed. A handy man may construct one of wood fairly inexpensively or assemble a precut one which is slightly higher in initial cost and upkeep. Perhaps cheapest in the long run is the popular steel shed which is dotting the landscape in ever-increasing numbers. It is attractive, practical, inexpensive and requires little attention. It provides good space without taking up too much of the yard area. Besides the aforementioned equipment, it may be necessary to store patio furniture and game equipment like badminton or croquet sets, the inevitable grill or hibachi, tree sprayers and ladders. The list expands with each subtle change in our technology.

Garage Storage Area

The garage is a logical place to find extra storage space. The first garages were built solely to house the automobile and protect it from the elements. Today that is only one of its functions. The newest ones are attached directly to the house or with an open or enclosed breezeway.

Garage storage is particularly desirable where there is no basement to solve many of the storage problems. Many garages built in the last decade or two are usually longer and wider than is necessary for car space, allowing for adequate traffic and movement around the car. This also provides space for closets and shelves along the sides.

MPS recommends that any mechanical equipment installed in the garage be located so as to eliminate possibility of damage by vehicles. When laundry equipment is installed in the garage, adequate space must be provided for its use. Other than laundry equipment, the most frequently stored mechanical equipment in a garage is the freezer. The above FHA-MPS caution should dictate its placement. Locating the laundry equipment in the garage is not uncommon and is usually done when the house is a small one. Climate must be favorable, of course. The housewife would hardly enjoy having her wash or washer freeze halfway through a cycle. It is a good idea to enclose the equipment within doors such as the shutter type which fold back and use minimum space. The garage should be easily accessible to the service area of the house so laundry and supplies can be carried directly to and from the kitchen.

The garage makes excellent storage for unused equipment, in or out of season garden tools, storm windows, ladders, a work bench, fireplace wood, bicycles and baby carriages. When there are rafters or some type of a ceiling, overhead garage storage space can be utilized. If the man of the house is a sailor, this is an excellent area for storing the mast or portable hull. It is not at all uncommon to find an entire boat stored in the garage or carport alongside the family car.

Other Storage Areas

There is a variety of other types of storage space. These usually reflect individual preferences and are unlike the usual storage areas, putting them in the frill or luxury class. Personal specifications make any generalization impossible except for the basics of accessibility, efficient planning and good lighting.

The cedar closet, while admittedly a luxury, is fast being phased out, even where money allows. Synthetic materials and fabrics comprise about 75 percent of our wardrobes and furnishings today. Futhermore, the moth has been conquered outside the environs of a cedar closet — even wool has been fabricated so as to be completely moth-proof. Rugs may require storage space, but pretreating with moth repellents eliminates any special storage requirements. Except for those purists who like the odor and the nostalgia of a cedar closet, the best storage-protection against moths is the average bedroom closet.

A wine cellar is the ultimate in storage considerations and is fast growing in popularity as the wine rage sweeps the country. The wine cellar may range from a very small closet to a small room. MPS has no basic requirements listed in its code. The "cellar" may not be in the cellar at all, but wherever its location, the prime concern should be steady temperature. The connoisseur will, in all likelihood, establish his own perfect temperature with some type of cooling equipment. Proper lighting arrangements are necessary for reading the year and other fine print on the labels.

The old root or vegetable cellar below ground or cellar level has all but disappeared. Some houses built in the 19th century or before still have them, but they are rarely used for food storage any more.

Garages and Carports

The automobile has revolutionized life in America in the past 50 years. A majority of families owning houses also own two or more cars. Americans like their cars and they like to keep them under cover in garages and in carports. Their desire to do this varies from area to area. In many northern parts of the country any house without a two-car garage is substandard. In parts of the West and South the demand for garages and carports is more flexible and often a one-car carport with an additional parking area for the second or third car is acceptable.

The garage that is built to FHA-MPS standards serves only to shelter the car. These standards call for a one-car garage or carport to be 10 feet wide and 20 feet long, measured from the inside of the studs and door to the edge of the opposite wall, stair platform or any obstruction, whichever is the narrowest dimension.

To build a garage to these minimum standards is false economy. For only a small additional amount of money, the garage can be built about 3 feet longer and wider. It then becomes a truly functional area of the house. Here is the cheapest and most convenient place to store all of the paraphernalia of outdoor living. If built even bigger, it can house the laundry and drying equipment and also serve as a workshop and a place for the children to play on cold or wet days.

The choice between a carport or a garage is mainly influenced by local custom and climate. Some carports are so elaborate that they cost almost as much as a garage. Others are very simple and consist only of a roof extended from the house supported on a few columns or a simple wall. A carport is a good choice when it fits into the neighborhood and the prime purpose will be to shelter the automobile from the sun and rain. It is also desirable when the house is so located on the site that a garage would tend to block the breeze or appear to crowd the house.

A garage has the advantage of providing shelter from the cold, ice and snow in the northern climates. Even in the South and West, ga-

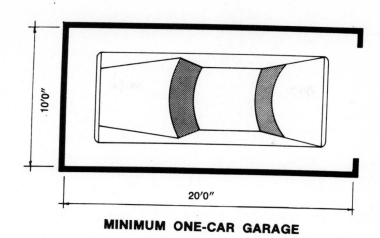

MINIMUM ONE-CAR GARAGE

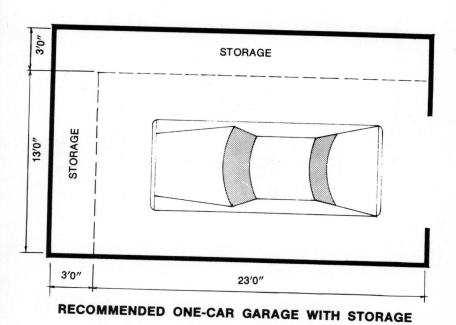

RECOMMENDED ONE-CAR GARAGE WITH STORAGE

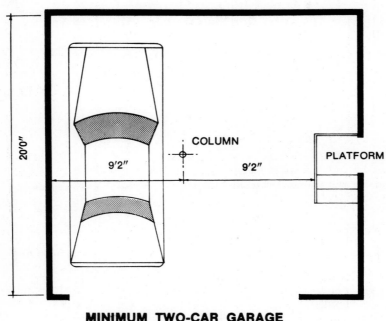

MINIMUM TWO-CAR GARAGE

20'0"

COLUMN

9'2"

9'2"

PLATFORM

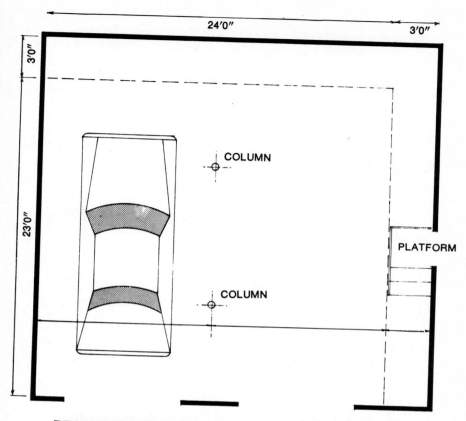

RECOMMENDED TWO-CAR GARAGE WITH STORAGE

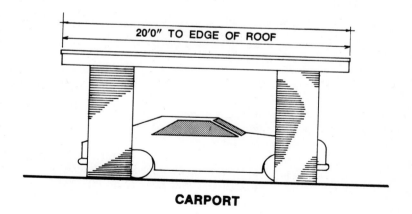

CARPORT

rages are selected by many who wish complete protection of their automobiles plus the other advantages of the garage.

There has been a definite trend away from the detached garage and towards an attached garage or carport or, if detached, connected to the house by a covered breezeway. Furthermore, as more and more activities center around the automobile, the entrance from the garage or carport has become the principal entrance to the house. Modern house design is taking this into consideration and making the entrance from the garage and carport into the house conveniently accessible to the various areas inside. An entrance that brings traffic through the work area of the kitchen or the living room is poorly located.

Often the site dictates where the attached garage or carport will be located. However, if there is a choice available it should not be on the south side blocking the sun; the west side is often a good location because the garage will protect the house and yard from winter winds.

Garage doors come in several types, the most popular being the overhead door which is raised and lowered on a track and is counterbalanced with a spring to make it easier to handle. The minimum acceptable width is 8 feet, but 9 feet is much better. On a double garage a 16-foot door is the minimum acceptable size. These double doors tend to be heavy and hard to open and close. Two single doors 8 or 9 feet wide are a much better arrangement. The minimum height for the door should be 6 feet, 4 inches. This will prove inadequate for a wagon with a rack of luggage or some recreation vehicles or trailers. A better height is 7 feet or more.

Automatic door openers and closers are very appealing. These devices open and close the door by power from an electric motor. They can be controlled with switches on the inside or outside of the garage or from a radio transmitter in the car. A small door from the garage to the back yard is a great convenience. At least one window is desirable for light and ventilation, especially if the garage is to be used for a work, play or laundry area. Some garages are built with some unfinished storage area under the roof eaves which can be reached via pull-down stairs or a ladder.

The wall between the house and an attached garage should have a one-hour fire resistance rating. This is required by many building codes.

In some of the northern parts of the country garages are heated. The system is usually designed to heat the garage to just above freezing rather than to 70 degrees. If the garage is insulated, a lot of heat is not required. Care must be taken to insulate all water pipes so they will not freeze. If a hot air system is used, no cold air return into the garage should be installed as this would provide an inlet for fumes into the rest of the heating system.

A good garage floor is a concrete slab. It should be one or two inches above the driveway level to prevent water from running into the garage. The floor should either be pitched to a central floor drain or to the driveway to allow the water from melting ice and snow to flow away.

The garage and carport should be lit and have electric outlets. The lights should be controlled from both inside the garage and inside the house. The garage light switch should be located at the front so it can be turned on without walking into the dark garage.

It is dangerous to have a driveway that enters the public street at a blind curve. A driveway that slopes upward to the street may also be dangerous, especially in the winter.

Although separate garages were once very common, it is now recognized as inconvenient to have a garage that is detached from the house with no connecting, sheltered accessway.

Chapter 4
Architectural Styles

It is not the purpose of this book to add wood to the fire of controversy that now rages about whether it is good or bad architecture to build houses using the traditional styles of the past.

It is only realistic to recognize that more than 50 million American homes reflect these traditional styles. They range from original Colonials and authentic Colonial reproductions and painstakingly restored Victorian villas to cheap tract speculation houses being constructed today that adopt one or more style features of the past.

Much confusion has existed as to the identification and origin of these styles. In this section an attempt has been made to illustrate many of the styles found throughout the country and to tell a little of their history and distinguishing characteristics.

The author realizes that the division among styles made here has been very arbitrary and that some of the styles are not nearly as simple to identify and describe as this book may seem to imply. However, keeping in mind the needs of the readers it was felt that an attempt should be made to provide basic knowledge on this subject. When a choice had to be made between absolute authenticity and clarity, it was made in the direction of clarity.

Brief History of Architectural Styles

In December, 1620, 102 Pilgrims landed at Plymouth. In spite of cold weather, they immediately started to build their first house. Shortly after it was completed, the house burned down forcing the Pilgrims to live much of that first winter in holes in the ground that were covered with timbers, canvas and sod.

By the middle 1600's there were thousands of houses in the Colonies. They ranged in type and style from small wood, thatched roof cottages to large stone, brick and frame houses. Many of these houses were constructed by craftsmen who formerly built houses in the large European cities. With this imported expertise and some imported materials they were able to construct houses that were equal in quality to those being built in Europe. Many of the Colonial styles developed in this early period of our history. At the same time the fad of copying styles from builders' handbooks and architectural design books was developing. At first, only portions of the exterior designs were taken from the books. Then, whole houses were constructed exactly like their European counterparts, built from the same books.

Different styles developed in different regions of the country. Many of the early colonists copied styles from their homelands. Others copied the classic Gothic, Greek and Roman designs while yet others developed truly indigenous American styles.

On each of the following pages is pictured a style still found today. Its key distinguishing characteristics, other characteristics and a brief history are outlined. The CTS System narrative, abbreviation and computer code are also given for each. See the Appendix for a complete discussion of the CTS System.

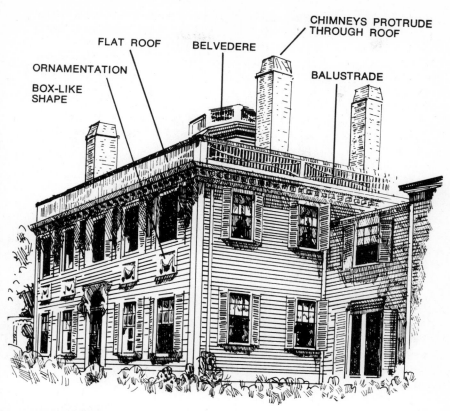

CHIMNEYS PROTRUDE
THROUGH ROOF

FLAT ROOF   BELVEDERE

ORNAMENTATION   BALUSTRADE

BOX-LIKE
SHAPE

Colonial American

FEDERAL (Federal - 101)

Key Distinguishing Characteristics

The Federal is a multi-story, symmetrical, box-shaped house with a flat roof.

Other Distinguishing Characteristics

Clapboard or brick exterior walls
One or more chimneys protruding through roof
Windows with small glass panes
Beautiful ornamentation
Belveder
Balustrade

History

Prevalent throughout the cities in the East in the 1700's, this style takes on many forms. Some consider this and the Adams style to be the same.

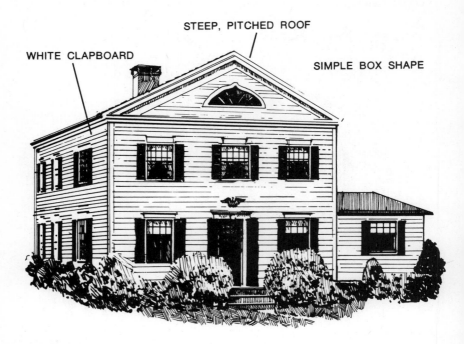

STEEP, PITCHED ROOF

WHITE CLAPBOARD

SIMPLE BOX SHAPE

Colonial American

NEW ENGLAND FARM HOUSE (N E Farm - 102)

Key Distinguishing Characteristics

This is a simple, box-shaped house. The traditional material for the exterior siding is white clapboard. A steep pitched roof is used to shed heavy snow.

Other Distinguishing Characteristics

Central chimney used to support frame
Two square rooms on each floor is typical layout
Interior layout poor by today's standards

History

These simple houses were built in the 1700's and 1800's throughout the New England countryside. They were the favorite style among the farmers who could not afford the more elaborate styles of the times. Simple to construct and maintain, they were designed to use the building materials and techniques of the period.

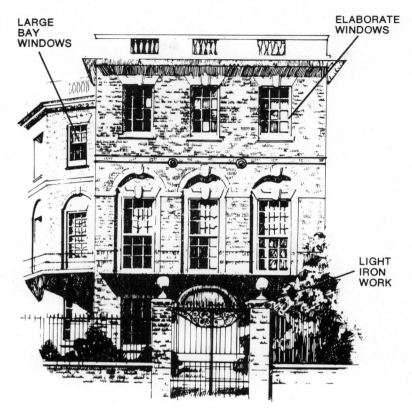

LARGE BAY WINDOWS

ELABORATE WINDOWS

LIGHT IRON WORK

Colonial American

ADAMS COLONIAL (Adams Co - 103)

Key Distinguishing Characteristics

A rectangular shaped, multi-story house of classic beauty, the Adams Colonial has large bay windows and a flat roof topped with a balustrade. The overall effect is one of lightness and delicacy.

Other Distinguishing Characteristics

Elaborate windows with small glass panes
Fantail windows on each side and above front entrance door
Chimneys protruding through roof
Carved steps with light iron rails

History

Introduced by Robert Adam in 1759 upon his return from a trip to Europe, the style was launched in New England when it was promoted by Charles Bulfinch and Samuel McIntire, two of America's famous early architects.

109

SHINGLES

GABLE ROOF

1½ STORIES

CENTRAL ENTRANCE

Colonial American

CAPE COD COLONIAL (Cape Cod - 104)

Key Distinguishing Characteristics

This is a small, symmetrical, 1½-story, compact house with a central entrance. The roof is the steep gable type covered with shingles. The authentic types have low central chimneys, but end chimneys are very common in the new versions. Bedrooms are on the first floor. The attic may be finished into additional bedrooms and a bath. A vine covered picket fence is traditional.

Other Distinguishing Characteristics

Traditionally, exterior walls were white clapboard, natural shingles or brick; modern versions have exterior walls of a wide variety of materials
Simple double-hung windows
Shutters same length as windows
Simple cornice with gutters immediately above first floor windows
Easy to build, maintain and heat
Inherent in the design is the necessity to walk through one room to reach another

History

The Cape Cod Colonial is the earliest dwelling type built by the American colonists that is still popular today. The early Cape Cods were very crude. They usually had one room on the first floor and a sleeping loft above. The modern version was the most popular house style in the U.S. from the 1920's through the 1950's.

110

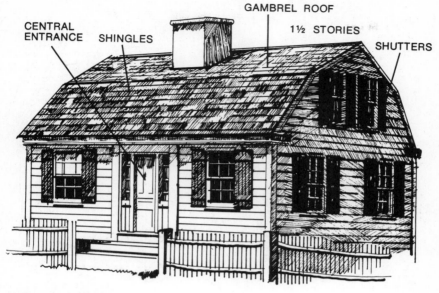

CENTRAL ENTRANCE · SHINGLES · GAMBREL ROOF · 1½ STORIES · SHUTTERS

Colonial American

CAPE ANN COLONIAL (Cape Ann - 105)

Key Distinguishing Characteristics

A small, symmetrical, 1½-story, compact house with a central entrance, the Cape Ann is distinguished from the Cape Cod style by its gambrel roof. The bedrooms are on the first floor. The attic may be finished into additional bedrooms and a bath.

Other Distinguishing Characteristics

Shingle covered roof
Traditionally, the exterior walls are white clapboard or natural shingles
Simple double-hung windows
Shutters the same length as the windows
Simple cornice with gutters immediately above the first floor windows
Easy to build, maintain and heat
Inherent in the design is the necessity to walk through one room to reach another

History

A variety of the Cape Cod that started in the Cape Ann area of Massachusetts, it is not nearly as popular as the gable-roofed Cape Cod style.

111

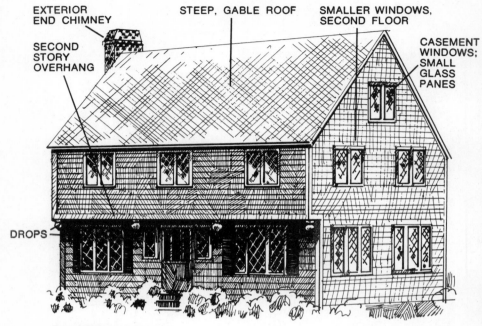

EXTERIOR END CHIMNEY

SECOND STORY OVERHANG

STEEP, GABLE ROOF

SMALLER WINDOWS, SECOND FLOOR

CASEMENT WINDOWS; SMALL GLASS PANES

DROPS

Colonial American

GARRISON COLONIAL (Garr Co - 106)

Key Distinguishing Characteristics

This is a 2½-story, symmetrical house with the second story overhang in the front. The traditional ornamentation is four carved drops (pineapple or acorn shape) below the overhang.

Other Distinguishing Characteristics

Exterior chimney at end
Older versions have casement windows with small
    panes of glass
Later versions have double-hung windows
Second story windows often smaller than those on first floor
Steep gable roof covered with shingles
Dormers often break through the cornice line

History

The Garrison Colonial is reputed to be patterned after colonial block-houses used to fend off Indians, but it is probably an outgrowth of the overhang style of the Elizabethan townhouse. The style probably has been perpetuated because of its pleasant appearance. The carved drops were the ends of the framing beams on early versions of the house. Later they were added on for decoration. Examples of this style built in the 1600's are still standing today.

112

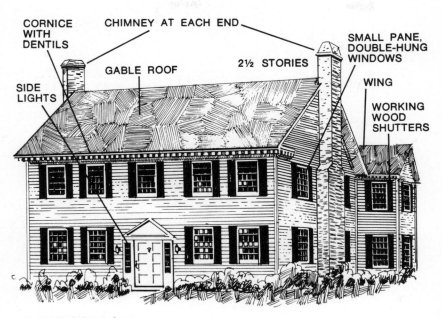

CORNICE WITH DENTILS — CHIMNEY AT EACH END — SMALL PANE, DOUBLE-HUNG WINDOWS

GABLE ROOF — 2½ STORIES — WING

SIDE LIGHTS — WORKING WOOD SHUTTERS

Colonial American

NEW ENGLAND COLONIAL (N E Col - 107)

Key Distinguishing Characteristics

The New England Colonial is a 2½-story, generally symmetrical, square or rectangular box-like house with side or rear wings. The traditional material is narrow clapboard siding. The roof is usually the gable type covered with shingles.

Other Distinguishing Characteristics

Originals had chimneys at each end for heating
Modern versions have only one chimney at end or in center
Windows are double-hung type with small glass panes
Shutters same size as windows
Central entrance door often has side lights and a fan light
Elaborate cornice with dentils
First floor has central hallway running from front to rear
Bedrooms are on the second floor

History

These large, roomy houses evolved from the Cape Cod style.

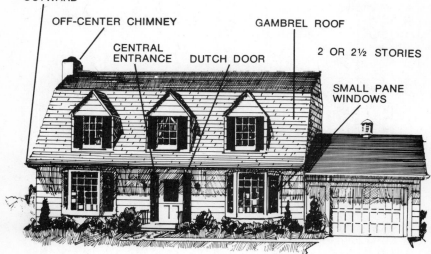

EAVES FLARE
OUTWARD

OFF-CENTER CHIMNEY

GAMBREL ROOF

CENTRAL
ENTRANCE   DUTCH DOOR

2 OR 2½ STORIES

SMALL PANE
WINDOWS

Colonial American

DUTCH COLONIAL (Dutch Co - 108)

Key Distinguishing Characteristics

The Dutch Colonial is a moderate sized, 2 to 2½-story house with a gambrel roof and eaves that flare outward.

Other Distinguishing Characteristics

Central entrance
Dutch entrance door
Double-hung windows with small panes of glass
Exterior may be made of a wide variety of material such as
    clapboard, shingles, cut stone, brick or stucco
Second story dormers through roof are common
Chimney rarely in the center

History

This is an indigenous American style that did not originate in Holland as is commonly believed. The Dutch settlers built them in Pennsylvania starting in the 1600's and soon after in New York.

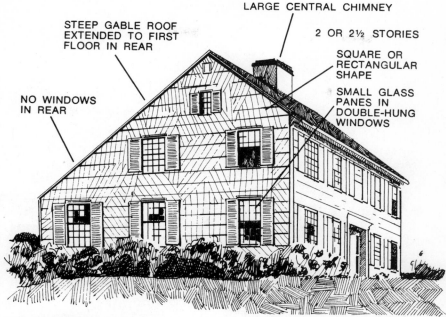

STEEP GABLE ROOF
EXTENDED TO FIRST
FLOOR IN REAR

LARGE CENTRAL CHIMNEY

2 OR 2½ STORIES

SQUARE OR
RECTANGULAR
SHAPE

SMALL GLASS
PANES IN
DOUBLE-HUNG
WINDOWS

NO WINDOWS
IN REAR

Colonial American

SALT BOX COLONIAL OR CATSLIDE (IN SOUTH)
(Salt Box or Catslide - 109)

Key Distinguishing Characteristics

This is 2 or 2½-story, square or rectangular house with a steep gable roof that extends down to the first floor in the rear.

Other Distinguishing Characteristics

Exterior walls usually clapboard or shingles
Large central chimney
Large double-hung windows with small panes of glass
No windows in rear
Shutters the same size as windows
In order to obtain traditional lean-to look, headroom in the rear
    must be sacrificed

History

The Salt Box is a Colonial New England farm house that grew in the form of a lean-to shed or room added to the rear of the house. The rear was oriented to the north to ward off cold winter winds. Its name comes from the resemblance to the salt box found in old country stores.

CLAPBOARD ON ADDITIONS

2½ STORIES

UNSUPPORTED HOOD OVER FRONT DOOR

PENT ROOF

GRAY LEDGE STONE WALLS

Colonial American

PENNSYLVANIA DUTCH COLONIAL OR PENNSYLVANIA GERMAN FARM HOUSE (Penn Dut - 110)

Key Distinguishing Characteristics

This is a massive, 2½-story, gray ledge stone house with a steep gable roof. An unsupported hood over the front entrance is traditional. When wings are added they are usually partly clapboard.

Other Distinguishing Characteristics

Shingle roof
Double-hung windows with small panes of glass
Ledge stone siding is long lasting and requires little care

History

The style was first built in the 1600's west of Philadelphia on the Main Line and in Lancaster, Pennsylvania. The name came from the mispronunciation of the word "Deutsch" since those who first built them were of German origin.

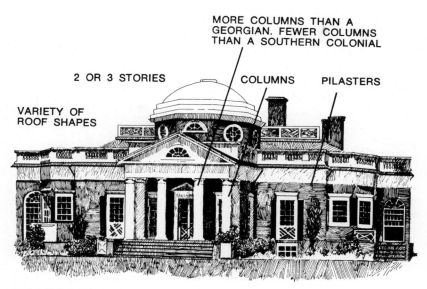

MORE COLUMNS THAN A
GEORGIAN. FEWER COLUMNS
THAN A SOUTHERN COLONIAL

2 OR 3 STORIES   COLUMNS   PILASTERS

VARIETY OF
ROOF SHAPES

Colonial American

CLASSIC COLONIAL (Classic - 111)

Key Distinguishing Characteristics

A large, impressive showplace, the Classic Colonial is a 2 or 3-story
house with a columned exterior. It has more columns than a Georgian
Colonial and fewer columns than a Southern Colonial.

Other Distinguishing Characteristics

Columns have Doric, Ionic or Corinthian capitals
Pilasters in the walls are common
Porticos at vehicular entrance with roof supported by columns
Variety of roof shapes on same house

History

Based on classic Greek architecture, the style was introduced by
Thomas Jefferson who built Monticello in 1772 and remodeled it
several times up to 1803. There is very little difference between a
Classic Colonial and a Greek Revival and many architectural historians
use the terms synonomously.

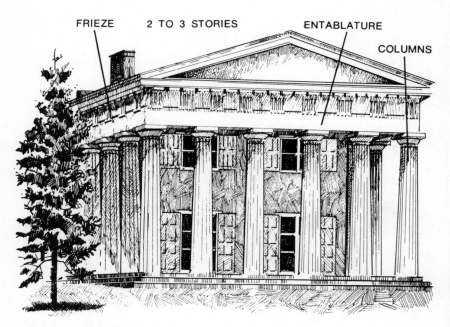

FRIEZE   2 TO 3 STORIES   ENTABLATURE

COLUMNS

Colonial American

GREEK REVIVAL (Greek - 112)

Key Distinguishing Characteristics

The Greek Revival style is a 2 or 3-story, symmetrical house that is a copy of a Greek temple complete with columns, architraves, friezes and cornices.

Other Distinguishing Characteristics

Windows small and hidden as they are not part of
     Greek temple architecture
The interior design is sacrificed for exterior design

History

There is little difference between a Greek Revival and a Classic Colonial. Many architectural historians consider them to be the same style. For clarity of identification, the houses that look like the traditional Greek temple are generally called Greek Revival. The style became popular in the U.S. in the early 1800's, especially for public buildings.

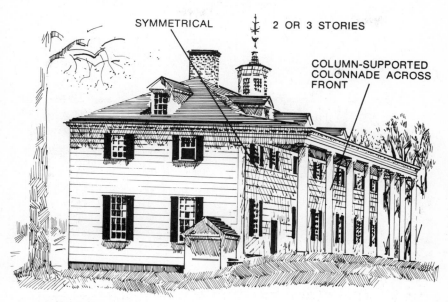

SYMMETRICAL 2 OR 3 STORIES

COLUMN-SUPPORTED COLONNADE ACROSS FRONT

Colonial American

SOUTHERN COLONIAL (South Co - 113)

Key Distinguishing Characteristics

The Southern Colonial is a large 2 or 3-story, frame house with a characteristic colonnade extending across the front. The roof extends over the colonnade.

Other Distinguishing Characteristics

Hip or gable roof covered with shingles
Symmetrical
Second-story balcony
Balustrade
Beleveder
Double-hung windows with small panes of glass
Shutters same size as windows
Cornices with dentils

History

The style is a southern conception of the New England Colonial and Georgian Colonial modified to suit the warmer climate. The classic example pictured here is Mount Vernon, built in the 1700's.

119

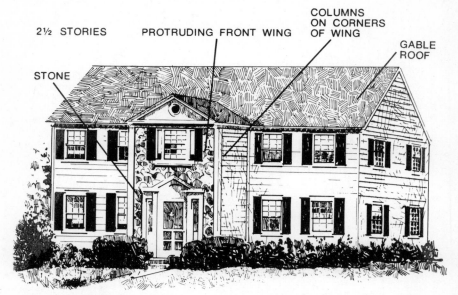

2½ STORIES — PROTRUDING FRONT WING — COLUMNS ON CORNERS OF WING — GABLE ROOF — STONE

Colonial American

FRONT GABLE NEW ENGLAND COLONIAL OR CHARLESTON COLONIAL OR ENGLISH COLONIAL
(F Gab NE or Charles or Eng Col - 114)

Key Distinguishing Characteristics

This is a 2½-story, symmetrical square or rectangular box-like house with a protruding front wing which distinguishes the style from a New England Colonial. The bedrooms are on the second floor.

Other Distinguishing Characteristics

House and wing have gable roofs
Basic house is usually clapboard
Wing front is often cut stone
Cornice on front or on all four sides
Double-hung windows with small panes of glass
Shutters same size as windows
Central entrance in front wing has side light and a fan light
Wing has stone columns on four corners
Central hallway runs from wing to rear of house
Bedrooms on second floor

History

Like the New England Colonial, this style evolved from the Cape Cod. The first houses of this style were built in the 1700's.

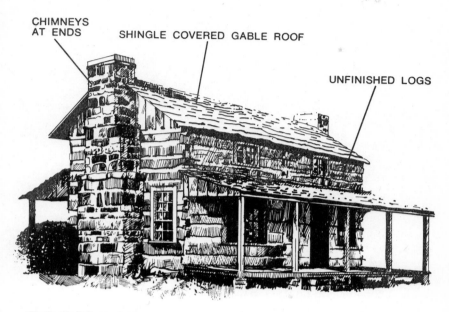

CHIMNEYS
AT ENDS

SHINGLE COVERED GABLE ROOF

UNFINISHED LOGS

Colonial American

LOG CABIN (Log Cab - 115)

Key Distinguishing Characteristics

This is a house built of unfinished logs.

Other Distinguishing Characteristics

Rectangular shape
Usually one story
Shingle-covered gable roof
Chimneys at ends
Few windows
Windows covered with oil soaked paper or small panes of glass

History

The first extensive use of the log cabin in America was by the Swedish immigrants. The log cabin eventually became the favorite house of the early pioneers. There are two basic types of log cabins. One is built with round logs like the traditional Lincoln cabin and the other is built with squared-off logs like the cabin pictured here. The illustration is of the Latta house in Arkansas. Several companies are producing reproductions today.

121

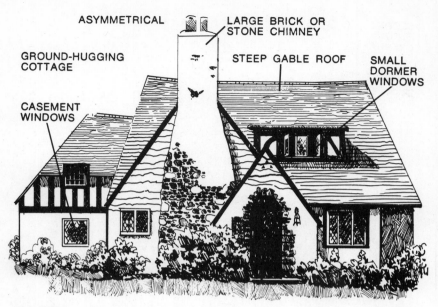

ASYMMETRICAL

LARGE BRICK OR STONE CHIMNEY

GROUND-HUGGING COTTAGE

STEEP GABLE ROOF

SMALL DORMER WINDOWS

CASEMENT WINDOWS

English

COTSWOLD COTTAGE (Cotscot - 201)

Key Distinguishing Characteristics

The smallest of the English styles, the Cotswold, is sometimes described as an Ann Hathaway or Hansel and Gretel Cottage. It is a ground-hugging, always asymmetrical style with a prominent brick or stone chimney in the front or on the side that appears to be very large in relation to the overall size of the house.

Other Distinguishing Characteristics

Walls of the originals were built of materials of the
    area — brick, stone, wood siding, half timbers
Steep gable roof with complex lines
Casement windows
Dormer windows smaller than other windows
Rooms tend to be small and irregularly shaped
Layout often necessitates walking through one room to
    get to another (even the bedrooms)
Upper story bedrooms have steep walls and need dormers

History

The Cotswold Cottage was first built in the Cotswold Hills of Gloucester, Hereford and Worcester, England starting about the time of the Norman conquest in 1066. The name is derived from "cot"-cottage and "wold"-wood — cottage in a wood. The romantic design was very popular throughout the U.S. in the 1920's and 1930's.

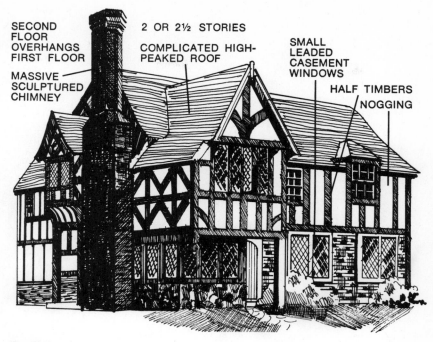

SECOND FLOOR OVERHANGS FIRST FLOOR

MASSIVE SCULPTURED CHIMNEY

2 OR 2½ STORIES

COMPLICATED HIGH-PEAKED ROOF

SMALL LEADED CASEMENT WINDOWS

HALF TIMBERS

NOGGING

English

ELIZABETHAN OR HALF TIMBER (Eliz or Halftim - 202)

Key Distinguishing Characteristics

This is a 2 or 2½-story house, often with part of the second story over-hanging the first. It has less stone work and is less fort-like than the Tudor. Stone and stucco walls with half timbers are most common.

Other Distinguishing Characteristics

Massive sculptured chimneys
Used brick between half timbers is called "nogging";
    when not covered with stucco it is "exposed nogging"
Complicated high peaked roofs which are expensive to
    install and subject to leaks because of complex valleys
Small leaded glass casement windows
Interior often has large halls and a spacious living room
    with a large fireplace and beamed ceilings
Bedrooms are on second floor

History

These houses were erected in England throughout the prosperous reign of Queen Elizabeth (1558-1603), particularly in the London area. The characteristic protruding second story supported by wooden brackets is a carry-over from narrow London lots where extra space could be obtained by overhanging the street.

123

FORT-LIKE APPEARANCE

2 TO 3 STORIES

SEMI-HEXAGONAL BAYS

BRICK OR STONE WALLS

HIGH CHIMNEY

CHIMNEY POTS

MOULDED STONE TRIM

STONE MULLIONS

English

TUDOR (Tudor - 203)

Key Distinguishing Characteristics

The Tudor is an imposing looking house with fortress-like lines. Siding is chiefly stone and brick with some stucco and half timbers. Windows and doors have moulded cement or stone trim around them.

Other Distinguishing Characteristics

Usually 2½ stories
Stone or cement window mullions and transoms
Casement windows with leaded glass (often diamond-shaped)
High, prominent chimneys with protruding chimney pots
Semi-hexagonal bays and turrets
Interior often layed out with odd-shaped rooms full of
    nooks and crannies, large fireplaces, beamed ceilings
    and rough plaster walls

History

The style started in England in the late 15th century during the reign of the House of Tudor. It did not become popular in the U. S. until the late 1800's.

124

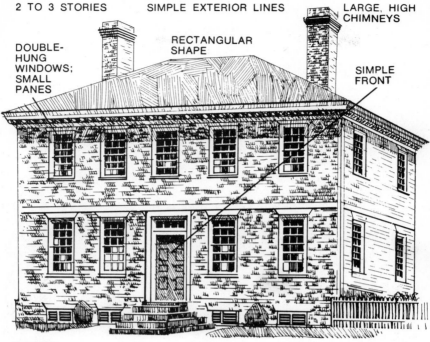

2 TO 3 STORIES   SIMPLE EXTERIOR LINES   LARGE, HIGH CHIMNEYS

DOUBLE-HUNG WINDOWS; SMALL PANES

RECTANGULAR SHAPE

SIMPLE FRONT

English

WILLIAMSBURG GEORGIAN OR EARLY GEORGIAN
(Williams or E Georg - 204)

Key Distinguishing Characteristics

The houses built in Williamsburg were representative of the early Georgian houses built in America throughout the early 1700's. They had simple exterior lines and generally fewer of the decorative devices characteristic of the later Georgian houses. Most were 2 or 3-story rectangular houses with two large chimneys rising high above the roof at each end.

Other Distinguishing Characteristics

Sliding double-hung windows with small panes
Simple front entrances

History

Williamsburg, Virginia was the cultural and political capital of the colonies during most of the 1700's. The houses here were based on the styles developed in England during the reign of the four King Georges. The Williamsburg houses were built in the early 1700's. With gifts of about $40 million, John D. Rockefeller, Jr. has restored the community. **125**

SIMPLE, INFORMAL STYLE

SMALL OCTAGONAL WINDOW

BRICK

HIP ROOF

DOUBLE-HUNG WINDOWS

1 CHIMNEY AT SIDE

English

REGENCY (Regency - 205)

Key Distinguishing Characteristics

This is a 2 or 3-story, symmetrical house with a hip roof. A small octagonal window over the front door is traditional.

Other Distinguishing Characteristics

Almost always brick, often painted white
One chimney on side
Double-hung windows
Shutters same size as windows
A simple informal style without the classic lines of the Georgian

History

The style reached its peak of popularity in England between 1810 and 1820. Many of these houses were built in the U. S. in the late 1800's, throughout the 1900's and are still being built today.

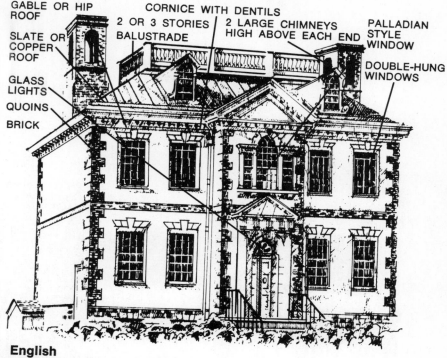

GABLE OR HIP ROOF

SLATE OR COPPER ROOF

GLASS LIGHTS

QUOINS

BRICK

CORNICE WITH DENTILS

2 OR 3 STORIES

BALUSTRADE

2 LARGE CHIMNEYS HIGH ABOVE EACH END

PALLADIAN STYLE WINDOW

DOUBLE-HUNG WINDOWS

English

GEORGIAN (George - 206)

Key Distinguishing Characteristics

The Georgian is a large, formal 2 or 3-story rectangular house characterized by its classic lines and ornamentation.

Other Distinguishing Characteristics

Traditional material is brick
Some made of masonry, clapboard or shingles
Gable or hip roof
Slate shingles or copper roofing
Two large chimneys rising high above the roof at each end
A Paladian style set of three windows on the second floor
    over the front entrance or at each end
Front entrance with Greek columns
Glass lights on side and above front entrance door

History

The style became popular in England during the reign of the four King Georges. They were built in large numbers here in the 1700's and 1800's. The house pictured is Mount Pleasant built in 1750 in Philadelphia. It contained a large number of the features that would be traditionally included in the Georgian style.

127

1½ TO 2 STORIES

DORMERS
BREAK
THROUGH
CORNICE — LARGE CHIMNEY

HIGH HIP OR
GABLE ROOF

VARIETY OF
BUILDING
MATERIALS

HALF
TIMBERS

French

FRENCH FARM HOUSE (Fr Farm - 301)

Key Distinguishing Characteristics

An informal 1½ to 2-story house made of a variety of building materials, this house has a high hip or gable type roof with steep slopes.

Other Distinguishing Characteristics

Large chimney
Half timbers used as decorative accent
Dormers that break through the cornice
Another common variation of this design (not pictured) is
   a house built around a courtyard with two symmetrical
   wings protruding towards the front.

History

The style is based on the farm houses found in the various, widespread provinces of France. Because of this, the style varies widely. They were most popular in America in the early 1900's.

128

1½ TO 2½ STORIES   PERFECTLY BALANCED

CURVE-HEADED UPPER WINDOWS THAT BREAK THROUGH CORNICE

BRICK

FORMAL LOOKING

HIGH, STEEP HIP ROOF

SOME HAVE 2 SYMMETRICAL 1-STORY WINGS

FRENCH WINDOWS AND SHUTTERS

French

FRENCH PROVINCIAL (Fr Prov - 302)

Key Distinguishing Characteristics

The French Provincial is a perfectly balanced, formal 1½ to 2½-story house with a high steep hip roof and curved-headed upper windows that break through the cornice.

Other Distinguishing Characteristics

French windows or shutters on first floor
Two symmetrical 1-story wings
Usually made of brick

History

The style originated in France during the reign of Louis XIV (1643-1715) by the rich who wanted showplace homes.

129

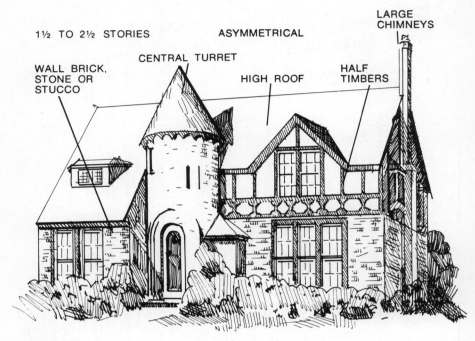

1½ TO 2½ STORIES   ASYMMETRICAL   LARGE CHIMNEYS

CENTRAL TURRET

WALL BRICK, STONE OR STUCCO   HIGH ROOF   HALF TIMBERS

French

FRENCH NORMANDY (Fr Norm - 303)

Key Distinguishing Characteristics

The main characteristic of this style is the central turret which is usually a staircase.

Other Distinguishing Characteristics

1½ to 2½ stories
Exterior walls usually brick, stone or stucco
Asymmetrical shape
Large chimneys
High, complicated roofs
Half timbers for decoration

History

This house style originated in the Normandy area of France where house and barn were one building. The turret was used for storage of grain or ensilage.

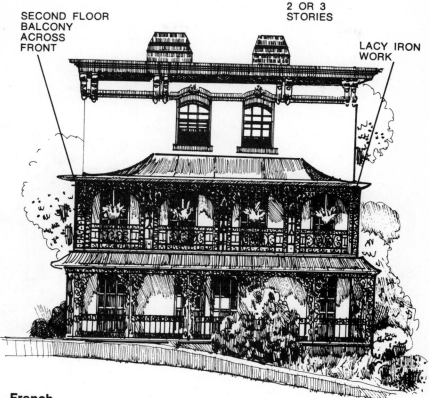

SYMMETRICAL

2 OR 3
STORIES

SECOND FLOOR
BALCONY
ACROSS
FRONT

LACY IRON
WORK

French

CREOLE OR LOUISIANA OR NEW ORLEANS
(Creole or Louisia or New Or - 304)

Key Distinguishing Characteristics

A balcony with extensive lacy iron work running across the entire front at the second story level distinguishes this style.

Other Distinguishing Characteristics

Symmetrical
2 to 3 stories

History

The style originated around New Orleans. Although it appears to have been influenced by French, Spanish and West Indian styles, it is not found in these places. Many of these houses were built in the late 1700's and 1800's.

131

1½ TO 2½ STORIES

GABLE ROOF

NATURAL WOOD LOOK

DECORATIVE WOOD WORK

LARGE GLASS WINDOWS

OPEN PORCHES

CURVED CORNICE

Swiss

SWISS CHALET (Swiss Ch - 401)

Key Distinguishing Characteristics

The Swiss Chalet is a 1½ to 2½-story, gable roof house with extensive natural decorative woodwork on the exterior.

Other Distinguishing Characteristics

Open porches
Large glass windows
Natural wood look
Curved cornice

History

The style is a direct copy of the mountain chalets of Switzerland. In this country too, they are built mostly in the mountain areas as ski lodges. The house pictured was designed and built by Paul Manchester.

1 TO 3 STORIES

PAINTED STUCCO
EXTERIOR WALLS

RED TILE ROOF

OVAL-TOP DOORS

Latin

SPANISH VILLA (Sp Villa - 501)

Key Distinguishing Characteristics

This is an asymmetrical, 1 to 3-story house with painted stucco exterior walls and red tile roof.

Other Distinguishing Characteristics

Oval top windows and doors
Wrought iron exterior decorations
Patio completely enclosed by exterior walls

History

American architecture has been influenced by Spanish styles since the days of the Spanish conquistadores. It is widespread throughout the warm parts of the U.S. and also appears in the north surprisingly frequently considering its basic unsuitability for cold climates.

133

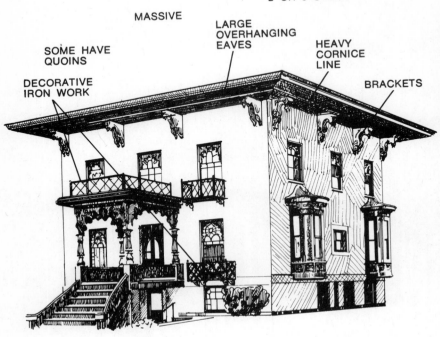

SOME HAVE QUOINS

MASSIVE

2 OR 3 STORY

LARGE OVERHANGING EAVES

HEAVY CORNICE LINE

BRACKETS

DECORATIVE IRON WORK

Latin

ITALIAN VILLA (It Villa - 502)

Key Distinguishing Characteristics

This is a massive, 2 or 3-story house of masonry with large overhanging eaves.

Other Distinguishing Characteristics

Many have square or octagonal towers
Quoins at corners
Heavy cornice line with supporting brackets
Decorative iron work

History

First built in the U.S. in 1837, this style was popularized by architect Andrew Jackson Downing in the middle 1800's. Architects Henry Austin and Richard Upjohn used the style extensively. The Bristol House pictured was designed by Austin and built in New Haven, Connecticut in 1845.

134

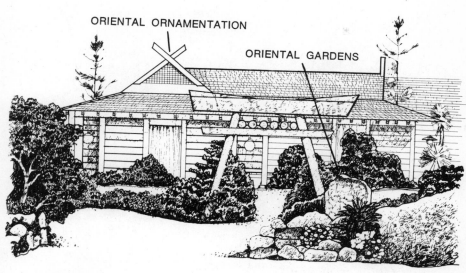

ORIENTAL ORNAMENTATION

ORIENTAL GARDENS

Oriental

JAPANESE (Japan - 601)

Key Distinguishing Characteristics

Exterior wall panels that slide to open (shoji)
Ornamental gardens often are enclosed behind bamboo fences or
walls designed for viewing from within the house.
Tile, thatch or wood shingle roof

Other Distinguishing Characteristics

Floors in a living area covered with tatami mats (always the same size,
about 3' x 6')
Interior walls with sliding panels (fusuma)
Modular type construction.

History

The Japanese style was first introduced into America in Hawaii and
California. It has been increasing in popularity in the last 10 years.
The classic house in Japan was built of wood, paper, tile, plaster and
stone. The classic Japanese house is surprisingly similar to our functional modern home.

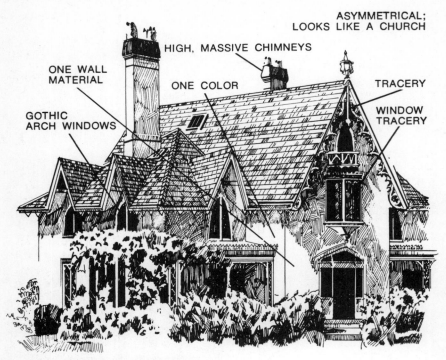

ASYMMETRICAL;
LOOKS LIKE A CHURCH

HIGH, MASSIVE CHIMNEYS

ONE WALL
MATERIAL

ONE COLOR

TRACERY

GOTHIC
ARCH WINDOWS

WINDOW
TRACERY

Nineteenth Century American

EARLY GOTHIC REVIVAL (E Goth - 701)

Key Distinguishing Characteristics

Gothic architecture can easily be identified by the pointed arch. The windows and doors are often this same pointed shape. Early Gothic revival might have only a few true Gothic characteristics. The house has a Gothic, church-like appearance and is asymmetrical. Usually only one color and one material are used.

Other Distinguishing Characteristics

Wood the most common material
Some built of stucco or brick
Fragile appearance, especially tracery and moulding
Pinnacles and battlements
High, massive chimneys

History

Gothic architecture was popular through Europe from the 12th to the 15th centuries. Many great Gothic churches are still standing today. It was one of the many styles copied in the U.S. as part of the fad of the 1700's and 1800's. The house pictured is the Rotch House in New Bedford, Massachusetts, designed by Alexander Jackson Davis in 1846.

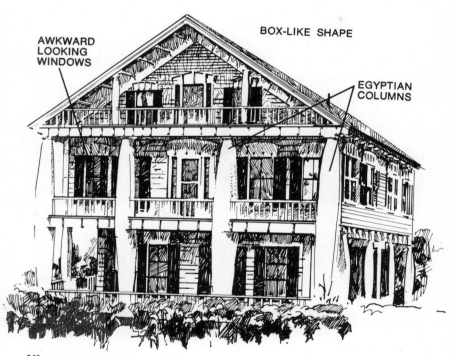

AWKWARD LOOKING WINDOWS

BOX-LIKE SHAPE

EGYPTIAN COLUMNS

Nineteenth Century American

EGYPTIAN REVIVAL (Egypt - 702)

Key Distinguishing Characteristics

This is a box-like house with Egyptian columns in the front.

Other Distinguishing Characteristics

Windows appear awkward since they are not part of authentic Egyptian architecture

History

Egyptian architecture was never very popular in America, even in the early 1800's when the fad of copying foreign styles was at its height. The best examples of Egyptian architecture are found in cemetery gates built during the same period.

137

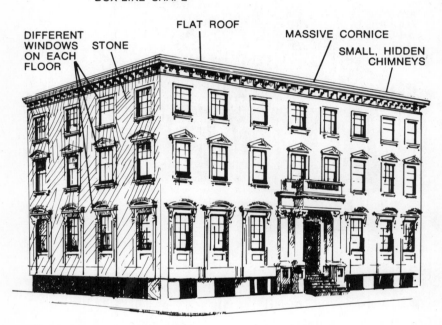

BOX-LIKE SHAPE

DIFFERENT
WINDOWS   STONE
ON EACH
FLOOR

FLAT ROOF

MASSIVE CORNICE

SMALL, HIDDEN
CHIMNEYS

Nineteenth Century American

ROMAN TUSCAN MODE (Ro Tusc - 703)

Key Distinguishing Characteristics

A house in this style is box-like and fills an entire lot, leaving no yard area. Windows on each floor are treated differently. The roof is flat with a massive cornice.

Other Distinguishing Characteristics

Chimneys small and hidden
Built of stone

History

The style was brought from Italy in the 1800's and used for several important public buildings and a few houses.

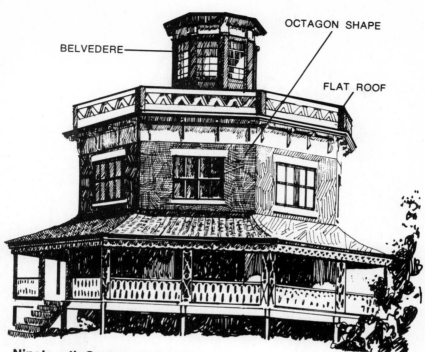

BELVEDERE—

OCTAGON SHAPE

FLAT ROOF

Nineteenth Century American

OCTAGON HOUSE (Octagon - 704)

Key Distinguishing Characteristics

The most distinguishing characteristic of this style is the octagonal shape.

Other Distinguishing Characteristics

Stone, wood or concrete walls
Flat roofs
Chimney or beleveder in the center
Variety of window and door types used

History

Most of these houses were built around 1850. They were a result of an idea put forth by Orson Squire Fowler, a popular author of the time who wrote about love, marital happiness, sex and phrenology. In several of his popular books he expounded the theory of a happy life in an octagon-shaped house. He thought that this design made for good interior circulation, heating and lighting. Octagon houses were built throughout the U.S. and several hundred are still lived in today. The house was explained in detail in his book, *A Home for All, or the Gravel Wall and Octagon Mode of Building, New, Cheap, Convenient, Superior and Adapted to Rich and Poor.*

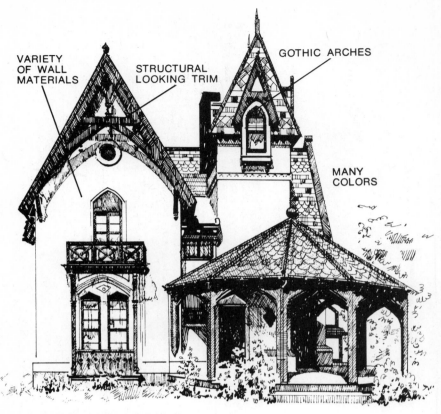

VARIETY
OF WALL
MATERIALS

STRUCTURAL
LOOKING TRIM

GOTHIC ARCHES

MANY
COLORS

Nineteenth Century American

HIGH VICTORIAN GOTHIC (Hi Goth - 705)

Key Distinguishing Characteristics

Like the early Gothic architecture, High Victorian Gothic can be identified as being Gothic by the pointed arch which is used extensively over the windows and doors. It is much more elaborate than the early Gothic with a great many colors and materials being visible. The decorative details give an appearance of solidity.

Other Distinguishing Characteristics

Multi-color effect of exterior walls obtained by the use
of stone, wood, brick
Exterior wood trim appears to be structural rather than merely trim

History

Most of the High Victorian Gothic houses were built after the Civil War and were based on European Gothic styles popular at the time. The house pictured is the Converse House in Norwich, Connecticut, built in 1870.

140

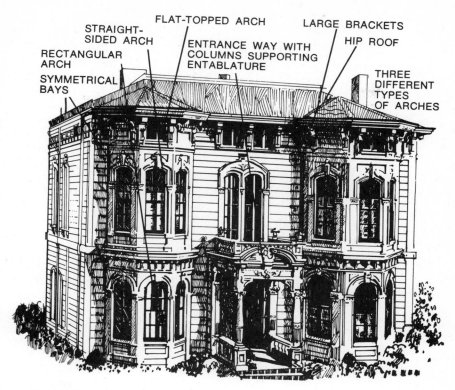

STRAIGHT-SIDED ARCH

FLAT-TOPPED ARCH

LARGE BRACKETS

HIP ROOF

RECTANGULAR ARCH

ENTRANCE WAY WITH COLUMNS SUPPORTING ENTABLATURE

SYMMETRICAL BAYS

THREE DIFFERENT TYPES OF ARCHES

Nineteenth Century American

HIGH VICTORIAN ITALIANATE (Vic Ital - 706)

Key Distinguishing Characteristics

The use of three different kinds of window arches is the primary distinguishing characteristic of this style. They are the straight-sided arch, flat-topped arch and rectangular arch.

Other Distinguishing Characteristics

Square shape
Symmetrical bays in the front
Hip roof
Small chimneys protruding through roof in irregular locations
Entrance-way has columns supporting an entablature
Cornice supported with over-size brackets
Exterior appears to be cut up and ornate

History

The style seems to have come to this country from Italy by way of England, where the style was popular in the early 1800's. The house pictured is in Portland, Oregon and was built by Mark Morris in 1882. **141**

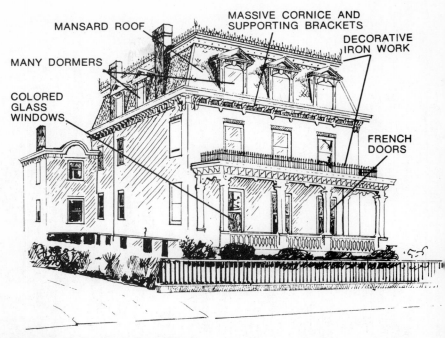

MANSARD ROOF

MASSIVE CORNICE AND
SUPPORTING BRACKETS

DECORATIVE
IRON WORK

MANY DORMERS

COLORED
GLASS
WINDOWS

FRENCH
DOORS

Nineteenth Century American

AMERICAN MANSARD OR SECOND EMPIRE STYLE
(Mansard or 2nd Emp - 707)

Key Distinguishing Characteristics

The main and distinguishing characteristic of this style is the roof design. The mansard roof slopes gently back from the wall line and then is topped with an invisible (from the street) section resembling a conventional hip roof. Through the roof protrude multiple dormers.

Other Distinguishing Characteristics

Decorative iron work
Decorative mouldings
Quoins on corners
Brackets supporting massive cornices
Colored glass in windows
Iron castings
French doors opening onto deep porches

History

The French Second Empire (the reign of Napoleon III from 1852 to 1870) was the period in which the style was popular in the U.S. The mansard roof was originally developed in France in the 1600's by Francois Mansart. It enjoyed wide popularity when revived in the 1800's and now in the 1970's.

142

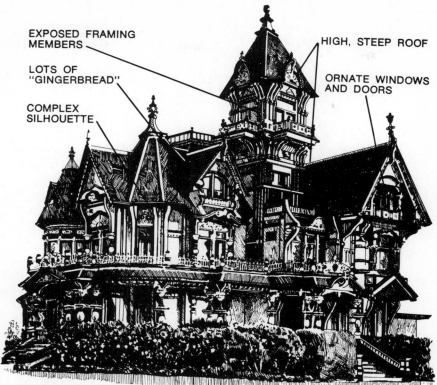

EXPOSED FRAMING
MEMBERS

LOTS OF
"GINGERBREAD"

COMPLEX
SILHOUETTE

HIGH, STEEP ROOF

ORNATE WINDOWS
AND DOORS

Nineteenth Century American

STICK STYLE OR CARPENTER GOTHIC
(Stick or C Goth - 708)

Key Distinguishing Characteristics

Exposed framing members, high steep roofs, complex silhouettes, diagonal braces and lots of gingerbread trim give this house style its characteristic look.

Other Distinguishing Characteristics

Clapboard siding overlayed with additional boards running
   horizontally, vertically and diagonally
Complex and ornate windows and doors

History

This style is the result of the advance in the technology of the 1800's. The public had become fascinated with what could be produced cheaply on the automatic bandsaw and they used these mill-cut ornate parts in great abundance on the houses of the period. The name applies to the many gingerbread scrolled houses of the 1800's. It is a very loosely defined style. The culmination of this fad took place in the construction of the house pictured, in Eureka, California, known as the Carson Mansion.

143

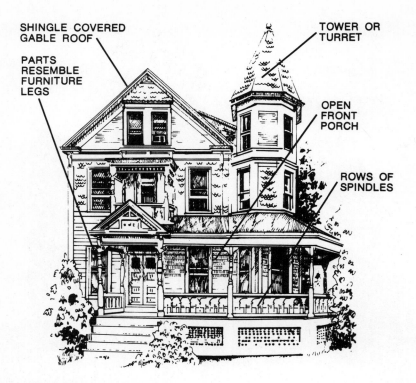

SHINGLE COVERED
GABLE ROOF

PARTS
RESEMBLE
FURNITURE
LEGS

TOWER OR
TURRET

OPEN
FRONT
PORCH

ROWS OF
SPINDLES

Nineteenth Century American

EASTLAKE (East L - 709)

Key Distinguishing Characteristics

The distinctive Eastlake type of ornamentation is the major characteristic of this style. Otherwise these houses look like a Queen Anne or
Carpenter Gothic. The ornamentation is very three-dimensional, having been made with a chisel, gouge and lathe rather than the scroll
saw. Many of the parts resemble furniture legs and knobs.

Other Distinguishing Characteristics

Rectangular shape
Open front porch
Tower or turret
Gable roof covered with shingles

History

Charles Lock Eastlake, an English architect, wrote a book, *Hints on
Household Taste,* first published here in Boston in 1872. The designs
in the very popular book became the basis of Eastlake style furniture
and houses. The houses, however, were not designed by Eastlake and
it was said that he did not like them.

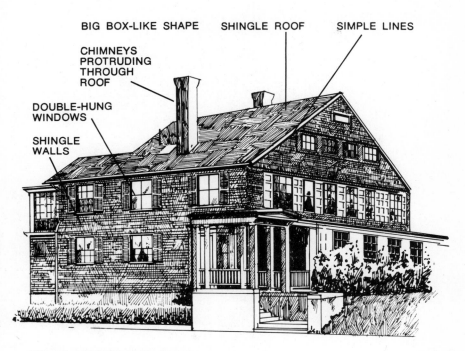

BIG BOX-LIKE SHAPE   SHINGLE ROOF   SIMPLE LINES

CHIMNEYS
PROTRUDING
THROUGH
ROOF

DOUBLE-HUNG
WINDOWS

SHINGLE
WALLS

Nineteenth Century American

SHINGLE STYLE (Shingle - 710)

Key Distinguishing Characteristics

The Shingle Style house is big and box-like with simple lines. The roof and exterior walls are covered with shingles usually stained or painted a dark color.

Other Distinguishing Characteristics

Gable roof
Double-hung windows
Shutters same size as windows
Several chimneys protruding through roof

History

This style developed in New England in the 1880's. It became very popular and many thousands were built. It eventually spread throughout the U.S. and many are in good condition and being lived in today.   **145**

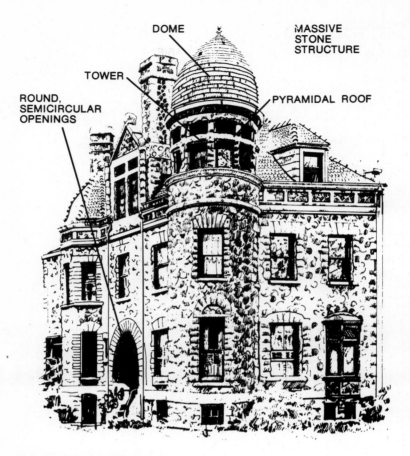

DOME

MASSIVE STONE STRUCTURE

TOWER

ROUND, SEMICIRCULAR OPENINGS

PYRAMIDAL ROOF

Nineteenth Century American

ROMANESQUE (Roman - 711)

Key Distinguishing Characteristics

This is a massive, stone structure with round, semicircular arched openings.

Other Distinguishing Characteristics

One or more towers topped with a dome
Pyramidal roof

History

This style was very popular in the 1800's for public buildings and churches. Only a few Romanesque houses were ever built in the U.S. Architect Henry Hobson Richardson used the style extensively. The house pictured is the Ayer House in Chicago which, unfortunately, was demolished in 1966.

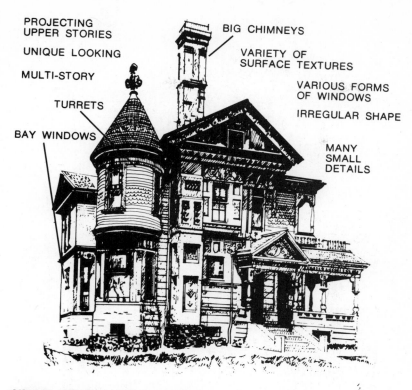

PROJECTING UPPER STORIES

UNIQUE LOOKING

MULTI-STORY

TURRETS

BAY WINDOWS

BIG CHIMNEYS

VARIETY OF SURFACE TEXTURES

VARIOUS FORMS OF WINDOWS

IRREGULAR SHAPE

MANY SMALL DETAILS

Nineteenth Century American

QUEEN ANNE (Q Anne - 712)

Key Distinguishing Characteristics

Queen Anne has come to be applied to any Victorian house that cannot be otherwise classified. They are all unique looking, multi-story houses, irregular in shape with a variety of surface textures, materials and colors.

Other Distinguishing Characteristics

Half timbering
Windows of various forms
Upper stories that project over the lower ones
Bay windows
Turrets
Big chimneys

History

First designed in England by Richard Norman Shaw, Queen Anne was started here as a style by architect Henry Hobson Richardson in the early 1870's.

147

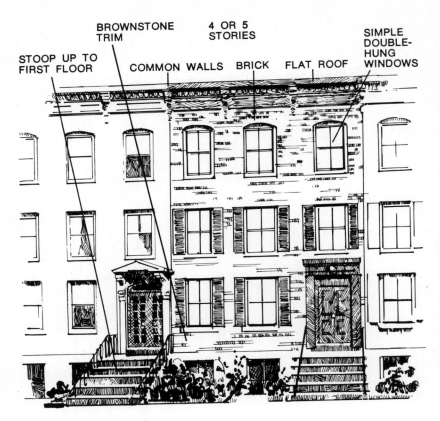

STOOP UP TO FIRST FLOOR

BROWNSTONE TRIM

4 OR 5 STORIES

COMMON WALLS   BRICK   FLAT ROOF

SIMPLE DOUBLE-HUNG WINDOWS

Nineteenth Century American

BROWNSTONE OR BRICK ROW HOUSE OR EASTERN TOWNHOUSE (Brown S or Br Row or E Town - 713)

Key Distinguishing Characteristics

These houses usually cover an entire city block. Most are 4 or 5 stories with a stoop leading up to the first floor. They have common side walls with the house on either side.

Other Distinguishing Characteristics

Built of brick
Often faced or trimmed with chocolate sandstone called brownstone
Flat roof
Simple double-hung windows

History

Brownstones became popular in the late 1800's in New York and other large eastern cities. They tend to develop their individual characteristics in each different city.

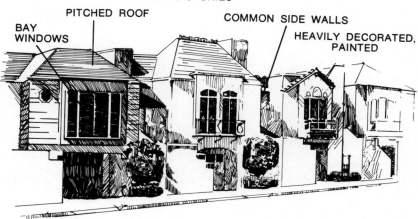

2 OR 3 STORIES

PITCHED ROOF

BAY WINDOWS

COMMON SIDE WALLS

HEAVILY DECORATED, PAINTED

Nineteenth Century American

WESTERN ROW HOUSE OR WESTERN TOWNHOUSE
(West Row - 714)

Key Distinguishing Characteristics

Like the eastern version of the townhouse, these houses are usually built to cover an entire street or block. They have common side walls with the house on either side.

Other Distinguishing Characteristics

2 or 3 stories
Exterior walls may be wood, stucco or brick
Bay windows
Heavily decorated and painted exteriors
Pitched roof

History

This style is the West's answer to the eastern row house. The main difference is that each unit has its own individual, unique design. They were built starting in the late 1800's and are still popular today. Thousands of these units have been built in San Francisco and other large western cities.

149

BALCONY ACROSS FRONT AT SECOND FLOOR

SHINGLE
ROOF

2 STORIES

RAIL SIMPLE
IRON OR WOOD

Nineteenth Century American

MONTEREY (Monterey - 715)

Key Distinguishing Characteristics

A 2-story house with a balcony across the front at the second floor level.

Other Distinguishing Characteristics

Asymmetrical shape
Balcony rail is simple iron work or wood as contrasted with lacy
iron work of Creole style.

History

Thomas Larkin, a Boston merchant, moved to Monterey, California and in 1835 built his version of a New England Colonial out of adobe brick. Most of the other houses in Monterey at that time were 1-story. In addition to the second story, his house also introduced to the area the second floor porch, roof shingles and the fireplace. The house was widely copied throughout the West and became known as the Monterey style.

150

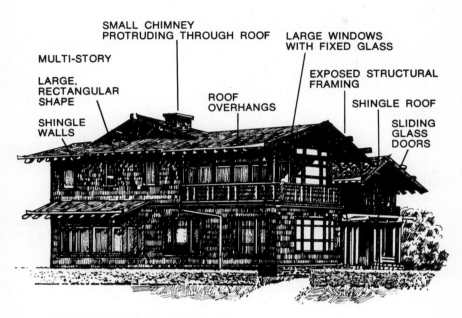

SMALL CHIMNEY
PROTRUDING THROUGH ROOF

LARGE WINDOWS
WITH FIXED GLASS

MULTI-STORY

LARGE,
RECTANGULAR
SHAPE

EXPOSED STRUCTURAL
FRAMING

ROOF
OVERHANGS

SHINGLE ROOF

SHINGLE
WALLS

SLIDING
GLASS
DOORS

Nineteenth Century American

WESTERN STICK (W Stick - 716)

Key Distinguishing Characteristics

The Western Stick is a large rectangular house with exposed structural framing.

Other Distinguishing Characteristics

Exterior walls and roof covered with shingles
Many windows have large fixed panes of glass
Roof overhangs the house to take advantage of the sun angles
Sliding glass doors -
Small chimney that protrudes through the roof at the end
of the house

History

This is a regional version of the Shingle Style that appeared in the late 1800's. One can detect both Oriental and Swiss influence in the design.   **151**

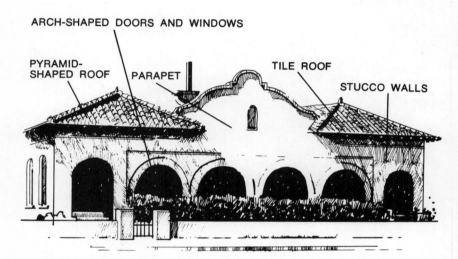

LOOKS LIKE OLD MISSION CHURCH

ARCH-SHAPED DOORS AND WINDOWS

PYRAMID-
SHAPED ROOF PARAPET

TILE ROOF

STUCCO WALLS

Nineteenth Century American

MISSION (Mission - 717)

Key Distinguishing Characteristics

These houses look like the old mission churches and houses of Southern California. The doors and windows are arch-shaped.

Other Distinguishing Characteristics

No sculptural ornamentation
Roof often hidden by the parapets
Towers and turrets with pyramid-shaped roofs
Exterior walls made of stucco

History

This is the style developed in the West in the 1800's when the people became aware of their architectural past. It was popularized as a reaction against the conventional eastern styles that were being built in the West at the time.

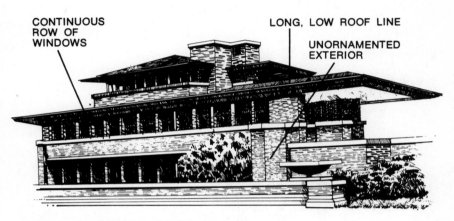

CONTINUOUS ROW OF WINDOWS

LONG, LOW ROOF LINE

UNORNAMENTED EXTERIOR

Early Twentieth Century American

PRAIRIE HOUSE (Prairie - 801)

Key Distinguishing Characteristics

A long, low roof line, continuous row of windows and an unornamented exterior distinguish this house. Designed to satisfy the physical and psychological needs of the inhabitants, it is unlike the traditional concept of a house that is a box subdivided into smaller boxes (rooms) each with some doors and windows.

Other Distinguishing Characteristics

Made up of individual areas in size, lighting and function, yet all under one roof

Many feel it is housing at its best

History

Frank Lloyd Wright, the designer of this house, dominated the architectural scene for the first half of the 20th century. His total impact is yet to be fully measured. Unlike many contemporary architects, he was vitally interested in home design. His principal concern was the organization of space. The Robie House pictured here was built in Chicago in 1909 and is one of the most famous Frank Lloyd Wright houses.

153

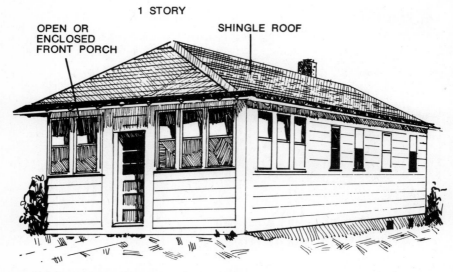

1 STORY

OPEN OR
ENCLOSED
FRONT PORCH

SHINGLE ROOF

Early Twentieth Century American

BUNGALOW (Bungalow - 802)

Key Distinguishing Characteristics

A Bungalow is a small, 1-story house that usually has an open or enclosed front porch.

Other Distinguishing Characteristics

Siding usually wood
Roof usually shingled
Many different regional types

History

Most Bungalows were built in the early 1900's and their popularity reached its peak around 1920. The terms, Bungalow and California Bungalow, were often used synonymously. Many Bungalow style houses are now called Ranch style. They are very popular in the shore communities of the east and west coasts.

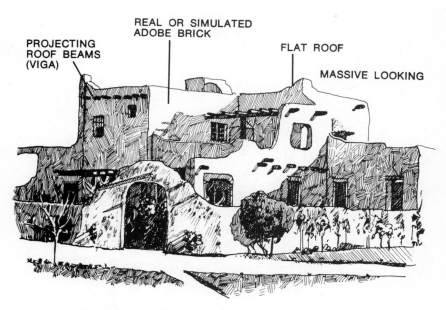

PROJECTING
ROOF BEAMS
(VIGA)

REAL OR SIMULATED
ADOBE BRICK

FLAT ROOF

MASSIVE LOOKING

Early Twentieth Century American

PUEBLO OR ADOBE (Pueblo or Adobe - 803)

Key Distinguishing Characteristics

This house is made of adobe brick or some other material that is made to look like adobe brick. The characteristic projecting roof beams are called viga.

Other Distinguishing Characteristics

The houses are massive looking
They rarely have arches
Roofs are flat
Long rain water gutters called "canales"

History

Based on the design of the houses of the Indians of New Mexico and Arizona, this became a popular style throughout the Southwest in the early 1900's.

CONTINUOUS WINDOWS

SIMPLE DESIGN, NO ORNAMENTATION

SMOOTH, UNIFORM WALL SURFACE

FLAT ROOF

Early Twentieth Century American

INTERNATIONAL (Internat - 804)

Key Distinguishing Characteristics

This design is very simple, with no ornamentation. The windows appear to be continuous rather than appearing to be holes in the walls.

Other Distinguishing Characteristics

Flat roofs
Smooth, uniform wall surfaces
Windows that turn the corner of the house.

History

The style was started in Europe in the 1920's by Walter Gropius, Ludwig Mies van der Rohe and the architects of the Bauhaus. It was introduced in the U.S. in the 1930's and is the basis of much modern architecture.

SMALL, COMPACT SHAPE   1 STORY

Early Twentieth Century American

CALIFORNIA BUNGALOW (Cal Bung - 805)

Key Distinguishing Characteristics

A 1-story, small compact house.

Other Distinguishing Characteristics

Usually made of wood
Traces of south seas, Spanish and Japanese influences on many

History

At the peak of the Bungalow's popularity, from 1900 to 1920, the name
Bungalow and California Bungalow were often used synonymously
although one writer of the time classified Bungalows into nine different types. The name comes from the Indian word "bangla" — of Bengal.   **157**

1 STORY　　LOW-PITCHED ROOF　　PICTURE WINDOWS　　GROUND-HUGGING　　SLIDING WINDOWS

Post World War II American

CALIFORNIA RANCH (C Ranch - 901)

Key Distinguishing Characteristics

A California Ranch is a 1-story, ground-hugging house with a low, pitched roof.

Other Distinguishing Characteristics

Large double-hung, sliding and picture windows
Sliding glass doors leading onto patios

History

The ranch house of the West has spread in popularity throughout the U. S. Today the term is commonly used to describe a wide variety of 1-story houses including many smaller ones which used to be called Bungalows.

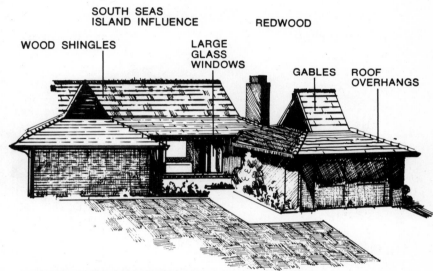

SOUTH SEAS
ISLAND INFLUENCE   REDWOOD

WOOD SHINGLES   LARGE
GLASS
WINDOWS   GABLES   ROOF
OVERHANGS

Post World War II American

NORTHWESTERN OR PUGET SOUND
(North W or P Sound - 902)

Key Distinguishing Characteristics

This is a low ranch type house with generous overhangs at eaves and gables. The exterior walls are often redwood.

Other Distinguishing Characteristics

    Roof shape reflects south sea island influence
    Large glass windows
    Wood shingle roof covering

History

The style was first developed in 1908 by Ellsworth Storey, a young architect who designed a group of these cottages. The style did not really become popular until after World War II.

159

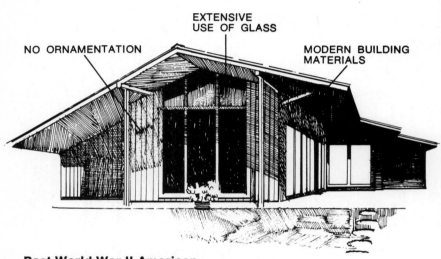

NO ORNAMENTATION

EXTENSIVE
USE OF GLASS

MODERN BUILDING
MATERIALS

Post World War II American

FUNCTIONAL MODERN OR CONTEMPORARY
(Fun Mod or Contemp - 903)

Key Distinguishing Characteristics

The exterior style of a contemporary house is an integral part of the overall design. Its function is to enclose some living areas with modern materials yet integrate the indoor and outdoor space into one unit.

Other Distinguishing Characteristics

Modern building materials
Extensive use of glass
Lack of ornamentation

History

The Contemporary is today's version of the revolution in house styles started by the great American architects Frank Lloyd Wright and Henry Hobson Richardson, the German architects of the Bauhaus and other architects around the world in the past 100 years.

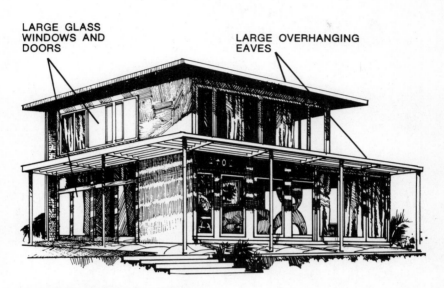

LARGE GLASS WINDOWS AND DOORS

LARGE OVERHANGING EAVES

Post World War II American

SOLAR HOUSE (Solar - 904)

Key Distinguishing Characteristics

This house has large overhanging eaves over the windows on both floors. A great portion of the exterior walls consists of large glass windows and doors. The house is positioned on the site to take advantage of the high summer sun and the low winter sun.

Other Distinguishing Characteristics

Some have eyebrow eaves that can be removed in the winter
Some experimental models have machinery that turns them so the front always faces the sun

History

The idea of a house that would utilize the changing position of the sun became popular after World War II. Several books and many articles were written on the subject. Many Solar Houses were built throughout the country and the principles of their design are still being used today by many architects.

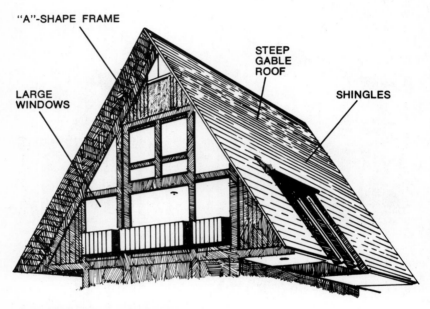

"A"-SHAPE FRAME

STEEP GABLE ROOF

LARGE WINDOWS

SHINGLES

Post World War II American

"A" FRAME (A Frame - 905)

Key Distinguishing Characteristics

In this style, the frame is the shape of one or more "A's".

Other Distinguishing Characteristics

Steep gable roof covered with shingles
Front and rear have large glass windows
Interior often only roughly finished
Sleeping area often is semi-enclosed or sometimes is a loft area

History

With the explosive growth in vacation homes since World War II, millions of families now own second homes that are used primarily for recreational purposes. The simple "A" frame style has been popular in the mountains and near lakes and oceans for this purpose.

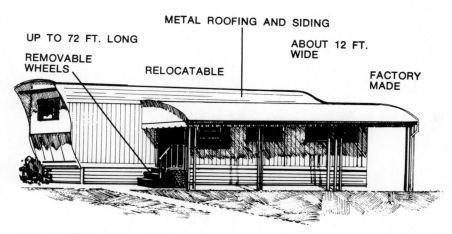

UP TO 72 FT. LONG

METAL ROOFING AND SIDING

ABOUT 12 FT. WIDE

REMOVABLE WHEELS

RELOCATABLE

FACTORY MADE

Post World War II American

MOBILE HOME (Mobile - 906)

Key Distinguishing Characteristics

This is a relocatable house that averages 12 feet wide and 60 feet long. It is produced in a factory and towed by truck or shipped by flat car to the site and often installed upon a prepared base.

Other Distinguishing Characteristics

Wheels and axles used to transport it are removed after installation
Siding and roof often metal

History

During the past 25 years the camping trailer has evolved into today's mobile home. The ongoing argument of whether the mobile home is a trailer or a house is still not settled. Whatever they are, their look is unique and they are considered here to be an architectural style. The American housing industry, for a variety of complex reasons, has failed to provide adequate housing for all Americans at a reasonable cost. This gap has been filled in part by the mobile home industry which had approximately 216,000 mobile home shipments in 1975.

163

PLASTIC EXTERIOR SIDING

CLEAN, MODERN
LINES

Post World War II American

PLASTIC HOUSE (Plastic - 907)

Key Distinguishing Characteristics

A major portion of the exterior siding of this house is fiberglass or other plastic material.

Other Distinguishing Characteristics

Moulded lines and new shapes
Use of new materials
Little or no ornamentation

History

The Plastic House is not an architectural style in the traditional concept. Many new styles may develop using plastic, but the experimental plastic houses being built today do look different from the traditional styles. This house is included to represent them all. The house pictured was designed by architect Valerie Batorewicz of New Haven, Connecticut.

Window, Roof and Ornamentation Styles

In addition to the basic house styles, individual parts of the house often have their own unique style characteristics. Basic types of windows can be shaped or combined into window styles with clearly identifiable characteristics.

Palladian windows, which are fashioned after those designed by Andrea Palladio, are characteristically found in Georgian homes, either in the end gables or above the front entrance in the second story. Gothic windows, characterized by their pointed arch, are found in many Victorian period houses. Circle head and fan windows are small ovals or circles used for decorative purposes, usually over a door. Bay or bow windows form a bay in a room which projects outward from the wall, usually on the first floor, and is supported by its own foundation. Oriel windows project from the outer face of the wall, especially on an upper story, and are supported by brackets or corbels.

Roof shapes also lend themselves to identifiable styles. A key distinguishing characteristic of some house styles is one particular roof style such as the mansard roof on the Second Empire Style and salt box roof on the Colonial Salt Box style. Other roof styles such as gable, gambrel, hip and flat roofs are found on many different styles of houses. Likewise, many house styles, such as the Georgian and Ranch, may have more than one roof style.

Greek columns, entablatures and ornamentation are an essential part of the Classic Colonial and Greek Revival styles. They are also often found on Georgian and other styles.

PALLADIAN WINDOW

WINDOW STYLES

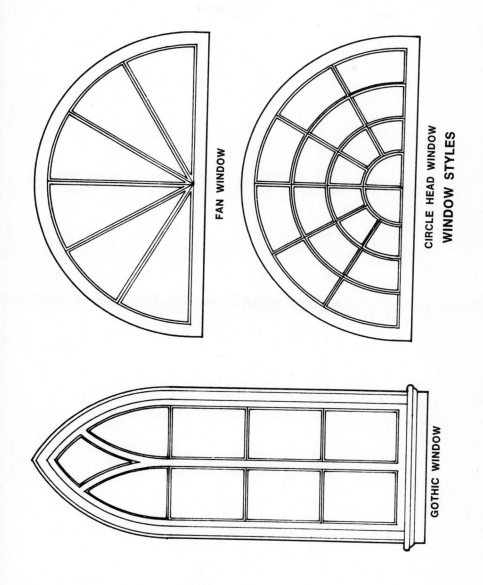

FAN WINDOW

CIRCLE HEAD WINDOW

WINDOW STYLES

GOTHIC WINDOW

WINDOW STYLES

ORIEL WINDOW

BAY WINDOW

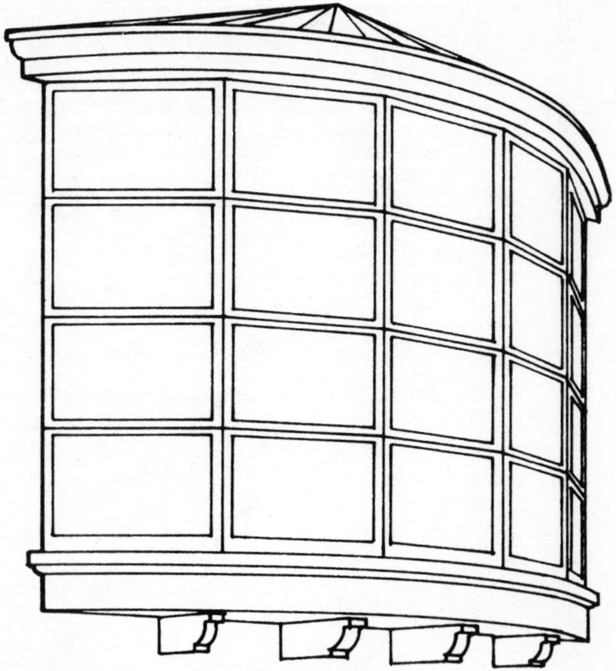

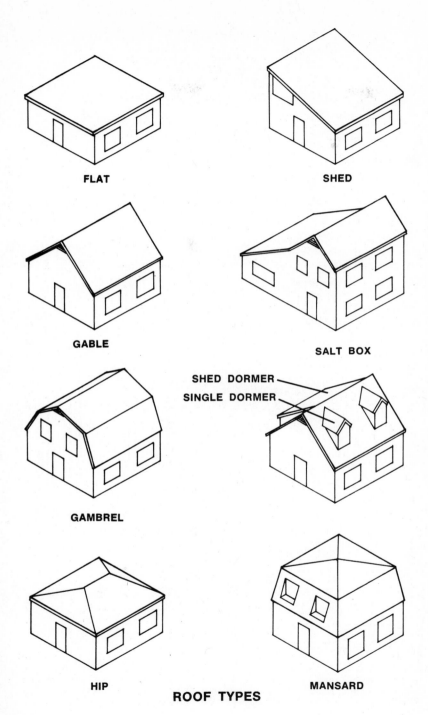

FLAT

SHED

GABLE

SALT BOX

GAMBREL

SHED DORMER

SINGLE DORMER

HIP

MANSARD

ROOF TYPES

"IT'S ALL GREEK TO ME"

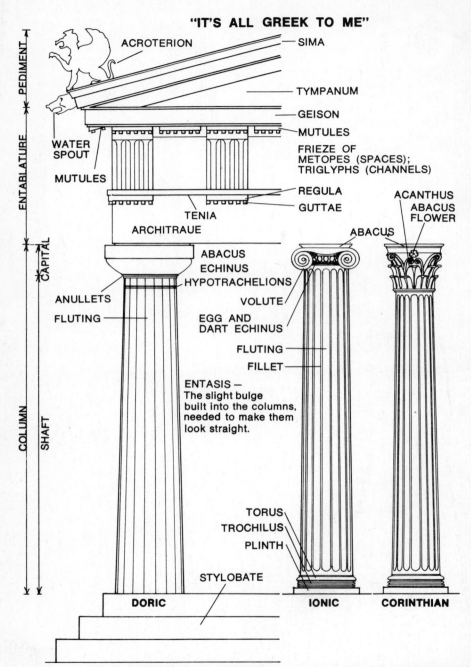

ACROTERION — SIMA

TYMPANUM

GEISON

MUTULES

FRIEZE OF METOPES (SPACES); TRIGLYPHS (CHANNELS)

WATER SPOUT

MUTULES

REGULA

GUTTAE

ACANTHUS
ABACUS
FLOWER

TENIA
ARCHITRAUE

ABACUS

ABACUS
ECHINUS
HYPOTRACHELIONS

VOLUTE

ANULLETS

FLUTING

EGG AND DART ECHINUS

FLUTING

FILLET

ENTASIS — The slight bulge built into the columns, needed to make them look straight.

TORUS
TROCHILUS
PLINTH

STYLOBATE

PEDIMENT

ENTABLATURE

CAPITAL

COLUMN

SHAFT

DORIC

IONIC

CORINTHIAN

A great many Classic Colonial, Greek Revival and other house styles exhibit Greek columns, capitals, entablatures and other Greek ornamentations. This illustration is a simple guide to Greek architecture and terminology.

Chapter 5
Basic
Construction

This section of the book attempts the impossible — to explain in a relatively few number of pages how a house is constructed. Though it may seem complex at first glance, the reader should not be scared away; he will be able to understand it. Most likely, he will find a great deal of the information familiar. Substantial reliance has been placed on the Minimum Property Standards established by the U.S. Department of Housing and Urban Development (FHA-MPS).

Substructures

Excavation

According to FHA-MPS, the objective of the excavation is to provide a safe and adequate support for footings and foundations. Adequate clearance is needed to assure protection against damage by decay or insect attack. Good excavation also provides drainage for and access to basementless space.

The process usually involves several steps. First, the location of the house is staked out by the builder or surveyor and the top soil is removed to a point where it is to be stored during the construction process and respread as the final step in the finish grading. Then the main hole is dug — first, usually with the help of a power shovel and then by hand to a depth 6 to 10 inches below the floor level. Finally, trenches are dug for the footings and service pipes, drains, drywells and septic tanks.

Excavations for footings and foundation walls should extend at least 6 inches into natural, undisturbed soil in order to provide adequate bearing except where bearing is on a stable rock formation and below the prevailing frost line, MPS specifies.

Footings and Foundation Walls

The objective of the footing, according to MPS, is to provide support for the dwelling without excessive differential or overall settlement or movement. It is the perimetric base of concrete that is laid by pouring the concrete into wooden forms set at a level below the frost line and on undisturbed earth. It should be noted that all substructures will settle to a certain extent unless they are located on solid bedrock and that excess shifting and settlement will cause cracks and leaks in the foundation wall and uneven floors in the house. Thus, local building codes specify the required depth for each region based on the local frost line depth.

Since the objective of the foundation is to provide construction which assures safe and adequate support for all vertical and lateral design loads, all foundation walls are poured or laid on top of the footings. Block walls must be properly laid and well mortared, then filled with concrete and made watertight with cement plaster or other waterproofing compounds. Cinder blocks are porous and are thus inferior to cement blocks for a solid foundation. Brick and tile, although good foundation materials, are costly and require substantial skill

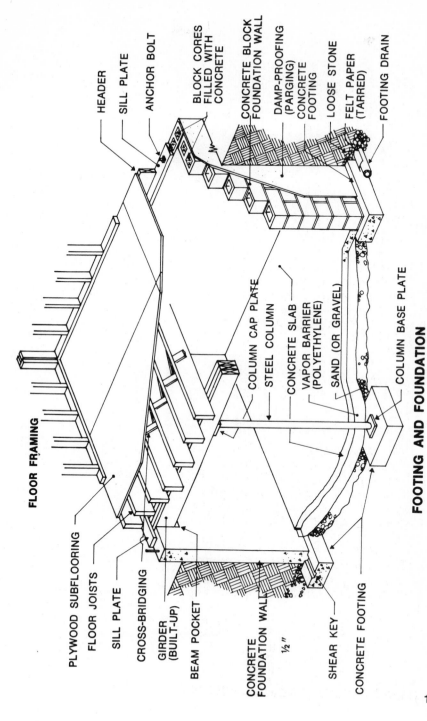

FLOOR FRAMING

HEADER
SILL PLATE
ANCHOR BOLT

BLOCK CORES FILLED WITH CONCRETE
CONCRETE BLOCK FOUNDATION WALL
DAMP-PROOFING (PARGING)
CONCRETE FOOTING
LOOSE STONE
FELT PAPER (TARRED)
FOOTING DRAIN

COLUMN CAP PLATE
STEEL COLUMN
CONCRETE SLAB
VAPOR BARRIER (POLYETHYLENE)
SAND (OR GRAVEL)
COLUMN BASE PLATE

PLYWOOD SUBFLOORING
FLOOR JOISTS
SILL PLATE
CROSS-BRIDGING
GIRDER (BUILT-UP)
BEAM POCKET

CONCRETE FOUNDATION WALL
½"
SHEAR KEY
CONCRETE FOOTING

FOOTING AND FOUNDATION

173

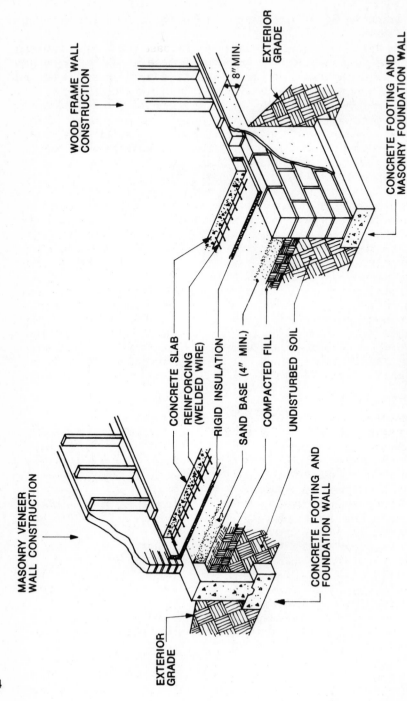

WOOD FRAME WALL CONSTRUCTION

8" MIN.

EXTERIOR GRADE

CONCRETE FOOTING AND MASONRY FOUNDATION WALL

MASONRY VENEER WALL CONSTRUCTION

EXTERIOR GRADE

CONCRETE SLAB

REINFORCING (WELDED WIRE)

RIGID INSULATION

SAND BASE (4" MIN.)

COMPACTED FILL

UNDISTURBED SOIL

CONCRETE FOOTING AND FOUNDATION WALL

STRUCTURAL SLAB ON GRADE

174

for proper laying, as does stone, which was once very popular in the Northeast.

The three basic forms of foundations are basements, crawl spaces and slab-on-ground. With the exception of those being constructed in the northern portions of the country, fewer and fewer houses are being built with basements. And, where basements are built, there is an increasing trend to gain additional living space by finishing portions into family rooms, utility areas, baths and lavatories, workrooms, kitchens and even bedrooms. In the event that the house has a basement, the height between the basement floor, which is constructed similarly to a slab, and the bottom of the joists is usually 7½ to 8 feet.

For basementless houses, the finish grade is a major factor in the choice between slab-on-ground or crawl space as a foundation. For slab-on-ground construction, it is important that the finished ground grade fall sharply away from the house in order to prevent flooding. Slabs are constructed by first building footings for support, although some slabs, known as "floating slabs," are built without them. The excavation is then covered with gravel and a vapor barrier and insulation is installed around the edge.

Crawl spaces, which provide flooding protection and also provide a convenient place to run heating ducts, plumbing pipes and wires that must be accessible for repairs, are constructed similarly to basements except that the distance from the floor to the joists is 3 to 4 feet. The floor can be concrete, as in a basement, or it can be dirt, often covered with a vapor barrier. In northern regions, crawl spaces must be insulated or heated to prevent pipe freezing and cold floors.

A Wet or Damp Basement

Dampness, of course, is the main problem with basements, since it damages wall and floor coverings, furniture, clothing and other possessions. It also poses a health hazard — especially when the basement is used for sleeping. Some of the causes of basement dampness which can be thwarted by the careful builder are poor foundation wall construction, excess ground water not properly carried away by ground tiles, poorly fitted windows or hatch, a poorly vented clothes dryer, gutters and downspouts spilling water too near the foundation wall and a rising water table in the ground.

A basement that is wet or damp only part of the year can usually be detected any time by careful inspection. All the walls should be checked for a powder-white mineral deposit a few inches off the floor. Only the most diligent cleaning will remove all these deposits after a basement has been flooded.

Stains along the lower edge of the walls and columns and on the furnace and hot water heater are indications of excessive dampness, as is mildew odor.

The causes of wet and damp basement are numerous. Some are easily corrected and others are almost impossible to correct. In areas where the soil drainage is poor or the water table is near the surface of the ground, it is necessary to have well constructed footing and foundation drains to maintain a dry basement. They should be installed when the house is constructed since it is a major expense to do so afterwards. The same is true of a vapor barrier under the basement floor, which is very easy to put down during construction but impossible afterwards.

Cracks in the floor and walls may be patched with various widely marketed compounds. A more drastic step is to dig down and repair the wall from the outside.

What appears at first to be a major water problem might be traced to a leak in a window or the hatch door. A simple caulking job will stop the water from coming in. Water will leak in through a window at the bottom of a well that does not drain properly in a heavy rain storm. Extending the drain line or deepening the dry well stops this problem.

The earth around the house should slope away from the foundation wall so ground water will not collect along the edge of the foundation. If there is an edge of the roof line without a gutter, water may be running off and collecting next to the foundation wall. The water that is collected by the gutter and flows into the leaders must be diverted away from the foundation wall. The leader should run into a sewer drain, dry well or splash pan—in that order of preference.

Dampness and mildew may also be caused by moisture condensing on the walls, ceiling and pipes. Proper ventilation eliminates this problem.

Main Bearing Beam and Columns

Since most houses are too large for the floor joists to be spanned from one foundation wall to the opposite foundation wall, one or more bearing beams resting on columns or piers are used to support the floor joists. If only one beam is required, it runs roughly down the center of the basement or crawl space.

Steel beams, because of their high strength, can span greater distances than wood beams of the same size. However, steel beams are subject to fire damage from relatively low heat. A steel beam will lose some of its strength at 500 degrees and at 1,000 degrees will buckle under a normal load. Beams that are covered with metal lath and plaster, on the other hand, will maintain their strength under much higher temperatures for long periods of time.

Wood beams, although not as strong as steel, are often used and are quite satisfactory. When a solid member is used, it is often 6 by 8 inches to 10 by 10 inches. Plank beams consist of several 2 by 6-inch to 2 by 10-inch planks placed side by side on end to achieve the desired thickness.

Most beams are supported by wood posts, brick or block piers or metal Lally columns which are concrete-filled steel cylinders. It is important that the post, pier or column rest on a footing, which should be at least 2 feet square and 1 foot thick. If brick and block piers are used, they should be at least 12 inches square but preferably 16 inches square. If wood posts are used, they should be set on a platform several inches off the floor so that any water on the floor from leaks will not rot them. Steel columns require caps and base plates.

Framing

Nine out of ten houses in this country are of wood frame construction. Many of them are covered with wood siding; others may be covered with wood shingles, composition shingles or siding, brick veneer or stucco. Regardless of the type of exterior covering, these houses are in the general classification of wood frame construction. However, there are three different types of wood frame construction: platform, balloon and plank and beam.

Platform Frame Construction

In platform frame construction, the subfloor extends to the outside edges of the building and provides a platform upon which exterior walls and interior partitions are erected. Platform construction is the type of framing most generally used for one-story houses. It is also used alone or in combination with balloon construction for two-story structures. Thus, because of its wide use, building techniques in some parts of the country have been developed almost exclusively around the platform system.

A platform-constructed house is easier than the others to erect because, at each floor level, a flat surface is provided on which to work. Moreover, it is also easily adapted to various methods of prefabrication. With a platform framing system, it is common practice to assemble the wall framing on the floor and then tilt the entire unit into place.

Balloon Frame Construction

In balloon frame construction, both studs and first-floor joists rest on the anchored sill. The second-floor joists bear on a 1 by 4-inch ribbon strip which has been let into the inside edges of the studs.

Balloon framing is a preferred type of construction for two-story buildings where the exterior covering is of brick or stone veneer since there is less likelihood of movement between the wood framing and the masonry veneer. Where exterior walls are of solid masonry, it is also desirable to use balloon framing for interior bearing partitions since this eliminates any variations in settlement which may occur between exterior wall and interior supports.

Plank and Beam Construction

In the plank and beam method of framing, beams of adequate size are spaced up to 8 feet apart and are covered with 2-inch planks which serve as the base for finish flooring or roof covering. The ends of the beams are supported on posts and the covering for exterior walls is attached to supplementary members set between the posts. Details for this method of construction are provided in "Wood Construction

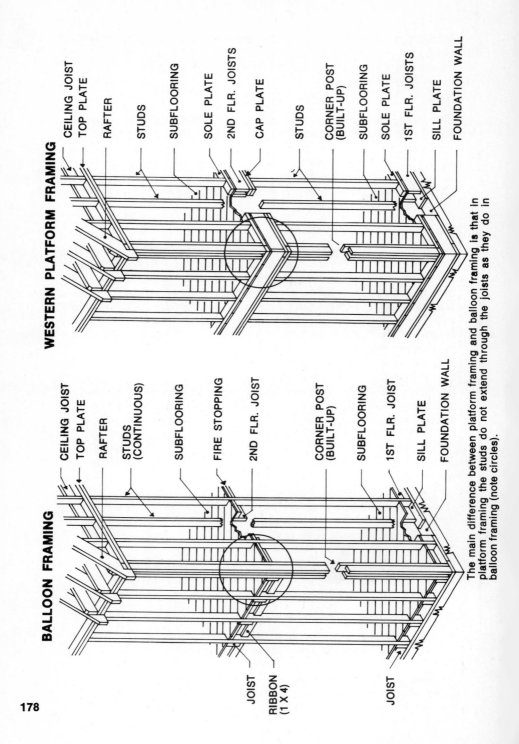

BALLOON FRAMING

CEILING JOIST
TOP PLATE
RAFTER
STUDS (CONTINUOUS)
SUBFLOORING
FIRE STOPPING
2ND FLR. JOIST
CORNER POST (BUILT-UP)
SUBFLOORING
1ST FLR. JOIST
SILL PLATE
FOUNDATION WALL

JOIST
RIBBON (1 X 4)

JOIST

WESTERN PLATFORM FRAMING

CEILING JOIST
TOP PLATE
RAFTER
STUDS
SUBFLOORING
SOLE PLATE
2ND FLR. JOISTS
CAP PLATE
STUDS
CORNER POST (BUILT-UP)
SUBFLOORING
SOLE PLATE
1ST FLR. JOISTS
SILL PLATE
FOUNDATION WALL

The main difference between platform framing and balloon framing is that in platform framing the studs do not extend through the joists as they do in balloon framing (note circles).

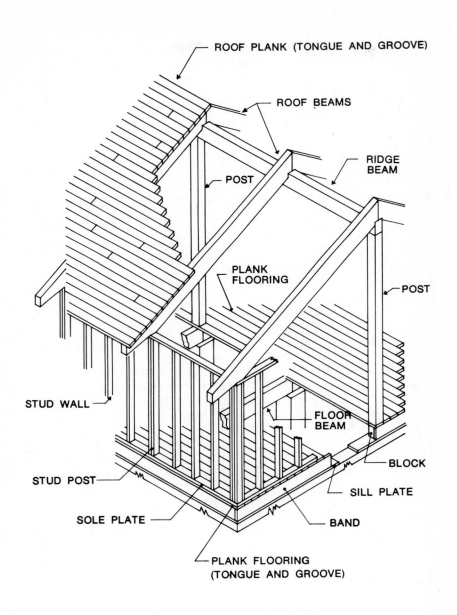

ROOF PLANK (TONGUE AND GROOVE)

ROOF BEAMS

RIDGE BEAM

POST

POST

PLANK FLOORING

STUD WALL

FLOOR BEAM

BLOCK

STUD POST

SILL PLATE

SOLE PLATE

BAND

PLANK FLOORING (TONGUE AND GROOVE)

PLANK AND BEAM FRAMING

179

Data No. 4" of the series published by the National Lumber Manufacturers Association.

One advantage of plank and beam framing is that it lends itself to use in the construction of contemporary houses where the planks and beams are used as exposed, finished materials.

Exterior Frame Walls

Framing

Exterior wall framing should be strong and stiff enough to support the vertical loads from floors and roof. Moreover, the walls should be able to resist the lateral loads resulting from winds and, in some areas, from earthquakes. Thus, the top plates should be doubled and overlapped at the wall and bearing partition intersection in order to tie the building together into a strong unit.

Studs, which in exterior walls are placed with the wide faces perpendicular to the direction of the wall, should be at least a minimum 2 by 4 inches for one and two-story buildings. In three-story buildings, studs in the bottom story should be at least 2 by 6 inches. In one-story buildings studs may be spaced 24 inches, on center, unless otherwise limited by the wall covering, while in multi-story buildings spacing should not exceed 16 inches on center. In all cases an arrangement of multiple studs is used at the corners to provide for ready attachment of exterior and interior surface materials.

Where doors or windows are to be located, provision must be made in the framing to carry the vertical load across the opening. Such provision is made by a header of adequate size, the ends of which may be supported either on studs (or by framing anchors when the span or opening does not exceed 3 feet in width).

If the builder chooses, a continuous header consisting of 2-inch members set on edge may be used instead of a double top plate. If a continuous header is used, the depth of the members must be the same as that required to span the largest opening and the joints in individual members should be staggered at least three stud spaces and should not occur over openings. Moreover, the members should be toe-nailed to studs and corners; intersections, with bearing partitions, should be lapped or tied with metal straps. Studs in gable ends should rest on wall plates with top notches to fit the end rafter to which they are nailed.

Defective House Framing

Once a house is a few years old, signs of defective framing can be detected visually. One sign is bulging exterior walls, which are best seen by standing at each corner of the house and looking along the wall. Another method is to make a plumb line out of a key and string and hold it against the wall. If the ridge line sags in the middle, trouble is developing.

Window sills that are not level are a sign of settling, defective framing or original sloppy carpentry. A careful house inspection should include the opening and closing of every window. Sticking windows may be a sign of settling or defective framing.

A sure sign of trouble is a large crack developing on the outside of the house between the chimney and the exterior wall. Other tip-offs to defective framing are cracks running outward at an angle from the upper corners of window and door frames.

Sagging and sloping floors may be detected visually or by putting a marble on the floor and watching to see if it rolls away. This may be a sign of defective framing.

Cracks in the walls other than those discussed above should be a cause of concern but in themselves are not conclusive evidence of framing problems. All houses settle unless built upon solid rock. Rare is the house that does not develop some wall and ceiling cracks. They should be of most concern when accompanied by some of the other signs of defective framing.

Sheathing

Exterior walls should be braced by a suitable sheathing applied horizontally or, preferably, diagonally to the framing. The diagonal method of placing sheathing is preferable to the horizontal because additional strength and stiffness may be provided by 1 by 4-inch members set into the outside face of the studs at an angle of 45 degrees and nailed to the top and bottom plates and studs. Moreover, where wood sheathing boards are applied diagonally, let-in braces are not necessary. In either case, sheathing should be nailed to sills, headers, studs, plates or continuous headers and to gable end rafters.

Wood sheathing is preferred by many builders because it provides a solid nailing base for applying exterior siding and finish. It usually is in square-edge, shiplap or tongued-and-grooved patterns. And when it is employed, joints should be made on studs unless end-matched boards are used, in which case each board should bear on at least two studs. Plywood, however, also makes an excellent sheathing material which can be nailed, stapled or glued to the frame. Other materials used as sheathing include fiberboard and specially fabricated gypsum panels which often have the sheathing paper and exterior finish incorporated into them.

Sheathing Paper

Weathertight walls are provided by covering the sheathing on the outside with sheathing paper which may be either asphalt-saturated felt weighing not less than 15 pounds per 108 square feet or any other impregnated paper having equivalent water-repellent properties but which will not act as a vapor barrier. Starting at the bottom of the wall, the sheathing paper should be applied by lapping it 4 inches at horizontal joints and 6 inches at vertical joints. Then, strips of sheathing paper about 6 inches wide should be installed behind all exterior trim and around all openings.

Flashing

Wherever joints occur in which dissimilar materials come together or wherever there is a possibility of leakage because of the type of construction, it is usually necessary to insert flashing — sheets or membranes of waterproof materials — to turn back the water.

Although flashing may be composed of impregnated felt or of combinations of felt and metal, it is generally composed of one of the following metals: copper, zinc, lead, leaded copper, tin, galvanized iron or steel, tin-plated steel, soft iron and copper-bearing or alloy steel. Of these, the pure metals are the best. And of the pure metals, copper, lead, zinc and leaded copper are both the most commonly used and the most effective because they corrode and deteriorate the least upon exposure to the weather. Coated sheets such as galvanized iron and tin-plated steel are satisfactory as long as the coating is unbroken. After that, however, corrosion may be accelerated by the electrolytic action set up between dissimilar metals.

No matter what the metal employed, the installation of flashing is practically identical. Thus, the following discussion, although based upon copper (the metal most widely employed for flashing), is applicable to all of the other metals it is possible to use.

When using metal exposed to changes in temperature found on building exteriors, the first point to remember is that metals have fairly high coefficients of thermal expansion and must be free to move with changes in temperature. If they are confined, the stresses set up in trying to change dimensions eventually cause fatigue cracks. Short, narrow strips, less than 10 to 12 inches wide, can be fastened along both edges without much danger; wider sheets, however, must not be confined along two opposite edges but must be free to move.

The second point to remember is that flashing must be installed in such a way that water is shed over any unsealed joints in it; i.e., joints must be so made that water could work through them only against the force of gravity. Furthermore, it must be impossible for driving winds to force water through. Thus it is usually necessary that the joint either provide a tortuous path in which the driving force of the wind is dissipated or that the laps in the joint be of sufficient length that water cannot possibly be driven through.

Copper flashing is generally 16-ounce "roofing temper" (R.T.), a soft and pliable material weighing 16 ounces per square foot. Heavier copper is required for heavy-duty flashing such as around certain kinds of tile roofs, whereas lighter copper (not less than 14-ounce) can be used in relatively protected points.

Finishes

Once framing, sheathing and flashing are dealt with, the matter of finishes for the exterior must be considered. Many types of siding

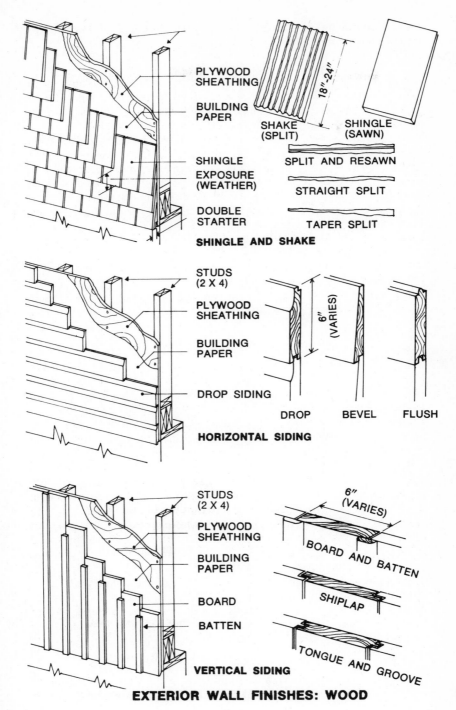

PLYWOOD SHEATHING

BUILDING PAPER

SHINGLE EXPOSURE (WEATHER)

DOUBLE STARTER

18"-24"

SHAKE (SPLIT)

SHINGLE (SAWN)

SPLIT AND RESAWN

STRAIGHT SPLIT

TAPER SPLIT

SHINGLE AND SHAKE

STUDS (2 X 4)

PLYWOOD SHEATHING

BUILDING PAPER

DROP SIDING

6" (VARIES)

DROP   BEVEL   FLUSH

HORIZONTAL SIDING

STUDS (2 X 4)

PLYWOOD SHEATHING

BUILDING PAPER

BOARD

BATTEN

6" (VARIES)

BOARD AND BATTEN

SHIPLAP

TONGUE AND GROOVE

VERTICAL SIDING

EXTERIOR WALL FINISHES: WOOD

183

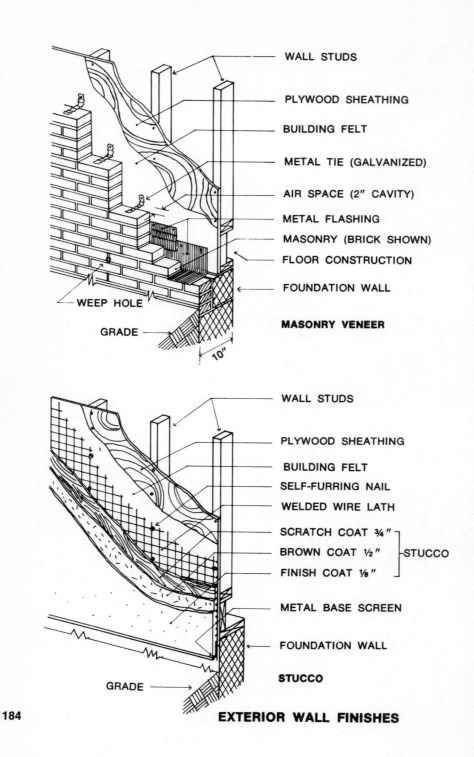

WALL STUDS

PLYWOOD SHEATHING

BUILDING FELT

METAL TIE (GALVANIZED)

AIR SPACE (2" CAVITY)

METAL FLASHING

MASONRY (BRICK SHOWN)

FLOOR CONSTRUCTION

FOUNDATION WALL

MASONRY VENEER

WEEP HOLE

GRADE

10"

WALL STUDS

PLYWOOD SHEATHING

BUILDING FELT

SELF-FURRING NAIL

WELDED WIRE LATH

SCRATCH COAT ¾"

BROWN COAT ½"   STUCCO

FINISH COAT ⅛"

METAL BASE SCREEN

FOUNDATION WALL

STUCCO

GRADE

EXTERIOR WALL FINISHES

and other exterior coverings are applied over wood framing. Often a house has more than one type of siding on it. Numerous patterns of wood siding are available. Names of different types, varying with the locality, include bevel, bungalow, colonial, rustic, shiplap and drop siding.

If bevel siding or square-inch boards are used, they should be applied horizontally and lapped 1 inch with nails driven just above the lap to permit possible movement due to change in moisture conditions. Moreover, they should be spaced so that the bottoms of the pieces coincide with the top of the trim over the door and window openings — an arrangement requiring careful planning by the carpenter before starting to apply the siding. Finally, when using bevel siding or square-inch boards, it is good practice to apply a liberal coating of water repellent to the end surfaces. No matter what the siding or exterior trim, it should be installed with corrosion-resistant nails, usually of galvanized steel or aluminum. Where wood sheathing is used, siding may be nailed at 24-inch intervals. Where other types of sheathing are used, nails should be driven through the sheathing into the studs at each bearing. Length of nail will vary with thickness of siding and types of sheathing.

Where shingles are installed in double courses (double layers), the butt of the exposed shingle extends about ¼ inch below the undercourse or layer in order to produce a shadow line. The undercourse should be attached to the sheathing with nails or staples and the outercourse attached with small headed nails driven approximately 2 inches above the butts and ¾ inch from edges.

In all cases, however, shingles should be nailed with corrosion-resistant nails of sufficient length to penetrate the sheathing, using two nails for shingles up to 8 inches wide and three nails for wider shingles. When shingles are installed in a single course or layer, the nails should be driven approximately 1 inch above the butt line of the following course.

Finally, it should be noted that when other than wood sheathing has been used, 1 by 3-inch horizontal nailing strips must be applied over the sheathing spaced to correspond with the weather exposure of the shingles.

Stucco is still popular in dry climates where it can be applied directly to the surface of a solid masonry wall. However, the application of stucco to a wood frame wall is generally more involved than that of other finishes. Thus, the high cost of labor for its application has reduced its popularity in the northern parts of the country.

When stucco, which is a type of plaster that can be variously patterned or colored, is used, however, its application involves the following steps. First, wood firing strips are nailed to the sheathing through the building paper. Then metal lathing is stretched onto the firing strips. Finally, the stucco is troweled onto the lathing.

A masonry veneer wall is really a frame wall with some variety of masonry siding, most commonly clay bricks, concrete bricks, split blocks or stone. In houses with masonry veneeer walls, all of the structural functions of the walls are performed by the framing and not by the one-unit thick masonry which is tied to the frame wall with rustproof metal ties spaced one tie for each 2 square feet of wall. When the walls are constructed, ¾ inch to 1 inch air space is left between the masonry and the sheathing and weep holes are installed at the base to let moisture escape.

A variety of other types of siding materials are available. These include aluminum, stone, hardboard, gypsum board, fiberglass and metals. Each of these require special techniques for proper installation. Instructions are provided by the manufacturers and should be carefully followed.

Solid Masonry Walls

Solid masonry walls, if well constructed, are very durable and easily maintained. However, they should be insulated and they do require a larger foundation than a wood frame wall. Such walls can be either one or two units thick. Single-unit-thick walls are most commonly made of 8-inch concrete.

While solid multiple-unit masonry walls were, until recently, the most common type of masonry wall, they are now mainly used only where required by local building codes. Where they are used, these walls are constructed either of two layers of brick, tile or cement block or of a combination of materials, with the higher grade on the outside and the cheaper unit as the back-up. In the latter case, masonry headers (bricks laid across two thicknesses of walls) or metal ties are used to tie the face units to the back-up units.

Hollow Masonry Walls

In the case of hollow masonry walls, a cavity masonry wall is built of two units that are separated into an inside and outside wall by between 2 to 4 inches of air space and bonded together with metal ties or joint reinforcement. The exterior wythe (thickness) is usually 4 inches and the interior wythe, 4 to 8 inches.

These cavity walls are used mainly in northern sections of the country for protection against severe outside temperatures and storms. They provide added protection from the elements when insulation is installed in the cavity as well as eliminating the need for furring on the inside wall since moisture penetration is almost impossible if cavity walls are properly constructed with flashing and weep holes (holes at the bottom to let any moisture escape).

Masonry bonded walls are similar to cavity walls except that the two wythes are joined by masonry header courses instead of by metal ties. Although they are economical to construct, their insulation qualities are inferior to cavity walls and they are, therefore, used mainly in the Southwest.

Clay Brick and Tile Installation

The primary factor in good brick and tile construction is proper installation. The major source of problems is partially filled mortar joints which will substantially reduce the strength of the wall and contribute to rapid disintegration and cracking. Water then penetrates the wall and expands during the freezing cycle further cracking the mortar.

Storage and Preparation of Materials on Site

Bricks and tiles should be stored off the ground and covered. High-suction brick when laid dried absorbs the water from the mortar; therefore it must be thoroughly soaked with water prior to installation. The brick or tile is then left to surface dry because surface water, by causing floating, will interfere with the bond between the brick or tile and the mortar.

Laying the Brick or Tile

This process begins by locating the corner and laying out the first layer of brick or tile to determine how much cutting or extending will be needed for an exact fit. A full, thick bed of mortar is spread for the first course. Several units are laid and then aligned with a level. After the first course is laid the corners are built up. A line is stretched from corner to corner for each course and the top outside edge of each unit is laid to this line. Head joints are completely filled with mortar on the end of the brick before placement. When the brick is shoved into place, mortar squeezes out of the sides and top, indicating that the head joint is completely filled. Alternate ways of filling the head joint are to throw a trowel of mortar on the end of the brick already in place or to place a full trowel of mortar on the wall and squeeze the brick into place, forcing the mortar up into the head joint.

Mortar Joints

There are two basic types of mortar joints—troweled joints and tooled joints. With troweled joints, after the brick or tile is laid in place the excess mortar squeezes out of the joint. If it is simply removed with a trowel and no further finishing done, it is a struck joint. If it is struck off parallel to the unit surface, it is a flush or plain cut joint. If it is struck on an angle with some of the joint removed, it is a struck joint. If some of the top of the joint is removed it is a weathered joint. For most residential purposes the weathered joint is preferred since it sheds water.

A tooled joint is a better joint which can be obtained by tooling it with a special tool which compresses and shapes the mortar in the joint. Common types of tooled joints are rodded joints, "V"-shaped joints and beaded joints.

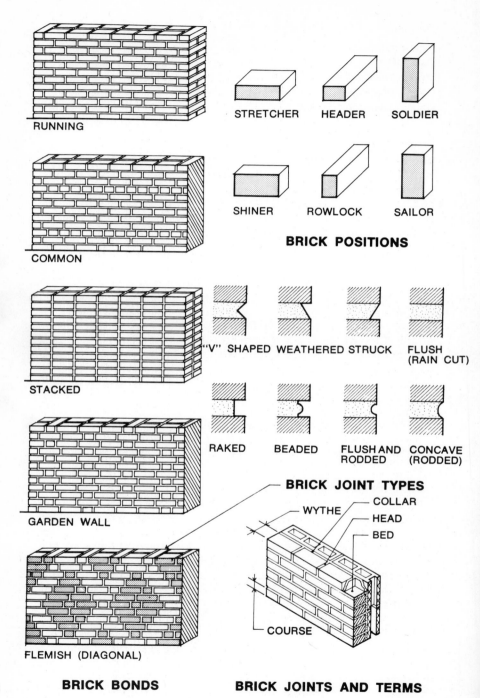

RUNNING

COMMON

STACKED

GARDEN WALL

FLEMISH (DIAGONAL)

STRETCHER   HEADER   SOLDIER

SHINER   ROWLOCK   SAILOR

BRICK POSITIONS

"V" SHAPED WEATHERED STRUCK FLUSH (RAIN CUT)

RAKED   BEADED   FLUSH AND RODDED   CONCAVE (RODDED)

BRICK JOINT TYPES

COLLAR

WYTHE

HEAD

BED

COURSE

188   **BRICK BONDS**   **BRICK JOINTS AND TERMS**

Parging

When more than one unit thick of brick or tile is used in the wall, the back of the face brick or tile may be covered with mortar. This parging will substantially increase the waterproofing of the wall. Basement walls are usually parged.

Weep Holes

By inserting short lengths of sash cord through the mortar at the bottom, a weep hole is made through which any water that gets inside a double unit wall may run out.

Cleaning

After the wall is completely laid and let stand for 48 hours its surface is cleaned from top to bottom with soap and water. Stains are removed with hydrochloric acid after the wall is seven days old.

Concrete Brick and Block Installation

A good, solid concrete brick or block wall depends upon good workmanship and proper installation. The joints must be well made and the unit carefully laid. The space between the bricks should be completely filled. Partially filled joints make a weak, leaky wall that is subject to cracking easily.

Storage and Preparation of Material on Site

The amount of moisture in concrete masonry is controlled to meet building specifications. Steps often taken to keep the units dry are putting them on planks to keep them off the ground and covering them with a waterproof cover. Concrete units are laid dry.

Laying the Brick or Blocks

The corner is located and the first layer of brick or block is laid out to determine if a cutting will be needed or if the desired wall length can be obtained by adjusting the size of the joints.

A full thick bed of mortar is spread for the first course and furrowed with a trowel to insure plenty of mortar along the bottom edge of the face shells of the block. Blocks are laid with their thicker side up. Several units are laid and then aligned with a level. After the first course is laid the corners are built up. A line is stretched from corner to corner for each course and the top outside edge of each unit is laid to this line. The balance of the laying process is generally the same as for clay brick and tile.

Insulation and Weatherproofing

Organic substances such as eelgrass, animal hair and sawdust were the first materials used in insulation. Production of manmade insulation material began in 1840 in Wales and in the U.S. in 1875, when "mineral wool" (inorganic fibrous insulation) was derived from iron slag. Pneumatic installation and blanket and bat production started in the 1920's followed by glass fiber insulation production in the mid-1930's.

However, prior to World War II, many houses were constructed with little or no insulation since their heavy building materials and tight-fitting window and door sashes gave them sufficient weather resistance. Newer houses, constructed of lighter materials and with less precise workmanship, require insulation to keep the heat and cold from penetrating.

Residential insulation, which is reasonable and easy to install as part of the initial construction, falls into the following five categories: loose, blankets and bats, foil, sprayed on and wallboard.

The most popular loose materials are rock wool, glass wool, slag wool, perlite, vermiculite, wood fiber, paper, cotton fiber paper and macerated paper. All of these products, which should be bugproofed and fireproofed before installation, may be blown or poured into the hollow spaces between the studs on the exterior walls and above the ceiling or below the roof rafters. The major disadvantage, however, is that they tend to settle, thus eventually leaving uninsulated spaces.

Blankets and bats are usally made of the same material as wallboard, loosely felted, glued between two sheets of treated paper or foil and quilted. When installed, they are stapled, clipped or nailed with lathing between the studs and under the rafters or over the ceiling. Here it should be noted that insulation under the rafters is called "cap insulation," while insulation in all the walls between the heated and the unheated portions of the house is "full insulation."

Sprayed-on insulation is a hot, viscous mixture that is sprayed onto the inside of the sheathing. When it cools and solidifies, it becomes a porous layer 1 to 2 inches thick. Foil, usually aluminum several thousandths of an inch thick, is another type of insulation which can be installed in up to four layers. It is especially effective in keeping heat out.

Wallboard insulation, which is used successfully under siding and roofs, is made of a variety of synthetic materials or wood and vegetable fibers, which are mechanically separated and then recombined by matting or felting to form many small air cells. The board is formed by putting the material under pressure. The more pressure used, the stronger and more rigid the board but the less effective its insulating qualities. Therefore, only moderate pressure is used, thus causing the boards to be much softer than regular wood.

The two primary benefits of insulation are fuel economy and occupant comfort. Its secondary benefits are the reduction of sound transmission and the reduction of the danger of fire spreading. Insulation keeps the heat inside when it is cold outside and the heat out when it is hot outside. Therefore, good insulation is important in most climates.

When examining the economic benefits of insulation, the first question that arises is, how much can be saved by the proper insulation of

a house? The answer depends upon many factors but mainly upon

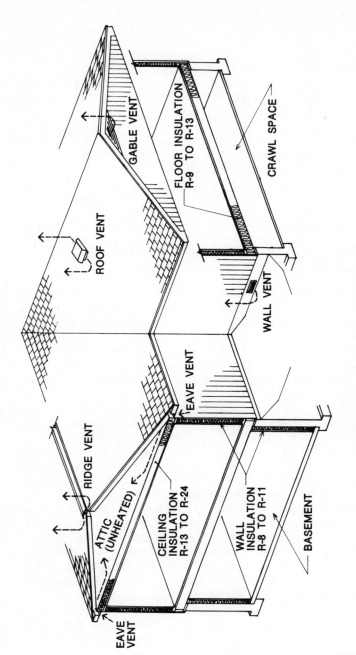

RIDGE VENT

EAVE VENT

ATTIC
(UNHEATED)

EAVE VENT

CEILING
INSULATION
R-13 TO R-24

WALL
INSULATION
R-8 TO R-11

BASEMENT

GABLE VENT

ROOF VENT

FLOOR INSULATION
R-9 TO R-13

CRAWL SPACE

WALL VENT

VENTILATION AND INSULATION

191

how weatherproof the windows and doors have been made and upon the quality both of the ceiling insulation and ventilation and of the wall insulation. The calculation of the heat losses and gains in terms of insulation is a subject best left to heating and cooling experts. However, local electric companies will supply complete information on how the losses and gains can be calculated.

Here are some rules of thumb for heating and cooling savings obtained by adding various types of weatherproofing and insulation to a frame, one-story building:

Insulation Description	Fuel Cost Savings
Weatherstripping all doors and windows	3-5%
Storm windows and doors on all openings	10-20%
Minimum attic or ceiling insulation	5-10%
Maximum attic or ceiling insulation	15-20%
Minimum wall insulation	5-15%
Maximum wall insulation	10-20%

The total difference between fuel costs for an uninsulated house and for an otherwise identical one with storm windows and doors and good insulation in the walls and ceiling can be 50 percent.

The standard measurement for the effectiveness of insulation is its "R" value (resistance to heat flow). The higher the "R" value the better the insulation. Thus, most brand name insulation products are marked with their "R" value.

Over-ceiling or under-roof insulation should have an "R" rating from R-13 in mild climates where there is no air-conditioning and gas or oil heat to R-24 or better in colder climates or hot climates or where there is electric heat or air-conditioning.

Exterior wall insulation requires an R-11 rating for hot or cold climates or electric heat or air-conditioning, down to R-8 for mild climates, no air-conditioning and gas and oil heat.

Floor insulation, if a house is built over a crawl area, should be at least R-9 and preferably R-13. When the house is built on a slab, only edge floor insulation is required. And when the house is built over a basement, no floor insulation is needed at all.

When examining the occupant comfort benefits of insulation the first to consider is reduction of the "cold wall" effect. The human body feels uncomfortable when it is losing heat too fast. And, even if a room itself feels warm, body heat will radiate to a nearby cold surface (wall, floor, ceiling) and produce a chilled feeling. In the summer, reverse conditions are in effect when excessively warm surfaces make it difficult for the body to maintain its normal temperature.

To compensate for the discomfort produced by heat radiation, most people will set the thermostat higher in the winter and lower in the summer, thereby increasing the fuel cost. But insulation helps to make a house comfortable without increased fuel costs because it helps make more uniform room-to-room and floor-to-ceiling air temperature differences. Moreover, it reduces drafts from convection currents which are generated by interior surface-air temperature differences.

Ventilation

According to MPS, a dwelling ideally should provide natural ventilation in areas such as attics and basementless spaces in order to minimize decay and deterioration of the house and to reduce attic heat in the summer. Thus, MPS makes the following recommendations:

1. Eight mesh per inch screening should be used to cover all exterior openings.
2. At least four foundation wall ventilators should be provided in basementless spaces or crawl spaces unless one side of such space is completely open to the basement.
3. Cross ventilation should be provided in an attic and in spaces between roofs and top floor ceilings by venting. All openings should be designed to prevent the entrance of rain or snow.

Interior Walls and Ceilings

Interior Masonry Walls

Interior masonry walls can be classified as either load bearing or non-load bearing. According to MPS, load bearing walls should be 6 inches thick when supporting not more than one floor and 8 inches thick when supporting more than one floor. Non-bearing partitions should be 3 inches thick and all masonry should be supported on masonry, concrete or steel. Wood framing should not be used for support and intersecting masonry walls should be bonded or anchored together.

Interior Wall Framing

An interior wall or partition is framed in generally the same way as an exterior wall. First, a shoe or sole plate is nailed to the sub-flooring. Then, to this plate are nailed studs either 2 by 4 inches for bearing walls or 2 by 3 inches for non-bearing walls. And finally, the top of the stud is nailed to a ceiling plate or cap.

Framing around door openings is also erected in the same manner as in exterior walls. Openings under 36 inches may be single or double framed (one or two studs) and jamb studs should extend in one piece from header to sole plate, according to MPS.

Ceiling Framing

The bottom side of the joists for the floor above act as the framing for the ceiling below. Finishes are applied to the ceiling joists similarly to the way in which they are attached to the wall studs. Usually ceiling joists should be a maximum of 16 inches on center.

Interior Wall and Ceiling Finishes

An MPS objective is to secure an interior wall and ceiling finish which will provide a suitable base for decorative finish, a waterproof finish in spaces subject to moisture and reasonable durability and economy of maintenance. The materials now most commonly used to fulfill this three-fold objective include plaster, gypsum, plywood, hardboard, fiberboard, ceramic tile and wood paneling.

Plaster Walls

Plaster walls are constructed by applying up to three coats of plaster over either metal lath, wire lath, wire fabric, gypsum lath, wood lath or fiberboard lath which has been attached to the studs or furring strips. Over metal lath, wire lath and wire fabric the plaster is usually applied in three coats. The first two are called the scratch coat and brown coat respectively. Together they are about $\frac{5}{8}$ inch thick. The final white coat is a maximum of $\frac{1}{16}$ inch thick. Over gypsum lath, wood lath, fiberboard lath or masonry, the plaster can be applied as above or in two coats to a minimum thickness of $\frac{1}{2}$ inch.

A well constructed plaster wall provides a high degree of sound-proofing. Its main disadvantages, however, are high cost and susceptibility to cracking.

Loose or Defective Interior Plastering

As long as cracked plaster is tight to the wall it may be sufficient to just patch and redecorate a crack.

Bulging plaster on the ceiling is dangerous and should be repaired. When this is suspected, the defect can often be detected by pressing a broom handle against the ceiling and feeling if there is any give in the plaster.

Gypsum Drywalls

Gypsum drywalls are constructed by nailing, gluing or screwing sheets of gypsum boards directly to the studs or masonry or to furring strips attached to the masonry wall.

As MPS prescribes, "When single $\frac{3}{8}$ inch thick sheets of gypsum board are used they should be nailed to studs 16 inches o.c., i.e., on center. When $\frac{1}{2}$ inch and $\frac{5}{8}$ inch thick sheets are used the studs may be 24 inches o.c. When two layers of $\frac{3}{8}$ inch thick or thicker gypsum board are used, the stud may be 24 inches o.c." Moreover, in the construction of gypsum drywalls, all exterior corners should be protected with metal corner beads, angles or wood moulding to prevent damage and all joints in wall board surfaces intended to receive paint or wallpaper finishes should be taped and cemented.

Gypsum drywalls eliminate the waste of time presented by the drying out period between coats connected with plaster walls. However, they carry with them the disadvantage of the possibility of nails popping out if there exists an improper moisture content in the studs.

Plywood

Plywood can be nailed or fastened directly to the studs. One-quarter-inch plywood requires studs 24 inches o.c. Plywood of any thickness can be nailed over gypsum board.

Hardboard

Like plywood, hardboard is nailed or fastened directly to the studs or furring strips. As MPS prescribes, "When applied alone it should be a minimum of ¼ inch thick with the studs 16 inches o.c."

Fiberboard

Also like plywood, fiberboard is nailed or fastened directly to the studs or furring strips. According to MPS, "When applied alone it should have a minimum of ½ inch thickness with studs 16 inches o.c. and ¾ inch thickness with studs 24 inches o.c." Although plywood, hardboard and fiberboard vary in cost according to thickness and quality, they often provide less expensive interior wall finishes than plaster or gypsum.

Ceramic Wall Tile — Cement Plaster Method

When ceramic wall tile is applied by the cement plaster method (also known as a mud job), a plaster wall is constructed by installing lath over water-resistant sheathing paper attached to the studs which should be firmly blocked to support the heavy weight of the tile. Next, two coats of plaster are applied to the lath. The first coat is the scratch coat that forms the base and the second is the mortar coat into which the tiles are embedded. The scratch coat is allowed to thoroughly harden and is then redampened. The tiles are then set into the freshly applied mortar coat by either "floating" or "buttering." "Floating" is placing the tile in the mortar with small, twisting motions, while "buttering" is spreading the mortar on each individual tile as butter is spread on bread. After the tiles are floated or buttered the final steps are to fill the joints between the tiles with a thin grout of white or gray portland cement and water and finally to wipe off all traces of cement on the surface.

Ceramic and Plastic Tile Wall — Adhesive Method

When a ceramic or plastic tile wall is used as an interior finish, first a wall of plaster, gypsum or other wallboard is constructed as previously described. Then, the entire surface to be tiled is sealed with a water-resistant sealer. Next, the tile adhesive is applied to the entire surface to be tiled with a notched spreader blade. Then the tile is set by the floating method. The buttering method of setting this tile should not be used as it does not make a good bond. Finally, the wall is finished in the same manner as in the plaster method.

Cracked, Loose or Leaking Tiles

The principal area where tile problems occur is around the tub, especially when there is a shower that splashes water on the tile wall. Defective grout will permit the water to seep behind the tile and loosen the glue. New types of waterproof adhesives help eliminate this. Tiles set in plaster also are less likely to present problems.

195

Here is an area where initial good workmanship will produce lasting, trouble-free results while shoddy work soon has to be redone. Often when repairs are undertaken it becomes necessary to replace the wallboard which has become damaged by the moisture. Special waterproof wallboards have now been developed to cut down this problem.

Wood Paneling

When wood paneling is used as an interior finish, MPS suggests that the "wood panel . . . be thoroughly seasoned and suitable for its intended use and applied over No. 15 asphalt-saturated felt or vapor barrier when application is direct to exterior wall framing or blocking" and that its maximum width be 12 inches, its minimum thickness ½ inch and the maximum spacing of supports 24 inches o.c.

Stairs

The objective, according to MPS, of the well-planned stairway is to provide safe ascent and descent and a design and arrangement of stairs which assures adequate headroom and space for moving furniture and equipment.

The major part of a staircase running from one floor to the next or from a floor to a landing is called a flight. Important components of a staircase are the stringers (also called carriages or horses) which run at an angle from one floor or landing to the adjoining floor or landing and which support the horizontal member or top of the stair called a tread and the vertical front of the stair called the riser. The approximately triangular shaped threads on a curve are winders. And the tip of the tread that extends beyond the riser is the nose. The width of the rise is called the run. Often under the nose is a piece of cove moulding.

The handrail can be attached to the wall with brackets or supported by posts called balusters. The end of the handrail that curves parallel to the floor is the easing and the top of the bottom baluster is a newel.

A complicated staircase must always be carefully planned. According to the Federal Housing Authority, a simple check for adequate design is to check the stairs on all of the following features: headroom, width clear of handrail, run, rise, winders, landings, handrail and railings.

Stair Construction

Depending on the custom of the area and facilities and skills of the builder, stairs are built in place or they are built as a unit in the shop or mill and delivered preassembled to the site to be set in place. Since construction of a spiral, dog-leg stairway requires special skills, most stairs built today are straight.

A built-in-place stair is made first by cutting the carriages and setting them in place with the top nailed to the header of the upper floor and the bottom resting on the lower floor. The treads and risers, which are tongued and grooved to fit together, are nailed to the stringers and

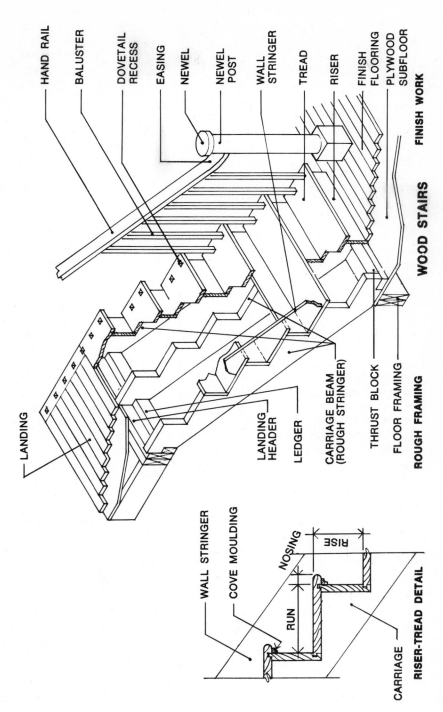

HAND RAIL

BALUSTER

DOVETAIL RECESS

EASING

NEWEL

NEWEL POST

WALL STRINGER

TREAD

RISER

FINISH FLOORING

PLYWOOD SUBFLOOR

FINISH WORK

LANDING

LANDING HEADER

LEDGER

CARRIAGE BEAM (ROUGH STRINGER)

THRUST BLOCK

FLOOR FRAMING

ROUGH FRAMING

WOOD STAIRS

WALL STRINGER

COVE MOULDING

NOSING

RISE

RUN

CARRIAGE

RISER-TREAD DETAIL

197

the cove moulding is nailed under the nose of the tread. Then, if a wall stringer is used, it is fitted into place.

Balusters can be attached to the treads by nailing (which, it should be noted, is the poorest method), by being inserted into drilled holes in the rail and tread (a more satisfactory method than nailing but one which carries the risk of the rail tending to loosen in time and turn) or by dovetailing the lower end of each baluster into the tread and fitting the upper end into holes bored into the lower side of the railing.

The rails are attached to the walls of enclosed stairs with wall brackets, which should not be more than 10 feet apart, and then securely screwed or lag-bolted to the studs or blocking. They should not, however, be screwed into the lath.

Interior Stairs

The first feature which must be taken into consideration with interior stairs is headroom which should be continuous and clear, measured vertically from the front edge of the nosing to a line parallel with the stair pitch. For main stairs there should be a minimum of 6 feet, 8 inches of headroom and for basement and service stairs a minimum of 6 feet, 4 inches. On interior stairs, the width clear of the handrail should be at least 2 feet, 8 inches for main stairs and at least 2 feet, 6 inches for basement or service stairs.

Main stairs with either a closed or open riser as well as basement stairs with a closed riser should have a run of at least 9 inches plus a 1⅛ inch nosing. Basement stairs with an open riser require a run of at least 9 inches plus a ½ inch nosing.

The maximum riser height condoned by the FHA for interior stairs is 8¼ inches. The winders should run at a point 18 inches from the converging end and should be no less than the run of the straight portion.

Landings of no less than 2 feet, 6 inches should be provided at the top of any stair run having a door which swings toward the stair. A continuous handrail should be installed on at least one side of each flight of stairs which exceeds three risers. Stairs which are open on both sides, including basement stairs, should have a continuous handrail on one side and a railing on the open portion on the other side. Railings should also be installed around the open sides of all other interior stairwells including those in attics.

Exterior Stairs

On all exterior stairs, with the exception of those running to the basement of the house, the FHA prescribes the following: (1) that the width be that of the walk but that it be no less than 3 feet, (2) that the run be no more than 11 inches, (3) that the rise be no more than 7½ inches and that all riser heights within a flight be uniform and (4) that a continuous handrail be installed on all open sides of stair flights to a platform more than four risers or 30 inches above finish grade.

Unprotected exterior stairs to the basement require the following, according to the standards set up by the FHA-MPS: (1) that the headroom be at least 6 feet, 4 inches, (2) that the width clear of handrail be at least 2 feet, 6 inches, (3) that the run be no more than 7½ inches and (4) that a handrail be installed on at least one side if the stairs exceed four risers.

For protected exterior stairs to the basement MPS prescribes the following: (1) that the headroom be at least 6 feet, 2 inches, (2) that the run be at least 8 inches plus a 1½ inch nosing and (3) that the rise be no more than 8¼ inches.

Roofs

Roof Framing

The roof must be constructed first so that it will support its own weight plus that of loads from snow, ice and wind and also so that it will act as a base for the application of the roof finish materials. The most common systems of roof construction used in houses are trusses, joists and rafters, joists alone, planks and beams and panelized construction.

A truss is made up of a number of individual boards (chords), usually 2 by 4 inches or 2 by 6 inches, arranged into a framework of triangles and connected by wood or metal gusset plates, metal gang-nail plates or metal ring connectors. Trusses are usually preassembled at a mill or on the ground. They are suspended between load-bearing walls spaced from 24 inches to 48 inches o.c. so that each truss acts as a unit to support the roof loads. Thus, they usually span the entire width of the house without additional support, allowing all the interior walls to be flexibly placed since they need not be load-bearing.

Joists are horizontal boards of from 2 by 4 inches to 2 by 8 inches which are suspended, with their narrow sides parallel to the ground, from exterior walls and often supported by interior bearing walls. Often when joists are used, rafters, usually of the same size, are erected as part of the framing system and run from the exterior bearing walls diagonally upward to a ridge board or hip rafter.

If sufficient joists are used, they can be suspended between the exterior bearing wall and a ridge beam supported by an interior bearing wall, thus eliminating the need for the rafters.

Plank and beam roofs consist of 2 by 6 inch to 2 by 8 inch boards usually connected by cutting them tongue and grooved so that they fit into a structural ridge beam and the exterior bearing wall beams. An alternate method used with plank and beam roofs is that of running them longitudinally, supported by the exterior wall and posts.

Panelized roof systems consist of preframed, precut and sheathed panels usually made up of 2 by 3 inch or 2 by 4 inch boards covered with plywood. In a stressed-skin roof panel the plywood skins are bonded to the framing members under heat and pressure by glue-nailing techniques or by special adhesives. A sandwich panel is sim-

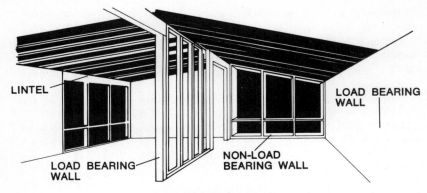

SLOPED ROOF JOISTS

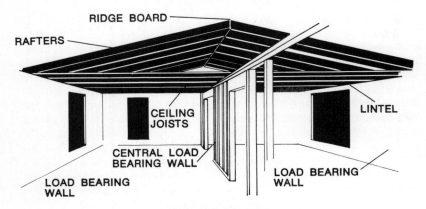

JOIST-AND-RAFTER

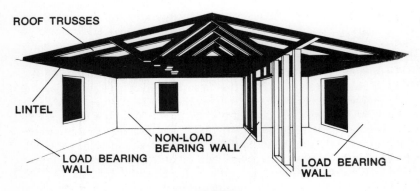

TRUSS

ROOF FRAMING SYSTEMS

ilarly constructed except that the faces are separated by weaker light-weight core materials instead of actual framing members.

Panels may be supported between transverse or longitudinal beams or can be centrally supported by bearing walls or beams as in joist roofs.

Roof Sheathing

According to MPS, the objective of roof sheathing is to provide safe support of roof loads without excessive deflection and to provide backing for the attachment of roofing materials. The most common types of sheathing are plywood, fiberboard and roof boards (planks).

Plywood sheathing should be from $\frac{5}{16}$ to $\frac{7}{8}$ inch thick. The thickness of the plywood chosen depends upon the type of plywood, the joist spacing and the roofing material to be used. The plywood is nailed or stapled to the frame.

There are several good fiberboard sheathing materials available that are installed similarly to plywood. When boards (planks) are used they should be at least $\frac{3}{4}$ inch thick and not more than 12 inches wide, tongue and grooved, shiplapped or square-edged. They are nailed either parallel to the rafters or diagonally across them. The roof boards should be nailed tightly to each other except in instances where shingle roofing is used in a climate where there is no snow. In these instances, the roof boards may be spaced the width of the board.

Roof Flashing

Whenever a roof is complicated by an intersection of the joining of two different roof slopes, adjoining walls or projections through the surface by chimneys or pipes or other protrusions, the joint must be flashed. Flashing is usually accomplished by first nailing metal strips across or under the point, then applying a waterproofing compound or cement and finally applying the roofing material over the edges to permanently hold it in place.

Roof Coverings

The objective of a roof covering is to prevent entrance of moisture and to provide reasonable durability and economy of maintenance, according to MPS.

Shingles and shakes made of wood, asphalt, asbestos, cement, slate or tile are used for the majority of house roofs. However, metal roof, clay tile and built-up or membrane roofs can also be found. Shingles are applied by nailing a double layer, known as a starter course, at the bottom of the roof and at the sides or rake of the roof so that they project at least 1 inch beyond the lower edge of the roof sheathing in order to enable rain to run clear of the roof. Each succeeding course or row of shingles is then nailed to the sheathing so that it covers the top of the row below, leaving part of the lower row exposed to the weather. The amount left exposed depends upon the type of shingle, its length and the effect desired. However, it is impor-

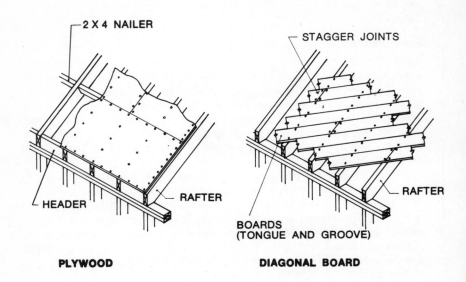

PLYWOOD

DIAGONAL BOARD

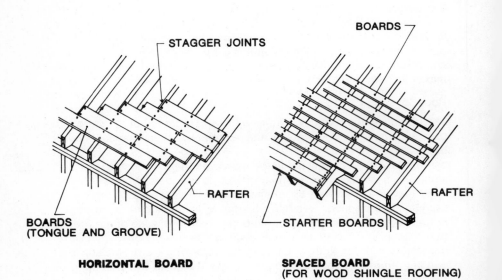

HORIZONTAL BOARD

SPACED BOARD
(FOR WOOD SHINGLE ROOFING)

ROOF SHEATHING

tant that the lines be kept straight and that the joints be staggered so that there is at least 1 or 2 inches of overlap to make the roof watertight.

There are various methods of joining the shingles and ridges, hips and valleys in conjunction with flashing materials to keep them watertight. Certain types of shingles, known as "diamond," "hexagonal," and "Dutch lap" are cut in special designs in order to reduce the amount of lap required and thereby save material. These shingles are installed according to individualized, special instructions.

Slate is laid over the sheathing covered with special impregnated slater's felt. A cant strip is nailed to the lower end of the eave and the starter course is nailed down with the long side parallel to the eaves. The second course, as well as each succeeding course, is nailed down with copper or zinc nails so that at least three inches overlap the course below it. Several different techniques are used with flashing to make the ridges, valleys, hips and rake waterproof. The application of tile shingles varies with the type of tile being used. However, some type of cant strip and roofing strip is required with almost every kind of tile shingle.

Shingle tile is applied to the roof in the same manner as slate. A cant strip is nailed to the edge of the eave and the tiles are nailed through predrilled holes in successive overlapping courses with copper nails. Special cut tiles are used for the ridge and for the hips and valleys.

Interlocking tiles known as French tile, Spanish tile, Mission tile, Roman tile and Greek tile are designed to provide maximum coverage with minimum material. The interlocking ridges on each side of the tile reduce the amount of lap needed for watertightness and hold the tile together.

The closest design to a flat tile is the French tile. The bottom row is laid on top of a cant strip. Specially shaped pieces are used at the hip, ridge and valleys. The tiles are nailed and cemented. Mission tiles and Spanish tiles are installed by turning each adjacent tile alternately concave ("pans") and convex ("covers") so that the covers fit down over the upturned edges of the pans. Then they are nailed with a single nail at the upper end. The convex tiles rest on 1 by 4-inch wood strips set on edge, running from eave to ridge. Specially shaped pieces are used at the ridge, rake, hips and valleys. Greek tile and Roman tile are both combinations of flat tile with upstanding outer edges and curved or angular covers which are tapered so that succeeding courses fit snugly.

Roll roofing is applied by nailing it down in strips that lap and then sealing the laps with roofer's cement. At the eaves and rakes the roofing is bent over the roof board, nailed down and bent over the top of the ridge. Sometimes when the roof is to be walked on, "duckboards" are laid over the roofing to protect it.

Defective Roofing

Water may leak through the roof for a variety of reasons. Asphalt shingle roofs may leak in a high wind if light grade shingles are used. **203**

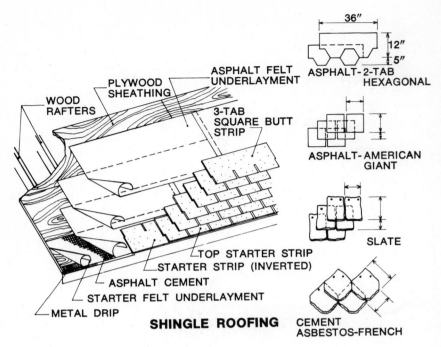

WOOD RAFTERS

PLYWOOD SHEATHING

ASPHALT FELT UNDERLAYMENT

3-TAB SQUARE BUTT STRIP

36″

12″
5″

ASPHALT-2-TAB HEXAGONAL

ASPHALT-AMERICAN GIANT

SLATE

TOP STARTER STRIP
STARTER STRIP (INVERTED)
ASPHALT CEMENT
STARTER FELT UNDERLAYMENT
METAL DRIP

SHINGLE ROOFING

CEMENT ASBESTOS-FRENCH

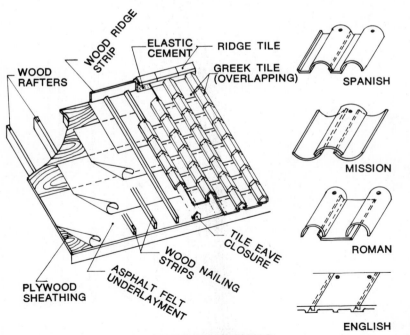

WOOD RIDGE STRIP

ELASTIC CEMENT

RIDGE TILE

GREEK TILE (OVERLAPPING)

WOOD RAFTERS

SPANISH

MISSION

ROMAN

TILE EAVE CLOSURE

WOOD NAILING STRIPS

PLYWOOD SHEATHING

ASPHALT FELT UNDERLAYMENT

ENGLISH

CLAY TILE ROOFING

204

As these shingles get older they curl, tear and become pierced with holes. Wood shingles may curl, split, become loose and broken and fall off the roof while asbestos shingles may crack and break. Metal roofs can rust, become bent and pierced with holes. Roll and built-up roofs may become loose, torn, patched and worn through.

Fireplaces

A fireplace with logs burning in the hearth still possesses a strong romantic appeal. Basically an amenity in most American homes rather than a heating system, the fireplace usually is constructed of masonry. There are many fireplace designs and variations, of which the simplest and most common is the single opening with a damper and hearth. Other designs, however, feature two, three or four openings.

The opening of the fireplace should be wider than it is high, but it should not be very deep because the shallower the fireplace, the more heat is reflected into the room. The inside walls should go back at an angle so that the rear inside wall is a least 1½ feet narrower than the front opening because square inside corners interfere with the air flow and cause smoking and poor combustion. The rear inside wall should slope forward to the back of the damper.

A damper is a desirable feature. It should be at least 8 inches above the top, but it should not be set directly on top of the fireplace opening because such a position would, at times, allow smoke to escape into the room. Another desirable feature is an ash dump connecting to an ash pit with an ash-cleanout door.

A well-designed smoke shelf, at least 4 inches wide, stops the cold air that flows down the inside of the chimney flue from blocking the flow of smoke upward. This is done by turning the cold air flow back upward, causing it to mix with the rising warm air and smoke.

The smoke chamber is the large space above the damper and smoke shelf. The back is straight, but the sides start at the ends of the damper and slope inward to the two inner sides of the flue. The rate of slope is about 7 inches inward to each 12 inches of height.

The inner hearth (really the back hearth under the fire), together with the cheeks and back of the fireplace, must be built of heat-resistant materials in order to withstand the intense heat of the fire. Thus, fire brick and fire clay are the most common materials used. A front hearth extending at least 16 inches from the front of the fireplace and at least 8 inches beyond each side is needed as a precaution against flying sparks. Since this hearth, which is usually supported by a trimmer arch or concrete slab, must be fireproof, it is often made of tile, brick or stone.

Chimneys and Vents

Chimneys and vents should be constructed and installed to be structurally safe, durable, smoketight and capable of withstanding action of flue gases.

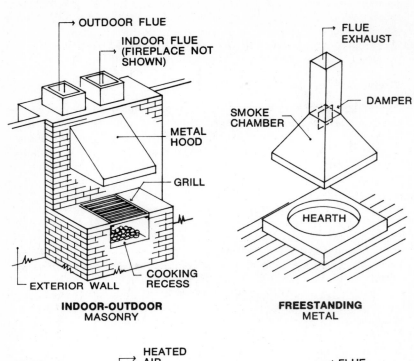

INDOOR-OUTDOOR
MASONRY

- OUTDOOR FLUE
- INDOOR FLUE (FIREPLACE NOT SHOWN)
- METAL HOOD
- GRILL
- COOKING RECESS
- EXTERIOR WALL

FREESTANDING
METAL

- FLUE EXHAUST
- DAMPER
- SMOKE CHAMBER
- HEARTH

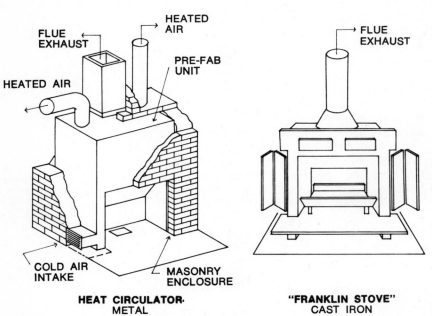

HEAT CIRCULATOR
METAL

- FLUE EXHAUST
- HEATED AIR
- HEATED AIR
- PRE-FAB UNIT
- COLD AIR INTAKE
- MASONRY ENCLOSURE

"FRANKLIN STOVE"
CAST IRON

- FLUE EXHAUST

FIREPLACE TYPES

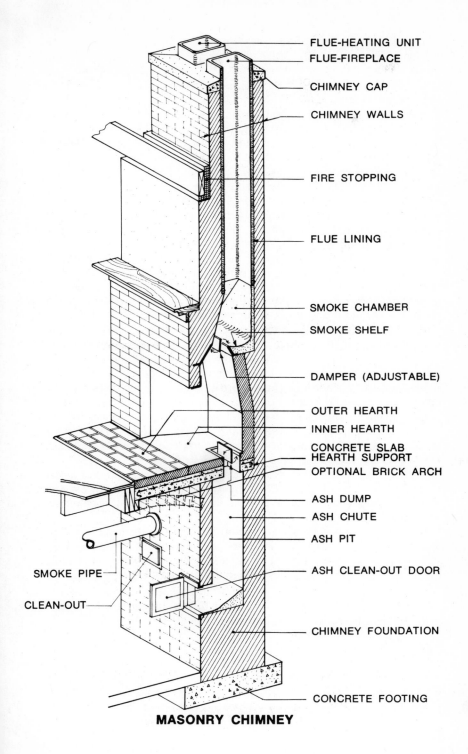

FLUE-HEATING UNIT

FLUE-FIREPLACE

CHIMNEY CAP

CHIMNEY WALLS

FIRE STOPPING

FLUE LINING

SMOKE CHAMBER

SMOKE SHELF

DAMPER (ADJUSTABLE)

OUTER HEARTH

INNER HEARTH

CONCRETE SLAB
HEARTH SUPPORT

OPTIONAL BRICK ARCH

ASH DUMP

ASH CHUTE

ASH PIT

ASH CLEAN-OUT DOOR

CHIMNEY FOUNDATION

CONCRETE FOOTING

SMOKE PIPE

CLEAN-OUT

MASONRY CHIMNEY

207

The efficiency of any heating system (except electric) depends upon the chimney or vent. Defective chimneys and vents may constitute serious fire hazards. A chimney may be a simple flue or an intricate masonry construction consisting of heater flues, ash pits, incinerators, ash chutes, fireplaces and fireplace flues.

Whatever its construction, the chimney is the heaviest portion of the house and it must be supported by its own concrete footings. Those footings must be so designed that they will not let the chimney settle faster than the rest of the building.

The masonry chimney walls should be 8 inches thick when they are exposed to the exterior of the house. The space between flues, when more than one flue occurs in a chimney, is called the wythe. In the best construction, this space is solidly filled with brick and the joints slushed full of mortar at least 4 inches thick. As MPS prescribes, "Masonry and factory-built chimneys shall extend at least 2 feet above any part of a roof, roof ridge or parapet wall within 10 feet of the chimney" and "masonry chimney walls shall be separated from combustible construction." A 2-inch air space filled with fireproof material is recommended. At the bottom of the chimney there should be an ash pit with a cleanout door into which run the flues from the fireplaces.

Flues from the furnace and hot water heater should not run into the ash pit because cold air below the smoke pipe connection will interfere with the draft in the flue. There should be a clean-out door close below the junction of the smoke pipe and the chimney.

The furnace and hot water heater are connected to the chimney by a smoke pipe through a metal or terra cotta collar built into the brickwork. The smoke pipe is slipped into the collar to a point where its end is just flush with the inside wall of the flue. If extended in error too far into the flue, the efficiency of the draft will be substantially reduced. For fire safety, smoke pipe should be at least 10 inches below the floor joists and the joists further protected with plaster or a shield of metal or asbestos. According to MPS, "The smoke pipe should not exceed 10 feet in length or 75 percent of the vertical height of the chimney, whichever is less."

The heart of the chimney is the vertical open shaft through which the smoke and gas pass from the fire to the outside air. The size of the flue required for best combustion efficiency depends upon the size of the fireplace or furnace, its design and the type of fuel being used.

The warm air or smoke rises up the flue. It ascends in spirals and occupies the greater part of the center of the flue. A rough surface retards this upward flow. The best way to overcome the roughness is to use a flue lining. The most common shapes for flue linings are circular, square and rectangular. A square or rectangular flue is less efficient than a round one. Its capacity is only the same as a round flue that would fit inside it. Thus, the main reason shapes other than circular are used is for ease of construction. It should be noted that a single flue should not be used for more than one heating device.

The flue should extend out of the top of the chimney wall a few inches. The top of the wall is capped with concrete, metal, stone or other noncombustible, waterproof material sloped from the flue to the outside edge. The cap should be at least 2 inches thick at the outside edge, according to MPS.

Another type of chimney is that made and assembled off the premises in a factory. Many of the prefabricated units consist of a flue liner encased in a concrete wall. The units should be Underwriters-approved and should be the proper size and design for the appliance to be used.

Gutters and Downspouts

Gutters and downspouts provide means for controlled water disposal from roofs to prevent damage to the property or to prevent unsightly appearance of walls when roof overhangs are not provided. MPS states that gutters are needed when either the soil is of such a nature that excessive erosion or expansion might occur or if the roof overhangs are less than 12 inches in width for one story or 24 inches in width for two stories. If gutters are not provided in the above instances, a diverter or other means must be provided to prevent water from roofs or valleys from draining on uncovered entrance platforms or steps.

Gutters or eavetroughs catch the rainwater as it reaches the edge of the roof and carries it to the downspouts or leaders. In northern climates it is important to install gutters below the slope line so that snow and ice can slide clear.

Metal gutters, which are attached to the house with various types of metal hangers, are the most common type now being made. Aluminum, copper, galvanized iron and other metals are used in the construction of these gutters, the size of which is determined by the size of the roof served and the rainfall of the area.

Wood gutters are still used, however, mostly in the Northeast. The most common types are cut from solid pieces of wood such as Douglas fir, cypress and redwood cedar. A wooden gutter is shaped with a semi-circular channel inside and an ornamental moulding (ogee) on the outside and it is sometimes completely or partially lined with metal. Care must be taken when two pieces of gutter are joined together to prevent the joint from leaking and rotting. Thus, the joint should be covered with a piece of metal embedded in elastic roofers cement.

Wood gutters are attached to the house with noncorroding screws bedded in elastic roofers cement to prevent leakage. Built-in gutters are made of metal and set into the deeply notched rafter a short distance up the roof from the eaves. Pole gutters consist of a wooden strip nailed perpendicularly to the roof and covered with sheet metal.

Downspouts or headers are vertical pipes that carry the water from the gutter to the ground and sometimes into sewers, dry wells, drain tiles or splash pans. Their most common shapes are rectangular, corrugated rectangular, corrugated round and plain round. They must be large enough to carry the water away as fast as they receive it.

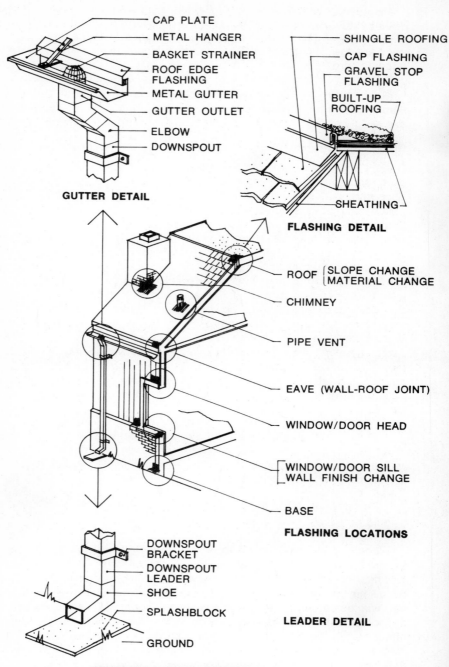

CAP PLATE
METAL HANGER
BASKET STRAINER
ROOF EDGE FLASHING
METAL GUTTER
GUTTER OUTLET
ELBOW
DOWNSPOUT

GUTTER DETAIL

SHINGLE ROOFING
CAP FLASHING
GRAVEL STOP FLASHING
BUILT-UP ROOFING

SHEATHING

FLASHING DETAIL

ROOF { SLOPE CHANGE / MATERIAL CHANGE
CHIMNEY
PIPE VENT
EAVE (WALL-ROOF JOINT)
WINDOW/DOOR HEAD
WINDOW/DOOR SILL / WALL FINISH CHANGE
BASE

FLASHING LOCATIONS

DOWNSPOUT BRACKET
DOWNSPOUT LEADER
SHOE
SPLASHBLOCK
GROUND

LEADER DETAIL

210

DRAINAGE AND FLASHING

The junction of the gutter and downspout should be covered with a basket strainer to hold back leaves and twigs, especially if the gutter is connected to a storm or sanitary sewer which might become clogged and difficult to clean out.

Floors

Floor Framing

The objective of wood floor framing is to provide floor construction which will assure safe and adequate support of all loads and also to eliminate objectionable vibrations, according to MPS.

Floor framing is relatively simple and the same regardless of the general framing method used for the rest of the house. Basically, floors consist of subflooring resting on joists (floor beams) stiffened by bridging; the joists are carried by girders, walls or loadbearing partitions.

In narrow houses the joists can run from exterior wall to exterior wall. Usually at least one girder is also required. Girders are generally supported by wooden posts, brick or block or poured concrete piers, hollow steel pipes (Lally columns) or steel pipes filled with concrete. All of these supports must rest on an adequate footing and have a suitable cap on which the girder rests.

Most house girders are made either of wood or steel "I" beams. Wood girders can be solid pieces of timber or built-up beams consisting of 2 or 3-inch boards standing on edge and spiked or lag screwed together to form a larger piece. The use of solid girders has substantially decreased in recent years because of their increase in cost and tendency to split (check) more easily than built-up ones. Another advantage with built-up girders is that, by spacing the joints, the beam can be made continuous and thus run unbroken from one end of the house to the other.

The use of at least one steel "I" beam is quite common. These beams are handled in the same manner as wood beams except that they usually come in sections and are bolted together. However, they should be covered with 1 inch of cement to protect them from heat in the event of a fire since it takes relatively little heat (only about 500 degrees to weaken and 1,000 degrees to buckle) to cause such a beam to lose its ability to hold the joists.

Subflooring

According to MPS, the objective in subflooring is to provide construction which will assure safe support of floor loads without excessive deflection and to provide adequate underlayment for the support and attachment of finish flooring materials.

Plywood is the most common material now being used for subflooring. When properly installed with the grain of the outer piles at right angles to joists and staggered so that the division between adjacent panels comes over different joists, it is as good as boards or planks.

Wood boards with a minimum thickness of ¾ inch and maximum width of 8 inches may be installed diagonally to or at right angles to joists. If installed at right angles to joists, finish floor should be installed across the subflooring.

Plank and beam floor systems are usually made with 2 by 6 or 2 by 8-inch tongue and grooved or splined planks spanning between beams generally spaced 4 to 8 feet on center. The planks serve as the subfloor and working platform and transmit the floor loads to fewer but larger members than in wood joist floor systems.

Panelized floor systems are increasing in popularity. Panels are either premanufactured in the mill or on the job site and set in place over the joists. Stressed-skin floor panels are made with framing members to which plywood skins are bonded either by the glue-nailing techniques or by adhesive applied under heat and pressure. Sandwich panels are similar, but the faces are glued to and separated by weaker lightweight core materials instead of actual framing members. A few houses are also made with steel and concrete subflooring systems.

Bridging

In order to stiffen the joists and prevent them from deflecting sideways, strips of wood or bands of metal are fastened crosswise between the joists and nailed top and bottom to form the bridging. There should be a line of bridging for each 6 to 8 feet of unsupported length of joist.

Floor Coverings

Carpeting, which can be installed over either finished flooring or subflooring, is rapidly gaining in popularity as a floor covering since manufacturers claim that it can be used in every room in the house. Thus, the FHA has reversed its former position on carpeting and now considers it part of the real estate.

Rubber-backed carpeting is glued with a special glue to the flooring or subflooring. Once down, it is hard to remove without damaging the carpet and the finished floor beneath. Jute-backed carpet, on the other hand, is installed over a carpet pad. It can be fastened down at the edges with tacks or by installing a tackless strip around the perimeter and attaching the carpet to it. Tackless strips can be glued to concrete and other hard surfaces. In either case, it is important to a good installation that the carpet be stretched in order to eliminate wrinkles or buckles.

Ceramic tile can be set in plaster. Over the subflooring is laid a waterproof building paper and then wire mesh. The setting bed of plaster must be 1¼ inches thick. (An alternate process is attaching the rough flooring to strips below the top of the joists which are beveled and extend into the plaster bed; the plaster bed must be 1¼ inches over the top of the joists.) On top of this comes a ⅛-inch skim coat.

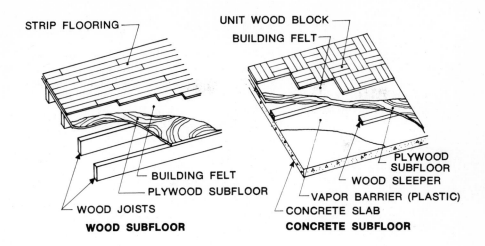

STRIP FLOORING

UNIT WOOD BLOCK
BUILDING FELT

PLYWOOD SUBFLOOR
WOOD SLEEPER

BUILDING FELT
PLYWOOD SUBFLOOR
WOOD JOISTS

VAPOR BARRIER (PLASTIC)
CONCRETE SLAB

WOOD SUBFLOOR　　　　**CONCRETE SUBFLOOR**

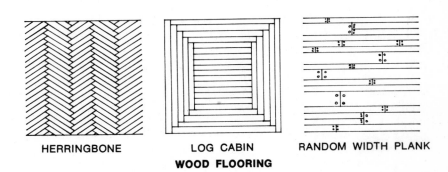

HERRINGBONE　　　LOG CABIN　　　RANDOM WIDTH PLANK

WOOD FLOORING

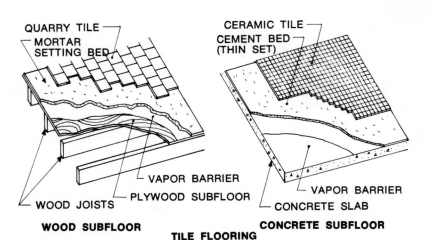

QUARRY TILE
MORTAR
SETTING BED

CERAMIC TILE
CEMENT BED
(THIN SET)

VAPOR BARRIER
PLYWOOD SUBFLOOR
WOOD JOISTS

VAPOR BARRIER
CONCRETE SLAB

WOOD SUBFLOOR　　　**CONCRETE SUBFLOOR**

TILE FLOORING

FLOORING

The tiles are soaked in water and then pressed firmly in place in the plastic setting bed. Mortar is compressed into joints, which are tooled the same day the tile is set, and then covered with waterproof paper and damp core.

Ceramic tile can also be attached with special adhesive to a smooth concrete floor or to a subfloor covered with a special adhesive or special underlayment material. The following steps are used for this method of installing ceramic tile. Before application of tile, the entire surface to be tiled is sealed with a suitable water-resistant sealer or a coat of tile adhesive. Then tile adhesive is applied to the surface with a notched spreader blade. The tile is set by the floating method, using a slight twisting motion. Before the grouting, time is allowed for the evaporation of volatiles from the adhesive. Grout is forced into joints, with care taken to leave no open joints. Then the joints are tooled.

Concrete slabs may be used for floor covering with no further treatment, painted with special concrete paint or covered with other flooring coverings. Resilient tiles are glued down with special adhesives according to the individual manufacturer's recommendation for his particular tile. They must not be installed directly over a board or plank subfloor. A suitable underlayment must first be installed.

Terrazzo flooring is made of colored marble chips mixed into cement. After being laid, it is ground down to a very smooth surface. There are two basic methods of installation: the unbonded method, where the cement is separated from the subfloor by an isolation membrane; and the bonded method, where the cement is bonded directly to the subflooring. In both methods, metal divider or expansion strips are installed to prevent cracking.

Wood block flooring may be installed over an underlayment or directly to most subfloorings. According to MPS, "The blocks are nailed with at least two nails in each tongued side (four per block), driving the nail at an angle of 40 to 50 degrees. They may also be installed by attaching them to a suitable underlayment with special adhesives. When installed over concrete the concrete should be first sealed with a primer compatible with the adhesive."

Wood strip flooring may be installed directly over the joists or over subflooring. When installed directly to the joist, hardwood must be $25/_{32}$ inches thick and not more than 2¼ inches wide. Softwood must always be at least $25/_{32}$ inches thick and installed at right angles to joists. Each flooring strip should bear on at least two joists. Approximately ½ inch clearance should be provided between flooring and wall to allow for expansion. The flooring is blind nailed to each joist with a threaded or screw type nail, driving the nail at an angle of approximately 50 degrees.

The boards over wood flooring are nailed at right angles to the subflooring except when the subflooring is plywood or is laid diagonally. It is best to put a layer of 15-pound asphalt impregnated felt or other suitable building paper on top of the subflooring in order to prevent drafts, dust and moisture from coming up through the strips. The flooring should be blind nailed, driving nails at an angle of 40 to 50 degrees.

Flooring must be kept dry at all times. It is often kiln-dried to a low moisture content of 6 to 8 percent and must be kept that way. Flooring that is moist when laid shrinks in the dry winter, leaving ugly joints. After the flooring is laid it is scraped, sanded and varnished.

Random width flooring of various lengths is laid to resemble old floors. In this process, the boards are blind nailed or pegged to simulate old floors. This flooring comes prefinished with nail holes to decrease the possibility of damaging it during installation. Plywood with special surface coatings and finishes is nailed down like any other flooring.

Other types of flooring are the many attractive vinyl tiles now available, rubber tiles, asphalt tile and rolled goods such as linoleum.

Weak and Defective Floors

Sagging and sloping floors can be caused by defective framing. There are, however, other reasons for floor troubles. The rest of the house framing may be fine and it may be just the floor joists that are too small or lack support because inadequate bridging is causing sagging or sloping. Poor carpentry in one area, however, is usually a tip-off to poor carpentry throughout the house.

Floors that have been exposed to water may warp and bulge upwards. Wide cracks between the floorboards are a sign of poor workmanship or shrinkage caused by wood that was improperly dried or not stored correctly at the time of installation. Fortunately, rough, stained, discolored, blemished, burned or gouged floors can usually be cured by refinishing.

Fire Protection and Safety

Fire Hazards

The following is a summary of the major items MPS requires a house to have in order to make it safe from fire. When houses are built to the lot line or with party walls such as exist in a townhouse, the party or lot line wall must extend the full height of the building without any openings and have not less than a two-hour fire resistance rating.

Fire stopping is required in concealed vertical spaces in walls and partitions at each floor level and at the ceiling of the uppermost story. Fire stopping should be of wood blocking, wood construction or noncombustible material. If wood, it should be a minimum thickness of 2 inches. If of a material other than wood, the fire stopping should provide equivalent protection.

Fireplaces should be supported on concrete or other masonry but not on wood framing. A separate flue is required for each fireplace. Where a lining of fire brick at least 2 inches thick is provided, the total thickness of the firebox wall, including lining, should be not less than 8 inches. Where fire brick lining is not provided, the thickness should be not less than 12 inches. And steel fireplace lining at least ¼ inch thick may be used in lieu of brick lining.

Fireplace walls and chimney should be separated from combustible construction by a 2-inch air space from all framing members; this airspace should be firestopped at floor level with an extension of ceiling finish, strips of asbestos board or other noncombustible material.

Every fireplace should have a damper which effectively closes the flue passage. Smoke chambers less than 8 inches thick should be parged with fire clay mortar (not masonry mortar) on all sides.

A hearth supported with noncombustible material made out of fire brick, brick, concrete, stone, tile or any other noncombustible heat resistant material should extend at least 16 inches in front of the fireplace opening and 8 inches on each side. Combustible material should not be placed within 3½ inches of the edges of a fireplace opening. Combustible material above and projecting more than 1½ inches in front of the fireplace opening should be placed at least 12 inches above the opening.

Factory-built fireplaces and their chimneys should be labeled and approved by Underwriters' Laboratories, Inc. and installed according to the conditions of the approval. The outer hearth for factory-built fireplaces should be of noncombustible material not less than ⅜ inch thick and may be placed upon the subfloor or upon the finish floor.

Chimneys should, of course, be constructed of fireproof materials. Smoke pipes should be constructed of a fireproof material equivalent to U.S. standard 24 gauge steel. Factory-built chimneys should be listed by Underwriters' Laboratories and installed in exact accordance with the listing. Chimneys should extend at least 2 feet above any part of a roof or roof ridge or parapet wall within 10 feet of the chimney.

Smoke pipes and flues should have a maximum length from appliance outlet to chimney of 10 feet or 75 percent of the chimney height, whichever is less. Fireclay flue lining should be installed in all masonry chimneys less than 8 inches thick. Heating equipment construction and installation should comply with published industry standards.

Domestic cooking ranges should be installed so that there is a minimum of 2 feet, 6 inches clearance between the top of the range and bottom of all unprotected wood or metal cabinets over the range. A 2-foot minimum clearance is acceptable when the bottom of the cabinet is protected with at least ½-inch asbestos mill board covered with not less than 28 gauge sheet metal (.015 stainless steel, .024 aluminum or .020 copper). The distance from the cooking surface to the bottom of the hood should be not less than the maximum projection of hood from the wall. The minimum distance from the edge of a burner to a wall or cabinet should be 10 inches.

Other Safety Hazards

The number of household accidents is staggering and unfortunately many of them result in death or serious injury.

Here are a few of the common causes of household accidents that can be eliminated:

1. Closets and cupboards that latch shut so they can only be opened from the outside and can trap children inside.
2. Doors that open out over stairs without a landing.
3. Steep, poorly lighted basement stairs without a handrail.
4. Bathroom light fixtures with pull strings or switches that can be reached from the tub or shower.
5. Swimming pools that are not 100 percent fenced in with at least a 4-foot fence and a gate that locks (the same goes for neighbors' pools, too).
6. No adequate convenient space to lock up dangerous cleaning products, medicines and other poisonous things so young children cannot get at them.
7. Too little headroom on stairs.
8. Porches, patios and stairwells without strong handrails around them.
9. Changes in floor levels in the house with only one or two steps difference.
10. Stair risers of unequal size.
11. Stairs without adequate lights that can be turned on and off from both the top and the bottom of the stairs.

Protection Against Termites and Decay

The subterranean termite is an insect which attacks in colonies and derives its nourishment from cellulose materials such as wood, fabrics, paper and fiber board. To obtain nourishment, the termite may attack wood structures above the ground by means of shelter tubes attached to foundation walls, piers and other construction members in contact with the ground. However, only under conditions which permit the insect to establish and maintain contact with soil moisture is a colony able to penetrate and consume wood in service. This requirement indicates that a barrier separating wood from earth, supplemented by inspection, is a practical and effective method for preventing damage by termites.

Protection of wood structures to provide maximum service-life involves three methods of control which can be handled by proper design and construction. One or more of the following methods may be employed: (1) controlling the moisture content of wood, (2) providing effective termite barriers, (3) using naturally durable or treated wood.

According to MPS, the objective is to provide protection of wood materials from damage by termites and decay by the application of suitable construction methods and control measures. It is possible by careful planning and attention to construction details to produce a frame house that will resist damage by subterranean termites and fungi which produce decay. Control of moisture content of wood is a practical and effective method for prevention of decay.

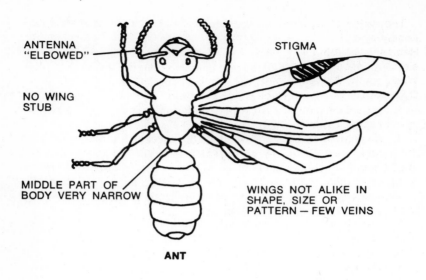

ANTENNA
"ELBOWED"

STIGMA

NO WING
STUB

MIDDLE PART OF
BODY VERY NARROW

WINGS NOT ALIKE IN
SHAPE, SIZE OR
PATTERN — FEW VEINS

ANT

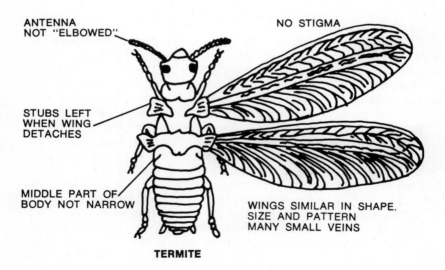

ANTENNA
NOT "ELBOWED"

NO STIGMA

STUBS LEFT
WHEN WING
DETACHES

MIDDLE PART OF
BODY NOT NARROW

WINGS SIMILAR IN SHAPE,
SIZE AND PATTERN
MANY SMALL VEINS

TERMITE

**DIFFERENCES BETWEEN WINGED ADULT
ANTS AND TERMITES**

The extent of termite control needed is determined by local conditions. Generally, it is required in all states except Alaska, Maine, New Hampshire, Vermont, Michigan (except very southern part), Wisconsin, Minnesota, North Dakota, South Dakota, Montana, Wyoming, Washington, Oregon, Idaho and the northern parts of Nevada, Utah, Colorado, Nebraska, Iowa and New York.

For effective termite control the following steps should be taken during construction:

1. All roots, wood forms and scraps of lumber should be removed from the immediate vicinity of the house before backfilling and before placing a floor slab. Particular care should be taken to remove all scraps of lumber from enclosed crawl spaces.

2. The building site should be graded to provide positive drainage away from foundation walls.

3. Provisions should be made for moisture-proofing the foundation walls in the manner described in the foundation section.

4. Adequate roof and wall flashing should be provided for in the manner described in the roof and wall section.

5. In crawl spaces, the clearance from the lowest wood joists or planks should be 18 inches and from the bottom of wood girders or wood posts, 12 inches.

6. In basements or cellars, wood posts that support floor framing should rest on concrete pedestals extending 2 inches above concrete floors and 6 inches above earth floors and separated by an impervious barrier.

7. Main beams or girders framing into masonry walls should have ½ inch of air space at top, end and sides.

8. Wood sills which rest on concrete or masonry exterior walls should be at least 8 inches above the exposed earth on the exterior of the building.

9. Wood siding and trim should be at least 6 inches above exposed earth on the exterior of a structure.

10. In exterior steps, the structural portions of wood stairs, such as stringers and posts, should be at least 6 inches above finish grade.

11. In porches and breezeways and on patios, the beams, headers and posts supporting the floor framing should be at least 12 inches above the ground. Floor joists should be at least 18 inches above the ground. Posts which rest on wood, concrete or masonry floors should be supported on pedestals extending at least 2 inches above the floor or at least 6 inches above exposed earth.

12. Planters, concrete steps or porch slabs resting on the ground should be below the top of the foundation or they should be separated from the wood in the main structure by at least 2 inches or otherwise protected from concealed termite penetration.

13. Ends of main structural members exposed to weather and supporting roofs or floors should rest on foundations which provide a clearance of at least 12 inches above the ground or 6 inches above concrete.

14. Shutters, window boxes and other decorative attachments should be separated from exterior siding to avoid trapping rain water.

15. Adequate ventilation should be provided to prevent condensation in enclosing spaces as described in the ventilation section.

Termite Barriers

Termite barriers are provided by any building material or component which can be made impenetrable to termites and which drives the insects into the open where their activities can be detected and eliminated. When there are adequate separation clearances as just described, termite barriers should not be needed except in very heavily infested areas.

The following are the five types of termite barriers employed when needed:

1. Preservative-treated lumber for all floor framing up to and including the subfloor. For this purpose, pressure treatment with an approved preservative is recommended.

2. Properly installed termite shields. These should be of not less than 26 gauge galvanized iron, or other suitable metal of proper thickness, installed in an approved manner on top of all foundation walls and piers and around all pipes leading from the ground. Longitudinal joints should be locked and soldered. Where masonry veneer is used, the shield should extend through the wall to the outside face of the veneer.

3. Chemical soil treatment. The following chemical formulations have been found to be successful and are recommended: aldrin, 0.5 percent in water emulsion or oil solution; benzene hexachloride, 0.8 percent gamma in water emulsion or oil solution; chlordane, 1 percent in water emulsion or oil solution; dieldrin, 0.5 percent in water emulsion or oil solution. In general, water emulsions are not injurious to plants. Applications of chemicals and precautions involved in their use should be according to manufacturers' recommendations.

4. Poured concrete foundations, provided no cracks greater than $\frac{1}{64}$ inch are present.

5. Poured, reinforced concrete caps, at least 4 inches thick, on unit masonry foundations, provided no cracks greater than $\frac{1}{64}$ inch occur.

Slab-on-ground construction requires special consideration in areas where the termite hazard is a significant problem. Concrete slabs vary in their susceptibility to penetration by termites; thus, they cannot be considered to provide adequate protection unless the slab and supporting foundation are poured integrally to avoid cracks or holes through which termites may enter.

Where other types of slab construction are used, termites may penetrate through joints between the slab and wall. They may also enter through expansion joints or openings made for plumbing or conduit. Thus it is necessary at these points to provide a barrier either by the use of termite shields, coal tar pitch or chemical soil treatment.

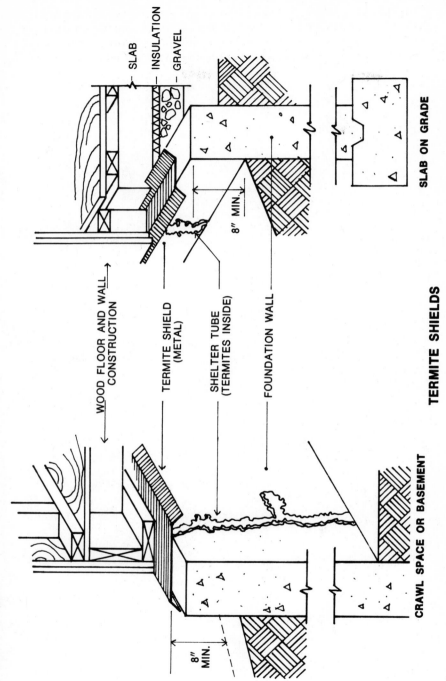

SLAB

INSULATION

GRAVEL

SLAB ON GRADE

WOOD FLOOR AND WALL CONSTRUCTION

TERMITE SHIELD (METAL)

SHELTER TUBE (TERMITES INSIDE)

FOUNDATION WALL

8" MIN.

8" MIN.

CRAWL SPACE OR BASEMENT

TERMITE SHIELDS

221

Masonry veneer in contact with the ground may provide access for termites in infested areas. For this reason the veneer should be kept at least 8 inches above the finished grade unless termite shields are installed in an approved manner or the soil on the exterior has received a chemical treatment.

Naturally durable or treated wood should be used when the member is so located that it cannot be maintained at a safe moisture content or where climatic or site conditions are such that construction practices alone are not sufficient for control of decay or termites. It should be used where wood is embedded in the ground, where it is resting on concrete which is in direct contact with earth or where it is not possible to maintain recommended separations between wood and earth.

Woods that are naturally decay-resistant are bald cypress (tidewater red), cedar, redwood, black locust and black walnut. Termite resistant species are redwood, bald cypress and eastern red cedar.

Wood may be treated by the pressure method in which it is impregnated with toxic chemicals at elevated pressures and temperatures in a retort. Wood may be preservatively treated by the non-pressure process of cold soaking, by the vacuum process or by the hot-and-cold-bath method. It is best not to cut the wood after it has been treated. However, if cutting is necessary, the new surfaces must be treated with a liberal application of the preservative used in the initial treatment.

The best way to check for termites is to hire a professional. The FHA, VA and other lending institutions require professional termite inspections in many areas of the country.

Termites work slowly. If they are caught in the early stages of infestation they may be stopped for a few hundred dollars. Damage done in the more advanced stages of infestation may cost thousands of dollars to repair.

Maintenance inspections of the house at least annually in the spring will detect problems at an early stage. Such inspections should concentrate on three specific areas: (1) foundation, including crawl spaces; (2) attic spaces and roof and (3) exterior surfaces, joints and architectural details. When termite shelter tubes are discovered, they should be destroyed and the ground below should be poisoned. When evidence of termites is discovered, it is best to use a professional exterminator to correct the problem. If dampness is noticed, proper clearances between wood and ground should be re-established and vents checked and repaired.

Painting and Decorating

In painting and decorating a house the objective should be to provide a coating which will give adequate resistance to weathering, protection from damage by corrosion, reasonable durability, economical maintenance and an attractive appearance.

All paints or other coatings should be standard commercial brands with a history of satisfactory use under conditions equal to or similar to the conditions present in the area concerned.

Application of paint or other coating should be in strict accordance with manufacturers' directions. Ready-mixed paint should not be thinned, except as permitted in the application instructions. Exterior painting should be done only in favorable weather. All surfaces should be free of dew or frost and must be dry to the touch except with certain masonry paints formulated for application to wet surfaces. Painting should not be done when the temperature is below 40 degrees. All surfaces to be finished should be clean and free of foreign material such as dirt, grease, asphalt or rust.

Application should be made in a workmanlike manner to provide a smooth surface. Additional coats may be required if the finish surface does not provide acceptable coverage or hiding. Certain pigments provide excellent hiding ability even when thinly applied. With paints of this type, care must be taken to obtain adequate coverage if the coating is to offer reasonable durability.

The following charts indicate the proper preparation, priming and finish coating recommended by the Federal Housing Administration for the various surfaces inside and outside a house.

Interior Wood, Metal and Concrete Surfaces

Type of Material	Preparation and Priming	Finish Coating
Windows Interior Doors Window Trim Door Trim Base Trim Paneling	If to be painted, fill and seal open grain wood to prevent raising; then prime with suitable primer compatible with finish coating.	One or more finish coats, same as for plaster.
Kitchen Cabinets Closet Shelving Closet Trim Miscellaneous Trim	If to be finished naturally, fill and seal open grain wood to prevent raising.	One or more coats of the following natural finishes[1]: Stain-wax. Stain followed by one or more coats of varnish. Shellac or lacquer with or without wiped paint undercoats of oil and wax finishes. Clear coats of varnish.
Plywood or Hardboard Interior Finish	None needed. If wallpapered, apply a smooth wall liner of blank wallpaper stock or smooth ¾-pound deadening felt to prevent grain showing through wallpaper.	Same as above for painting and natural finishes.
Structural Steel	Prime with one coat of red or blue lead or zinc chromate.	When exposed, paint with one or more coats of paint compatible with primer.
Other Interior Metals[2]	Same as exterior metal.	Same as exterior metal.

Interior Wood Floors

Sand floors.
When open grained, apply one coat of filler, wipe off excess.
Prefinished floors require no preparation.
Floors covered with wall to wall carpet should be rough sanded to smooth joints and edges.

Finish floors with one of the following:
One coat sealer and two coats of wax.
Stain and two coats of wax.
One coat shellac, varnish or lacquer and coat of wax.
Two coats of varnish or shellac.
Two coats floor and deck paint and one coat of wax.
Flooring prefinished at the factory requires no in-home finishing.

Interior Concrete Floors[3]

When oil paint is used, neutralize surface before painting.
For other finishes, clean before applying finish.

Finish floors with one of the following:
Two coats of resin emulsion paint.
Two coats solvent rubber paint.
Two coats floor deck enamel.
Two coats oil base paint.
Apply coat of wax over all finishes.

[1]When used in kitchen or bathroom, must provide a durable, waterproof finish.
[2]May be left unfinished.
[3]May be left unfinished in uninhabitable rooms and their closets and when covered with other floor material.

Interior Walls and Ceilings

Type of Material	Preparation and Priming	Finish Coating
Bathroom and Kitchen Walls with Plaster Surfaces	When to be painted, prime with a suitable primer compatible with finishing paint unless finish paint is self-priming type; then no primer is needed.	One coat of the following: Casein paint (water thinned). Enamel (solvent thinned). Latex paint (PVA, acrylic, styrene, butadiene — water thinned). Oil paint (solvent thinned). Resin-emulsion paint (solvent thinned).
	When to be papered, plaster should be thoroughly dry, neutralized with zinc sulfate solution, if necessary, and coated with a coat of wall sizing.	One layer of sunfast and waterfast colored wallpaper that is either waterproof or waterproofed.
Paster Surfaces covered with tile or other approved finish material	Prepare according to finish materials manufacturers' instructions.	None needed.
Bathroom and Kitchen Walls with gypsum wallboard surfaces	Tape and cement joints in accordance with manufacturers' instructions. When to be painted, apply coat of wallboard sealer (not needed if finish coats are self-sealing types). When to be papered, seal entire surface with a coat of latex, oil or varnish base sealer.	Two coats of the same as for bathrooms and kitchens with plaster surfaces. Same as bathrooms and kitchens with plaster surfaces.
Gypsum Wallboard Surfaces covered with tile or other approved finish material	Prepare according to finish materials manufacturers' instructions.	None needed. Prefinished gypsum wallboard may be used except in bathrooms and kitchens.
Masonry Surfaces	None needed when left unfinished. Prime with primer compatible with finish paint.	May be left unfinished. If painted, same as plaster.

Exterior Wood Surfaces

Type of Material	Preparation and Priming	Finish Coating
Wood Siding[1,2] Millwork[1,2] Trim[1,2]	Fill knotholes and cracks with putty. Seal with knot sealer or aluminum paint. Prime with primer or initial coat of finishing paint.	Apply two coats of finish paint to obtain coating 4.5 to 5.0 mil thickness.
Top and Bottom of: Exterior Wood Doors Casement Sash Awning Sash Double-Hung Window Sash	Same as above	Same as above
Wood Shingles[1,2] Wood Shakes[1,2] Roughsawn Siding[1,2]	None required	Two coats of oil stain, pigmented stain or oil shingle paint.
Exterior Plywood	Seal all edges prior to erection with a heavy coat of primer, aluminum paint or white lead linseed oil paste. Seal surfaces same as wood siding.	Apply two coats to sanded surfaces to obtain 4.0 mil thickness. On unsanded surfaces (grooved-siding) apply two coats of opaque penetrating finish such as pigmented oil stain. Do not use ordinary house paint or clear finish.
Wood Porch Floors Wood Porch Decks	Caulk joints and seal surfaces and prime same as wood siding.	One or more coats of exterior type floor and neck enamel.

[1]Factory-applied finishes of comparable quality also acceptable.

[2]When made of edge grain redwood or red cedar, may be left unfinished.

Caution: Natural finishes or surface coated finishes (for example, varnishes and synthetics) should be used only on surfaces which are not in direct exposure to rain or sunlight. Varnish, if used, should be spar varnish formulated for exterior use; not less than two coats should be applied. Penetrating finishes such as sealers, oils, pigmented oils and water repellents should be used on surfaces exposed to sunlight or rain.

Exterior Metal

Type of Material	Preparation and Priming	Finish Coating
Galvanized Steel Galvanized Iron	Clean with turpentine or mineral spirits and wipe dry. Prime with a coat of zinc dust or zinc oxide primer.	Coat of exterior house paint or another coat of zinc dust, zinc oxide primer.
Copper[1]	Prime with one coat of a primer compatible with finish coat.	Coat of exterior house paint.
Steel	Remove all dirt, grease, rust, loose scale, etc. Prime with at least one coat of red or blue lead or zinc chromate primer; or Factory-applied rust inhibitive phosphate coating.	Coat of exterior house paint or exterior aluminum paint.
Iron	Same as steel	Same as steel
Stainless Steel[1]	Same as steel	Same as steel
Aluminum[1]	Clean surface with aluminum cleaner. Prime with coat of zinc chromate primer.	One or more coats of exterior house paint or exterior aluminum paint.
Zinc-Copper Alloys[1]	Same as galvanized steel	Same as galvanized steel
Terne Plate	Prime with coat of primer compatible with lead-tin alloy coating.	Same as steel

[1]May be left unfinished except when located so as to cause staining of surfaces below.

Exterior Masonry Units

Type of Material	Preparation and Priming	Finish Coating
Foundation Walls	None needed	None needed
High Density Concrete Brick	None needed	None needed
Solid Split Concrete Block	None needed	None needed
Other Concrete Masonry Units	Brush surface with one of the following to fill pours: Portland cement grout. Coat of cement water paint. Grout made of equal parts of portland cement, fine sand and water.	Two coats of one of the following: Cement-water paint (water thinned). Latex paint (PVA, acrylic, styrene, butadiene — water thinned). Rubber paint (synthetic or natural — solvent thinned). Oil paint (solvent thinned).
Clay Brick	None needed	Same as other concrete masonry units.[1,2]
Clay Tile	None needed	Same as other concrete masonry units.[1,2]
Concrete or Stucco Finish	None needed	Same as other concrete masonry units except that paints formulated especially for alkaline surfaces preferred.[1]

[1]May be left unfinished.
[2]Latex paint not to be used on glazed brick or tile.

229

Chapter 6
Building Materials

A myriad of building materials goes into a house. And each type of material—from wood to steel—can be produced with a variety of characteristics and qualities. Choosing the proper material for a particular purpose and then picking the most desirable grade of that material are extremely important factors in good house construction.

Wood

General Information and Properties

Wood is the basic material of which most houses in this country have been made in the past, are being made today and apparently will be made in the foreseeable future. Fortunately, because of wise forest management programs we are growing more timber each year than we are cutting down and losing through natural causes. However, certain species are being depleted faster than they are being replaced and conservation work is still needed.

Wood's popularity as a house building material is due to its universal availability, low price, high compression and tension strength, resistance to bending and resistance to impact. It can be easily worked with simple tools and is highly durable, as evidenced by the many 300-year-old wood houses still standing today. It also has excellent insulating qualities and its natural grain is very decorative and appealing. Four basic building materials are made from raw wood: lumber, plywood, laminated timber and fiberboard.

Wood is divided into two classes: softwood and hardwood. This division is based on a botanical difference and not necessarily on the actual hardness or softness of the wood. Softwood comes from coniferous (cone bearing) trees, most of which have needles rather than leaves. It is used for framing lumber, sheathing boards, roofing, subflooring, siding, some flooring, some moulding and some paneling. Hardwood comes from deciduous (leaf shedding) trees with broad leaves. It is used for hardwood floors, some interior paneling, cabinets, furniture and some moulding.

Lumber

After a log is felled it is sawed or planed into various sizes and shapes called lumber. Lumber is divided into the broad classifications of boards (1 to 1½ inches thick, 2 inches and wider), timbers (5 inches and thicker and 5 inches and wider) and specialty products (various sizes). The vast majority of lumber used in residential construction is 2 inches thick or less.

Plywood

Softwood plywood is used extensively in house construction. The most common size sheet is 4 by 8 feet, but lengths up to 16 feet or longer and widths up to 10 feet are also available. The number of plies is always odd and the minimum number is three, which, together, run from ⅛ inch (special use only) to ⅜ inch thick; five plies run from ½ to ¾ inch thick and seven plies run from ⅞ to 1⅛ inch thick.

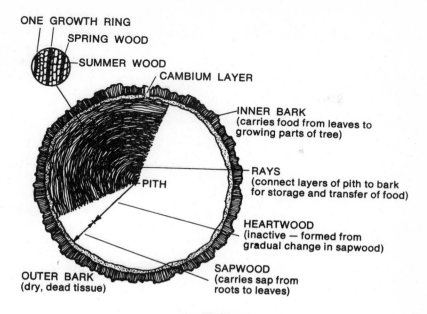

ONE GROWTH RING

SPRING WOOD

SUMMER WOOD

CAMBIUM LAYER

INNER BARK
(carries food from leaves to
growing parts of tree)

PITH

RAYS
(connect layers of pith to bark
for storage and transfer of food)

HEARTWOOD
(inactive — formed from
gradual change in sapwood)

OUTER BARK
(dry, dead tissue)

SAPWOOD
(carries sap from
roots to leaves)

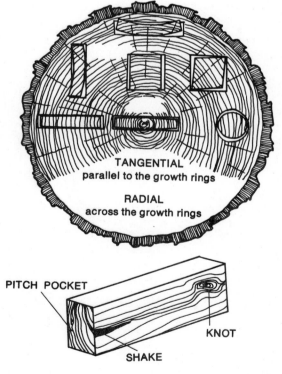

TANGENTIAL
parallel to the growth rings

RADIAL
across the growth rings

PITCH POCKET

KNOT

SHAKE

RAW WOOD   233

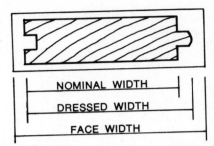

NOMINAL, DRESSED AND FACE SIZES OF LUMBER

NOMINAL SIZE OF LUMBER IS ITS UNFINISHED SIZE, KNOWN COMMERCIALLY AS 1X6, 2X4, ETC. SEASONING AND PLANING BRING IT TO ITS **DRESSED SIZE.** THE **FACE SIZE** IS THAT PORTION OF PIECE EXPOSED TO VIEW WHEN IN PLACE.

DRESSED AND MATCHED (D & M)
USED FOR SUBFLOORING, FLOORING SHEATHING OR WHERE TIGHT JOINTS ARE REQUIRED BETWEEN BOARDS

SHIPLAP
USED FOR SUBFLOORING, WALL AND ROOF SHEATHING OR LOW-COST SIDING

VERTICAL SIDING   DROP SIDING   MOULDINGS
PATTERNED

TYPICAL TYPES OF WORKED LUMBER

234

Two types of softwood plywood are available. The exterior type is used whenever the wood will be used where it will get wet, such as for fences, sheathing and siding. It is glued with adhesives that are insoluble in water. Interior plywood is made with less expensive water soluble adhesives; therefore it must not be used where it will be exposed to water or high humidity.

Softwood plywood is graded by appearance. Poor appearance is caused by defects in the face sheets of veneer. The grades range down from the best "N" Grade used only for special work; "A" Grade, good for the great majority of exposed painted surfaces; "B" Grade (in which defects have been repaired), good for concrete forms where a smooth surface, but not appearance, is important; "C" Grade which contains light unrepaired knots and is often used for sheathing; to "D" Grade (with larger knotholes and defects) which is unsuitable for most house construction.

There are also three additional softwood plywood grades called engineered grades. Standard Grade is designed for use as floor, wall and roof sheathing. Structural I and II Grades are unsanded and made with waterproof adhesives and are used in making stressed skin panels, box beams and trusses.

Hardwood plywood is used as a decorative material for wall paneling, doors and block flooring. It takes its name from whatever species is used for the face veneer, i.e., mahogany plywood, maple plywood, etc.

Another major difference in the manufacture of softwood and hardwood plywood is the use of a solid "core" or extra-thick middle ply in many hardwood panels. Lumber-core plywood contains a core made up of narrow strips of solid wood. It is often used for cabinets and doors.

Particle-board core plywood contains a core made of wood particles bonded together with a resin binder. It has great dimensional stability and is good for counter and table tops. Mineral core plywood is used for fire-resistant panel construction. Veneers are bound to a core of hard, non-combustible material.

The most common size hardwood panel is 4 by 8 feet but, like softwood plywood, larger sizes are available on special order. The number of plies is always odd and the minimum number is three, which, together, run from ⅛ to ¼ inch thick; five plies from ¼ to ⅝ inch thick, seven plies from ⅝ to ¾ inch thick and nine plies from ¾ to 1 inch thick.

There are four types of plywood made, depending upon the kind of adhesive used. Type I is made with waterproof adhesive for use where the plywood will come in contact with water; Type II is made with water resistant adhesive for use in areas of high humidity and dampness and Type III is made with moisture resistant adhesive for use in areas with some humidity. The Technical Grade is a special product that provides equal stiffness in both directions.

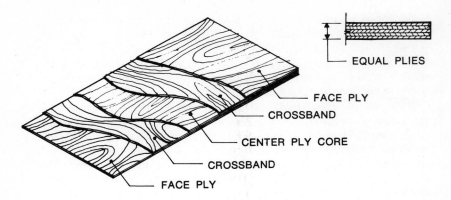

EQUAL PLIES

FACE PLY

CROSSBAND

CENTER PLY CORE

CROSSBAND

FACE PLY

SOFTWOOD VENEER

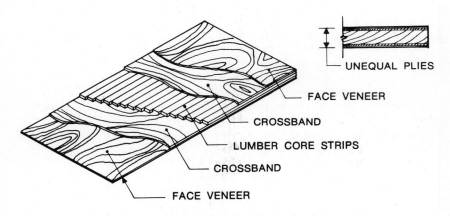

UNEQUAL PLIES

FACE VENEER

CROSSBAND

LUMBER CORE STRIPS

CROSSBAND

FACE VENEER

HARDWOOD LUMBER CORE

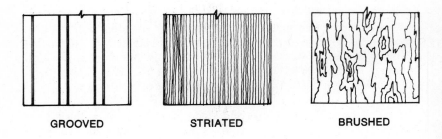

GROOVED

STRIATED

BRUSHED

PLYWOOD

Hardwood is also graded to indicate quality of the face veneer. The grades in order of descending quality are Premium, Good Grade (#2), Utility Grade (#3), Backing Grade (#4) and Specialty Grade (SP). Hardwood panels are primarily used for their decorative quality. Their appearance is affected by tree species (any of almost 250 different species are used to make hardwood plywood). The part of the tree the wood is cut from affects the look of the grain, as does the method of cutting. Finally, the manner in which the pieces of veneer are applied and matched produces still more variation in the appearance.

A multitude of special manufacturing processes and treatments make possible an almost unlimited variety of plywood finishes and characteristics. Many manufacturers, in addition to making the standard panels, also make a line of their own special panels. Some of the common things done to alter the appearance are to groove the surface, striate it, brush it or emboss it.

The use of plywood as a building material increases yearly. Plywood makes a highly acceptable structural material that lends itself to precutting, prefabrication and modular home construction. The hardwood plywoods offer an almost unlimited number of decorative possibilities and are easy and economical to install and maintain.

Shingles and Shakes

The handsplit shingle or shake was one of the first exterior covering materials used in this country, dating back to early colonial days. The colonists first made them from white pine, often smoothing the split surfaces on a "shaving horse." The early settlers in the Pacific Northwest found the coastal natives living in wood "long houses." Cabins made by these early settlers are still intact today.

What made them so durable was western red cedar wood which is highly resistant to rot and decay. Most of the wood shingles and shakes made today in this country are western red cedar. Other desirable characteristics of this wood are its fine, even grain, exceptional strength, low weight, low rate of expansion and contraction with changes in moisture content and high impermeability to water.

The difference between a shingle and a shake is that a shingle is sawed and a shake is split. A shake may have a sawed back and still be called a shake.

Shingles that are properly applied to walls should last the life of the structure. When applied to roofs their life expectancy decreases as the slope of the roof decreases and the grade of the shingle decreases. A premium shingle on a normal house roof with a 45 degree slope should last 20 to 25 years in most parts of the country.

Wood Flooring

Wood flooring can be made out of either hardwood or softwood. Its longstanding popularity comes from natural appearance, durability, comfort underfoot and ease of installation and maintenance.

Although the first settlers' homes probably had dirt floors, it wasn't long before wood floors became a standard feature of the colonial house. Those first floors were made of wide softwood planks. Gradually, as manufacturing processes improved, the narrow hardwood strip floors became the most popular. However, there has been a recent increase in demand for the old random length, wide plank flooring. When more formal houses were desired, small pieces set in intricate designs called parquet floors were installed. Wood blocks are also used for flooring.

The bulk of wood flooring is made from 12 different tree species. The most popular hardwood used for house floors is oak, followed far behind by the other hardwoods—maple, birch and beech. The two most popular softwoods are southern pine and then Douglas fir. Far behind them come western hemlock, eastern white pine, ponderosa pine and half a dozen other woods.

More than 70 percent of all wood flooring is from about 20 species of oak, which is divided into the two broad classifications of white oak and red oak. With the exception of appearance there is little difference between them for flooring purposes. For floors that take an extra amount of abuse, sugar maple makes a good flooring since it is the hardest and most wear resistant.

Installation

Wood strip flooring may be installed over wood subflooring covered with polyethylene film on top of which are set 1 by 2-inch wood strips 16 inches on center.

The handling of the flooring material on the site is very important. If properly shipped it will still be at the 6 to 8 percent moisture content that was obtained by the kiln drying process. When exposed to moisture the flooring will quickly absorb it, swell, twist and otherwise get out of shape and will not fit well together. Furthermore, flooring that is laid moist will shrink in the dry winter inside air and leave open joints between adjacent strips.

The laying process is started by nailing down a strip next to the wall. The nails are driven through the top of the board close to the wall so they will be covered by the baseboard. Subsequent strips are fitted lightly against each other and blind nailed to the subflooring or 1 by 2-inch wood strips.

Strip flooring can be laid in various patterns. The simplest and most common method is to run all the strips in the same direction. Another method is to lay the boards parallel to all four sides of the room. This pattern is called "log cabin." Parquet floor designs are made by cutting the strips into small pieces and laying them in a geometrical pattern. It is possible to make a very complicated design by using small pieces and woods of different colors.

If the flooring is not prefinished it must be scraped by hand (a process that some think brings out the best look in the wood) or sanded with a machine.

Wood blocks can be installed similarly to strips or attached with an adhesive. Wood planks are sometimes laid to simulate old plank floors by using pieces of random widths and lengths. The floor may be blind nailed or pegged.

Concrete

After wood, the most important building material in house construction is concrete. It is used for foundation footings, cellar floors and walls, sidewalks, driveways, garage floors and septic tanks. Concrete has four desirable qualities that make it a good building material: it is strong and will support heavy loads; it is watertight and therefore is good for subsurface installation; it is extremely durable, which makes it ideal for footings and foundations that must last the entire life of the house, and it is workable and may be poured into forms and troweled to obtain a smooth or designed surface.

Concrete is a mixture of portland cement, water, fine aggregate and coarse aggregate and admixtures. When portland cement is mixed with water a chemical reaction takes place. The mixture sets and then hardens into a solid mass that will stay solid forever. Detailed information on concrete manufacture is included in Chapter 8.

White portland cement is used for special purposes where the white color is desirable, such as in precast walls, terrazzo, stucco and tile grout. It is also used if a color is to be added to the mixture. There are also several other types that are used for non-residential construction purposes.

Masonry

A brick made of clay is one of the oldest building materials. There is little recorded history that does not mention brick and other masonry materials for building. Today it is widely used for wall construction in houses in almost every part of the country and for many miscellaneous purposes ranging from patios to walls to fireplaces. The cement block has also developed into a major residential building material. Primarily used for basement walls, it is also used for complete wall construction or covered with a veneer of stucco, brick or wood.

Masonry used for exterior walls must as a minimum be strong, watertight and durable. Its appearance varies widely in color, texture and pattern. Masonry is broken down into the two broad classifications of clay masonry units and concrete masonry units.

Clay Brick and Tiles

Solid clay masonry units are called bricks and hollow units tiles. They are made of surface clays, shales and fire clays. All three are compounds of silica, alumina and varying amounts of metal oxides and other impurities. Metal oxides affect the final color of the brick.

Surface clays are sediments that are found near the earth's surface. Shales are a type of clay that has naturally been subject to high pressure and has become quite hard. Fire clays are found deep within the earth and have to be mined out. They have refractory qualities (resistance to heat).

Clay Brick and Tile Classification and Properties

The term "brick" means a solid masonry unit. Cored brick may have from three to more than 20 core holes, taking up to 25 percent of the total brick volume, and still classify as a solid unit. Bricks may be custom made but most are produced in standard sizes and shapes.

The properties of the brick depend upon the characteristics of the clay used and the manufacturing process. Bricks with the highest compression strength and lowest absorption rate are usually made by the stiff mud process and burned at high temperatures. In general, the higher the burning temperature the higher the compression rate. Well-made bricks are very durable. The only action of weathering that has any significant effect on them is alternate freezing and thawing in the presence of moisture.

The natural color of brick depends upon the chemical composition of the clay used, the burning temperatures and amount of oxygen used in the burning process and the length of time the brick is burned. Colors range from light grays and creams to dark reds and blacks. Underburning tends to produce lighter colors and overburning darker colors. However, underburned bricks are softer and more absorptive and have decreased compressive strength. Overburning produces a dark clinker brick that is hard and less absorbent and has increased compressive strength.

Textures range from very smooth, resulting from the pressure exerted by the steel dies, through fine and medium to those which have been marked by cutting, scratching, rolling, brushing or otherwise roughing up the surface as the clay leaves the die.

Fire shrinkage increases with higher temperatures at an uneven rate resulting in bricks of uneven sizes. This problem is well acknowledged and builders accept bricks with size variations.

Hollow Masonry Tile Classification and Properties

The term "hollow tile" means a hollow masonry unit with the core holes exceeding 25 percent of the total tile volume.

Structural clay tile is produced as load bearing and non-load bearing. When designed to be laid with the cells in a horizontal plane, the unit is called side construction or horizontal cell tile. When it is designed to be laid with the cells vertically, the unit is called end construction or vertical cell tile.

Structural facing tile is produced glazed or unglazed. Glazed tile is produced from a high grade, light burning fire clay to which a ceramic glaze is applied that fuses into a glass-like coating on the tile during the burning process. Unglazed tile is produced from light or dark-burning clays and shales and may have smooth or tough textural finishes.

Architectural terra cotta is a made-to-order decorative glazed tile used as a veneer to make a multi-colored interior or exterior wall.

Concrete Brick and Block

Solid concrete masonry units are called bricks and hollow masonry units blocks. They are made of a relatively dry mixture of cementitious material, aggregates, water and admixtures. The cementitious material is usually portland cement (either the normal or high-early-strength type), portland blast furnace slag cements, fly ash, silica flour or other pozzolanic (ash) materials.

The primary factor in the selection of aggregates is their availability locally. They are classified according to their weight as dense or normal. These classes include sand, gravel, crushed limestone and air-cooled slag. Lightweight aggregate includes expanded shale or clay, expanded slag, coal cinders, pumice and scoria.

A few admixtures have been found to be beneficial in concrete brick and block manufacture. Air-entraining agents increase the plasticity and workability and distribute minute air bubbles uniformly throughout the hardened concrete, increasing its ability to withstand frost action. They also improve the appearance and reduce the amount of breakage of freshly molded units. Water-repelling agents such as metallic stearates reduce the rate of absorption and the capillarity of the units. Accelerators (usually calcium chloride) speed up the hardening during cold weather. Workability agents are used to compensate for poorly graded aggregates.

Concrete Brick and Block Classification and Properties

Many of the physical properties of concrete blocks are similar to regular concrete, previously discussed. However, there are some important differences. Concrete bricks and blocks are made with a substantially lower percentage of concrete in the mixture and a much lower water-to-cement ratio than regular concrete. Fine aggregate is used and is often made of porous or lightweight materials. Unlike structural concrete, blocks and bricks contain a relatively large volume of inter-particle void spaces between the pieces of aggregate that are not filled with cement.

The compressive strength of the units is highest and water absorption rate lowest when they are made of sand and gravel, limestone or air-cooled aggregate, in that order. It is less for units made of expanded shale, expanded slag, cinders, pumice or scoria, in that order.

Bricks and blocks are made in a wide variety of surface textures for aesthetic and structural reasons. The texture is varied by controlling the gradation of aggregates, the amount of water used and the degree of compaction at the time of molding. The surface may be ground to obtain an exact size. The face contour is controlled with the use of a face mold.

Most concrete blocks made with portland cement range in color from white to black and tan to brown. Other colors can be obtained by the addition of mineral oxide pigments to the concrete mixture before it is cured.

Mortars

Mortar is a mixture of cementitious materials (portland cement, lime, masonry cement, sand and water). It is used to bind the individual masonry units together and to the other building elements. It also seals the spaces between the masonry units, compensates for the size variations, binds the units to the metal ties and reinforcing (if any) and improves the appearance of the wall aesthetically by creating shadow lines and/or color effects.

Mortar Properties

Mortar is laid while it is soft; later it hardens. In its soft state it must have good workability. To be workable the particles must not separate, must spread easily and hold the weight of the units firmly enough, without sagging, drooping or smearing, to permit their alignment.

The workability of mortar is affected by its water retention capacity. It must be sufficient to prevent the moisture from bleeding into the dry masonry units which tend to attract it. When this happens a thin layer of water forms between the mortar and the masonry unit and the bond is substantially weakened. When the mortar hardens, its most important quality is its bond strength. Bond strength is affected by the type of mortar mixture used, type of unit being bonded, workmanship and curing.

The durability of mortar is measured by its ability to resist damage caused by the natural cycle of heating and thawing. Correctly installed mortar should last 35 years or more before requiring any maintenance. The addition of an air-entraining agent to the mixture substantially improves the durability of mortar.

The compressive strength of mortar is controlled by the amount of portland cement in the mixture. The more cement, the higher the compressive strength; the more water, the less the compressive strength. This is rarely a problem in residential construction, however.

Cracked or Loose Mortar Joints

Mortar that is not perfectly installed will in time become soft, crack and fall out. This will in turn weaken the wall and allow water to seep through. The joints can be inspected visually and by poking them with a pointed instrument. Defective joints should be repointed.

Iron

Iron and steel products are found throughout a house. Nails, beams, plumbing fixtures, hardware, appliances, cabinets, electrical components and heating systems all contain iron or steel.

Iron is an element found all over the earth. In fact, about 5 percent of the earth's crust is iron and possibly even a higher percentage of the earth's molten core is iron. It is rarely found in its metallic form, however. Instead it is usually combined with oxygen and mixed with other minerals. Iron ore is a mixture of iron, oxygen and other minerals in a combination containing enough iron to make it economically feasible to extract the iron.

Pure iron is soft and has little usefulness. It is hardened by adding small amounts of carbon and other elements. Steel is a mixture of iron and about 1.2 percent carbon. Cast iron is a mixture of iron and substantially less carbon, while wrought iron is a mixture of iron and substantially more carbon.

Materials

Iron ore is the primary raw material. It is mined from either underground mines or open pit mines. The largest mines in the U.S. are in the Lake Superior area. Good iron ore may contain as much as 70 percent iron. Fuel is necessary to produce iron from the ore. Coal, oil and natural gas are all used as fuels.

Fluxes are materials used to separate the impurities. Basic fluxes are limestone and dolomite which are mined. Acid fluxes are sand, gravel and quartz rock which are taken from quarries. Fluorspar, a mineral that is also mined, is a neutral flux used to make the slag more liquid.

Refractories, like fluxes, are basic or acid. They have high heat resistance and are used to line the steel-making furnaces. Ganister, which is a quartzite rock, is an acid refractory. Magnesia, a basic refractory, is a mineral that is extracted from sea water

Classification of Irons

Cast Iron: By combining selected pig irons, scrap metals and alloying elements, cast iron can be made with increased hardness, toughness, corrosion and wear-resistance. This permits the casting of complex shapes difficult to produce by machining or rolling. Bath tubs, pipe and radiators are often made of cast iron.

Malleable Cast Iron: This type of iron contains a mixture of raw materials that improves its breakage resistance by increasing the toughness and ductility. It is used for builders' hardware.

High Strength and Nodular Cast Iron: By adding molybdenum, chromium and nickel to the blend, a high strength cast iron is produced that is used for special purposes. By adding magnesium, nodular cast iron is produced which has high strength, ductility, toughness, machinability and corrosion resistance.

Wrought Iron: Mechanically mixing iron-silicate with the pig iron and rolling it results in an iron with packed and elongated particles of iron and slag, giving it a fibrous grain structure. The iron becomes tough, ductile and easy to weld. It has low compressive and tensile strength but high rust resistance and is used for exterior rails, ornaments and pipes.

Steel

Materials

The basic materials of steel are pig iron, carbon, phosphorus, sulfur, manganese and silicon. These materials, when present in the pig iron as it comes from the blast furnace, are called residual elements. When they are added they are called alloying elements.

Properties

The properties of steel can be changed by adding various alloys to it, such as aluminum, copper, lead, chromium, nickel, molybdenum, tungsten, vanadium, tellurium, titanium and cobalt.

The American Iron and Steel Institute (AISI) has a letter and numerical designation system for identifying the alloys that have been added to a type of steel and how it was made.

Alloy steels are high in cost and are used only where carbon steel of similar properties is not satisfactory. They are used for motor parts such as those found in vacuum cleaners, washing machines, tools and appliances.

Stainless steel has high resistance to corrosion. It obtains this characteristic from the addition of chromium. When chromium oxidizes a thin, airtight layer of chromium oxide forms over its surface which prevents corrosion. Stainless steel is classified by the AISI as martensitic steel when it contains 12 to 18 percent chromium; ferritic steel when it has 18 to 27 percent chromium; austenitic steel when it has 16 to 26 percent plus 3.5 to 22 percent nickel and sometimes 5.5 to 10 percent manganese.

Martensitic chromium steel has moderate corrosion resistance, high strength hardness and abrasion resistance. It is used for cutlery and bearings. Ferritic chromium steels have high corrosion resistance and are used as trim on such items as automobiles, appliances, gutters and downspouts, range hoods and range tops.

Austenitic nickel-chromium steels are the most common stainless steels. They have excellent corrosion resistance and high strength and ductility and are chemically resistant. They are used for kitchen sinks, counter tops and appliances, kitchenware and utensils.

Heat resisting steel obtains its characteristics with the addition of 4 to 12 percent chromium. It retains its physical and mechanical properties at temperatures up to 1,100 degrees.

Aluminum

Aluminum is a modern building material that was unknown 200 years ago and not commonly used except for jewelry and tableware until the 20th century. Its use has doubled in the past 15 years.

Materials

Aluminum is the most plentiful metal, but it is never found free like gold or silver. It is always combined chemically with other elements. Almost all rocks have some aluminum in them. However, it is not economically profitable to extract it unless the ore has an aluminum oxide content of at least 45 percent. Aluminum ores are called bauxite. Almost all American bauxite is in Arkansas, which unfortunately supplies us with only about a quarter of our need. The rest is imported.

Products Made from Aluminum Parts

The principal aluminum products used in residential construction are doors and windows, gutters and downspouts, siding and insulation. Each of these products is covered in its respective section in this book.

Sheet aluminum is used for many other household building products. Among them are air-conditioning and heating ducts, awnings, flashing, garage doors, range hoods, letters and numbers, shingles, termite shields, terrazzo strips, venetian blinds and weatherstripping. Extrusion aluminum products used in houses include builders' hardware, fascia plate, gravel stops, railings and thresholds. Bar aluminum products include grating and screws. Wire aluminum is used for nails and screens. Aluminum foil is used for insulation and vapor barriers. Cast aluminum is used for hardware, grilles, letters and sculptural relief panels.

Glass

The use of glass goes back to prehistoric times, although the exact place and date of its origin are unknown. The oldest known specimens of glass were from Egypt where around 2000 B.C. the glassmaking industry was well established. Glass was used for Roman window panes at the time of Christ. The process for casting glass was invented in France in 1688.

Types of Window Glass

The most common window glass used in houses is standard sheet glass. In the building trade the thickness of glass is known as its strength. The most common types are single strength and double strength.

Single strength is used only in small lights (panes) and in areas of low wind velocity. MPS states that nothing less than single strength glass should be used. The strength of the glass should be increased as the size of the window increases and in areas of high wind velocity. Double strength is suitable for most regular window glass use.

Plate glass tends to distort the view less than standard glass and is used where good vision is important, as in a picture window. It is more expensive than standard or double glass.

To obtain better insulation, two pieces of single or double strength glass may be hermetically sealed together with an air space between them. This is known as insulating glass. It is often used in electrically heated houses. Each light of glass comes with a label that shows the manufacturer's name, the quality and the thickness.

Glass that is tinted or bronzed reflects heat and has some insulation qualities. By reflection it reduces the ability to see inside the house from the exterior.

Types of Door Glass

According to MPS, single strength glass should only be used in doors when the area of the pane is not more than 6 square feet and the glass is not less than 10 inches above the floor. Double strength or stronger annealed glass is required when the size of the pane is increased. When the door does not have bars, special safety glass should be used. Safety glass may be made by laminating two sheets of single or double strength glass to a tough plastic sheet. Shower and tub enclosures should also be safety glass.

Fully tempered glass is often used in doors which might be hazardous if a child ran into them. This glass breaks up upon impact into small pieces rather than sharp-edged pieces. Tempered and laminated glass may be identified by the permanent label on it showing the manufacturer's name, the type of glass and the thickness, which should be left visible on the lower side when the glass is installed. Glass which meets rigid safety standards may also have an Underwriters' Laboratories label permanently on it.

Resilient Flooring Products

Resilient flooring products have dense, nonabsorbent surfaces and the ability to spring back into place. Linoleum was the first resilient flooring. It was discovered in England more than 100 years ago. After World War I, asphalt tile evolved from a troweled-on mastic flooring. After World War II, vinyl asbestos and vinyl flooring products were introduced. They come in a wide variety of textures and colors. New vinyl products are continuously being introduced and they appear to be, along with carpeting, the floor covering of the future. Resilient flooring is manufactured in sheet and tile form. Wall base, thresholds, stair treads and feature strips are available to make complete matching flooring installations possible.

Linoleum

The term "inlaid linoleum" is used for all types of true linoleum products, including plain, marbleized, jasper and molded types, to distinguish them from a thin-faced or enameled felt sheet product typically made in rug sizes that is often mistakenly referred to as linoleum. This inexpensive product is not suitable for permanently adhered installation and is often found as a temporary covering over damaged floors in older houses.

Linoleum has good grease and burn resistance, average durability and ease of maintenance, fair resistance to stains, good resilience and some soundproof quality. However, it has poor resistance to alkalies and indentation and it should not be installed on floors that are on or below grade.

Asphalt Tile

Asphalt tiles traditionally have been classed into several groups. The least expensive group is Group B which is quite dark in color. Group C is of medium darkness and medium price and Group D is light-colored and more expensive. Group K is most expensive and includes many special colors and patterns.

Asphalt tile has good resistance to alkalies, fair resistance to burns and indentation, fair durability and ease of maintenance but poor grease and stain resistance (except for an uncommon grease-resistant type) and poor resilience and quietness. A major advantage is that it may be installed below grade, on grade or above grade and is often used over cement basement floors. It is also expensive.

Rubber Tile

Rubber tile is flexible and has a high-gloss surface. It is very resistant to burns and stains, has good resilience, sound absorption and durability. It has fair resistance to grease, alkalies and indentation. Like asphalt tile, rubber tile may be installed below grade, on grade or above grade. In many ways it is superior to asphalt tiles but is more expensive for comparable colors.

Cork Tile

Plain cork tile has an attractive appearance. It is very resilient and makes a very quiet floor covering. However, its resistance to grease, alkalies, stains, burns and indentations is poor. It is also difficult to maintain and has poor durability. A clear sheet of vinyl is sometimes fused to the surface. This substantially improves the ease of maintenance and grease and stain resistance. However, it decreases the resiliency and soundproof qualities and changes the appearance. Both plain and vinyl cork tile may be installed above grade and on grade but should not be installed below grade.

Vinyl Asbestos Tile

Vinyl asbestos tile has good resistance to grease, alkalies and burns. It is easy to maintain and quite durable. This tile has poor stain and indentation resistance and poor resilience and quietness, however. It may be installed anywhere, which is a major advantage, and it represents a good value for the price.

Solid Vinyl Tiles

Solid vinyl tile comes in a wide variety of colors. It is very durable, is quiet and has good resistance to grease, alkalies and stain. It also has fair resilience and resistance to burning and indentation. Solid vinyl tiles are relatively easy to maintain and may be installed anywhere. Usually vinyl is more expensive than vinyl asbestos tile of similar color and design.

Asbestos and Rag-Backed Vinyl Tiles and Sheets

In general these products have good durability and are easy to maintain. They have good resistance to grease, alkalies and stain and are fairly resilient and quiet. However, they are susceptible to damage by burning and indentation. Solid and asbestos-backed vinyl flooring may be installed anywhere, but rag felt-backed vinyl flooring should be installed only above grade.

Selection Criteria

Once the decision has been made to use a resilient floor covering rather than wood, carpet or concrete, a decision must also be made as to which type of tile to use. The first consideration is that of moisture. Ground moisture generally affects tiles laid on floors that are below ground level or on concrete slabs that are not completely protected by a vapor barrier. Wood subfloors over crawl spaces protected by a vapor barrier are generally free of ground moisture, as are floors over basements and second or higher floors.

Linoleum, cork and rag-backed vinyl are all attacked by ground moisture and should not be installed where it is present. Concrete releases moisture as it cures and care must be taken not to install linoleum, cork or rag-backed vinyl on any concrete floor no matter where it is located unless it is fully cured.

Surface moisture coming from spilled water or floor mopping, common in halls, laundry rooms, bathrooms and kitchens, tends to seep through the seams and affect the adhesives and subflooring. Sheet products tend to reduce this problem and special water-resistant adhesives should be used wherever surface moisture may be present.

Floors laid in kitchens are often subject to grease and alkalies which are found in many cleansers. For this reason asphalt and cork tiles generally should not be used in the kitchen. Linoleum has a high grease resistance but a low alkali resistance and must be kept waxed when installed in the kitchen.

Durability is the ability of the flooring to be serviceable and maintain an attractive appearance over a long period of time. Cork has poor durability and should be used only when appearance and quiet are the primary considerations of choice. Because linoleum has low durability it is used when price is a factor and in low-wear areas. Generally, the vinyl tiles and sheets are the most attractive and have the most desired features. Unfortunately, they are also the most expensive and the final selection is usually based on a weighing of the poorer characteristics of the other products against their lower cost.

Carpeting

Carpets were woven in Babylon 5,000 years ago. Egyptian pharaohs carpeted palace floors around 1500 B.C. Oriental carpet weavers returned with the crusaders to 14th century Europe and did a lively business.

Carpeting continues to grow in popularity as a floor covering for all rooms in the house. Since World War II American carpet production is up more than 400 percent. The FHA has reversed a longstanding policy and now accepts wall-to-wall carpeting over subflooring as an approved floor material and has established acceptable minimum standards. This wide acceptance of carpet can primarily be attributed to technological advances in carpet production. Most people choose carpet for its comfort and decorative value. As a bonus they get its excellent sound reduction characteristics.

Raw Materials

Backing yarns hold the pile tufts in place and provide stiffness to help the carpet lie flat. The most common materials used are jute (sisal fiber), kraftcord (wood pulp fiber), cotton, nylon and rayon.

Wool is the traditional pile fiber. Its good performance has been proven over many years and it is the standard by which the other fibers are judged. Most carpet wool is imported. Wools from various regions have different characteristics. Woolen yarn, which most carpets are made of, is composed of interlocked long and short fibers from 3½ to 5½ inches long.

Nylon, a synthetic fiber, accounts for about half of the residential carpet now being made. There are two basic types of nylon, known as Type 6 and Type 66, which differ chemically and in their color retention characteristics. Otherwise they are very similar. Staple nylon fiber is composed of 1½ to 6 inch fibers spun into yarn. Continuous filament nylon consists of bundles of continuous fibers which are formed into yarn without spinning.

Acrylics and modacrylics (modified acrylics) are synthetic fibers. Acrylic is a polymer composed of at least 85 percent acrylonitrile. Modacrylic is a polymer composed of between 35 to 85 percent acrylonitrile. Polypropylene is a synthetic fiber mostly produced in bulked continuous filament yarn.

Cotton, rayon and acetate have not proven very satisfactory for carpet construction but are still being used occasionally.

Performance Characteristics

A major factor in carpet performance is the fiber used. Pile thickness, weight and density are also important factors, plus the overall construction of the carpet. Durability is the ability of the carpet to resist destruction or loss of the fibers and is the overall measure of its serviceable life. The carpet must resist wearing away from foot traffic, which causes the fibers to break.

Alkalies are found in foods and cleaning materials. Acids are also common in food. Carpeting that will be exposed to either of these must have a high resistance.

Insects and fungi attack natural fibers but do not attack synthetic fibers. Wool carpeting, however, can be treated during manufacture, thereby greatly reducing its vulnerability.

Cigarette and other burns are a constant hazard many carpets are exposed to. Synthetic fabrics tend to melt and the heated area is very visible afterwards. Wool burns, but more slowly than synthetic fabrics.

Attractiveness of the carpet is decreased when the carpet compresses under a heavy load. This tends to be related more to the carpet's construction than to the type of fiber. The ability of the carpet to recover from being compacted by a load is called crush resistance. Pile density, spacing of the tufts and tightness of yarns affect crush resistance more than the type of fiber. The same is true of the ability of the carpet to retain the surface texture imparted during manufacture.

Much has been written about the ease of maintaining carpeting as compared with other floor coverings. About the best that can be said, however, is that no form of flooring is clearly superior under all conditions. Carpeting has proved to be easy to maintain in many household situations but difficult if not impossible in others. Stains and grime sometimes are impossible to remove.

Wool has excellent texture retention, durability, appearance and ease of maintenance. It has excellent crush resistance and when chemically treated it is resistant to insects and fungi. It has good resistance to abrasion, burns, compression, staining and soiling, but it should be avoided in kitchens and bathrooms since it has a low resistance to alkalies and acids and does not easily wet-clean. In the winter, wool carpeting builds up static electricity and gives electric shocks. It is also expensive compared to other carpet fibers.

Nylon now accounts for more than half the fiber used in residential carpeting. It is low in cost and comes in many bright, popular colors. It has excellent durability, appearance retention and texture retention as well as being easy to wet-clean and generally to maintain. Nylon also has excellent resistance to abrasion, alkalies, acids, staining, insects and fungi and makes a good kitchen, bathroom or family room carpet. It also has good compression and crush resistance but does soil very easily and, like the other synthetics, melts when burned and builds up static electricity.

Acrylic has most of the good features of nylon. It is somewhat less resistant to acids but has better general soil resistance. The appearance and high durability of acrylic compare favorably with wool and it is much less expensive. Modacrylic has about the same characteristics as acrylic.

Polypropylene is the newest of the synthetic fibers and is gaining popularity. Its principal use is for indoor-outdoor and kitchen carpeting. It has low moisture absorbency and high stain resistance. The colors are locked in chemically and are very fade resistant. Although it is easy to wet-clean, polypropylene is not very resilient and its resistance to crushing and compression is poor.

Mixtures and blends of the various fibers are common. In general, a carpet must consist of about a third of a fiber before it takes on the characteristics of that fiber. As the various synthetic fibers are improved there seems to be less blending being done.

Standards of Quality

To date, the carpet industry still has failed to adopt industry-wide standards of quality. Several of the major fiber manufacturers and the Wool Institute have established standards for use of their respective fibers and their label of approval is a good starting point upon which to make a selection.

The FHA has issued a set of minimum standards which may become the basis of an overall industry standard. Few carpets indicate whether or not they meet the FHA minimum standards. However, many dealers have the information and will issue a certificate for those products that comply. Federal law requires that all carpets be identified with a label that states the fiber content.

Carpeting that has been correctly selected for the area in which it is being used and is well cared for should last from 7 to 15 years.

Cushioning

All carpeting should be laid with cushioning that is either bonded to the backing or installed separately. Cushioning substantially increases resilience and durability. Separate cushioning is called padding. With loomed carpeting, foam or sponge rubber is often bonded directly to the back of the carpet at the mill.

Felted hair makes the most economical conventional padding. It is made with a waffle design that provides a skidproof surface and improves resiliency. Sometimes it is reinforced with a jute backing or a burlap center. Another type of padding is made of felted hair or jute and coated on one or both sides with rubber to hold the fibers together and to provide additional resilience.

Rubber padding, in addition to being bonded directly to many carpets, is also made in separate sheets. A layer of burlap is usually bonded to one side to facilitate installation by permitting a smooth, even stretch of the carpet, which is difficult to accomplish on top of plain rubber. Rubber padding in general is superior to felted hair, but it is more expensive and should not be used on radiant-heated floors.

Miscellaneous Roofing and Siding Products

Asphalt Roofing and Siding

Asphalt roofing and siding products are classified into saturated felt products, roll roofing and siding and shingles.

Saturated felt is impregnated with asphalt but is otherwise untreated. It is used for shingle underlayment, sheathing paper and lamination strips for built-up roofing. It comes in 15-pound and 30-pound weight per square (the amount of material required to cover 100 square feet of roof).

Roll roofing and siding come both smooth and coated with mineral aggregates. Some are in split rolls which, when applied, give a patterned edge.

Shingles are classified as either strip shingles or single shingles. **251**

Standards of Quality

A leaky roof means rotting attic timber, stained ceilings and walls and damage to the rest of the house and its contents. A roof lasts for substantially less time than the whole house and the quality of the shingles will determine how long the roof will last before it will have to be repaired or replaced. Roofing is damaged fastest by wind and sun; therefore, in the hot climates and in high wind areas of the country quality becomes even more important.

Fortunately, the Underwriters' Laboratory (U.L.) tests and rates most of the roofing being produced today for fire resistance, wind resistance, quantity of saturant and efficiency of saturation, thickness and distribution of coating asphalts, adhesion and distribution of granules, weight count, size, coloration and other characteristics of finished products before and after packing.

In high wind areas, a minimum of U.L. Class C wind resistant roofing should be used. More fire protection and usually longer trouble-free life is obtained from wind resistant Class B and Class A shingles. In low wind areas, wind resistant shingles are not necessary, but a minimum of Class C should be used to provide fire protection and satisfactory life. Classes B and A are still better quality.

Shingles are sold by weight per square. The higher the weight the longer the shingle will last and the more it will cost. However, as labor is a big factor in roofing cost, it pays to buy heavyweight shingles since their total cost in the long run will be less. A heavy shingle weighs more than 30 pounds per square.

Slate Shingles

Slate is a natural mineral that is taken from a quarry. A standard roofing slate is $\frac{3}{16}$ inch thick and may be 10 to 24 inches in length and 8 to 14 inches wide. It makes a durable roofing material that may last throughout the life of the house if cracked and missing slates are replaced on a regular basis. Because they are heavy compared to other roofing materials the roof framing must be especially strong to support a slate roof. The other principal disadvantage is that they are expensive compared to other roofing materials. They are often used on houses of the English and French styles.

Tile Shingles

Roof tiles are manufactured the same way that wall tiles are. They come in a variety of styles and are used on houses done in Spanish, Italian and Greek styles. Tile makes an excellent roofing material, especially in climates where the temperature does not fall below freezing.

Gypsum Products

Gypsum (hydrous calcium sulfate) is a common, rocklike mineral found throughout the world, usually combined with clay, limestone and iron oxides. It is unique in that when it is intensely heated (calcined) it gives up some of its water content and becomes powdered. When cooled, gypsum can be restored to its original rocklike state by adding water. This mixture remains plastic for awhile and then sets. During this plastic state it can be easily shaped and worked. Gypsum has been used for 4,000 years for construction. Plaster of Paris, an early form, received its name from the mines under Paris where gypsum was found.

Gypsum also has unique fire resistant qualities. When exposed to heat, the surface releases its water in the form of steam. This absorbs some of the heat and delays the raising of the gypsum's temperature to above 212 degrees. When the water is released, the surface becomes calcined and protects the underlying gypsum from the heat. This dual action makes gypsum an excellent fire resistant material. Other materials used to manufacture gypsum products are aggregates, mineral and organic fibers and lime.

Classification of Gypsum Board Products

Regular wallboard is surfaced with gray liner on one side and calendared manila paper on the other side. Many are produced with tapered edges. Predecorated wallboard has a decorative finish on one side. Insulating wallboard is regular wallboard with a bright finish aluminum foil bonded to the back. It often eliminates the need for a separate vapor barrier. Fire resistant (Type X) wallboard has a core especially formulated with additives and glass fibers for greater fire resistance. Water resistant wallboard is made by using a multi-layered covering of chemically treated paper and a gypsum core formulated with asphaltic adhesives.

Backing board serves as a base to attach other types of wallboards or tiles. It differs from other types of wallboard in that both surfaces are covered with a gray liner paper not suitable for decorating. It comes in its regular form, fire resistant (Type X) or with insulating foil. Backing board is made with a highly water resistant vinyl surface for use as a base for tile in areas such as stall showers which are subject to direct wetting.

Gypsum sheathing is used as a base for other external sidings. It is usually made of a water resistant gypsum core completely enclosed with a firmly bonded water repellent paper which eliminates the need for sheathing paper.

Formboards are used as permanent forms for gypsum concrete roof decking. They are surfaced on an exposed face and longitudinal edges with calendared manila paper treated to resist fungus growth. A vinyl-faced board, which provides a white, highly reflective and durable surface, is also available.

Gypsum lath boards are used as a base for plastering. They consist of a gypsum core enclosed by a multi-layered fibrous paper covering designed to assure good bond with gypsum plaster. The main types are plain, perforated, insulating and fire resistant (Type X).

Classification of Plaster Products

Neat gypsum base coat plaster does not contain any aggregates. It comes in varieties with fiber already added. It is usually combined on the job with the aggregates and is used for base coats over metal or gypsum lath, gypsum and clay tile and concrete block and brick. It is also used as a scratch coat over metal lath and, with sand added to it, as a brown coat.

Wood-fibered base coat plaster consists of calcined gypsum integrally mixed with selected coarse cellulose fibers which provide bulk and good coverage. It is used over the same bases as neat plaster, but it provides greater fire resistance than neat and other sanded basecoat plasters.

Ready-mixed base coat plaster is neat plaster to which aggregates and all other additives except water are added at the mill.

Bond base coat plaster, in addition to plaster, contains lime and other chemical additives which improve bond with concrete and other nonporous surfaces.

Gauging finish coat plaster consists of coarsely ground gypsum of low consistency. It is generally used without aggregates. However, the addition of fine silica sand increases its crack resistance when it is used over base coat plasters containing lightweight aggregates.

Keene's cement is similar to gypsum gauging plaster except that it is burned dry in a kiln instead of a calciner or kettle. The result is a denser material with greater impact, abrasion and moisture resistance.

Other types of plaster used in house construction are gypsum-lime putty-trowel plaster and Keene's cement-lime putty-trowel plaster, both of which are made by the addition of lime. Keene's cement-lime-sand-float plaster is used where a minimum of cracking is desirable. Prepared gypsum finish plaster is available in a variety of mixtures, all of which require only water to be added on the job. Veneer (thin coat) plaster is a high strength plaster for very thin applications. Molding plaster is used to make intricate surfaces with close tolerances. Acoustical plasters contain chemicals that form small air bubbles in the plaster to increase its sound absorption properties.

Mineral Fiber Products

Fiber products are made by combining asbestos fibers, portland cement and water. The mixture is formed under pressure into thin, hard, rigid sheets or pipes. Roof shingles of this material were produced at the turn of the century. The material is completely inert, durable, fire resistant and maintenance free. Its main drawback is low resistance to impact.

Classification of Products

Wall siding and shingles and roofing shingles are made in a wide variety of styles, shapes, textures and colors.

Flat sheets are made in sizes from 4 by 8 feet to 4 by 12 feet and ⅛ to 2 inches thick. They are used for both interior and exterior applications where a high amount of fire resistance is needed, such as near fireplaces, ranges, ovens and furnaces. They are made with smooth or textured surfaces, with finishes of natural gray, preprimed for painting or with pigmented plastic coatings.

Insulated panels are made in the factory by assembling layers of insulating core material between flat sheets. These sandwich panels are used for exterior walls, interior partitions and roof decking.

All of the products should conform to the standards set by the American Society for Testing and Materials and bear a label to that effect.

Cabinets

The history of bathroom and kitchen cabinets goes back several hundred years in this country. They were first called ambries and cupboards and were used to store food. The walls in the kitchen between the chimney and the side wall were filled with open and closed shelves.

Prior to World War II most cabinets were made of wood. Many were built in the mills, while some were constructed on the site by skilled cabinet makers. After the war, huge mass production cabinet plants took over the bulk of cabinet production. Most of what can be said about kitchen cabinets is also true of bathroom cabinets, so they are covered together here.

Cabinet Types, Styles, Finishes

Kitchen cabinets are classified by their function as base cabinets, wall cabinets or miscellaneous cabinets which includes broom cabinets, oven cabinets, lazy Susans, fruit and vegetable bins and tray-holders. Bathroom cabinets, also known as vanities, are similar to kitchen base cabinets, differing mainly in height and depth.

Both kitchen and bathroom cabinets are manufactured in a variety of styles, the most popular of which are colonial, contemporary, traditional, Mediterranean and Spanish. Woods used in cabinet manufacture are ash, oak, birch, maple, walnut, ponderosa pine, Douglas fir and western hemlock. A variety of finishes such as walnut, ebony, white, nutmeg, fruitwood, cherry, pecan, birch and oak are used.

Drawer construction varies according to cabinet quality and is a good way to judge the overall quality of the cabinet. In the good grade cabinets the drawer sides are connected to the front and back with multiple dovetail joints. Cheaper cabinets have drawers with lock shouldered and square shouldered joints.

Factory made metal cabinets are often used today. Drawers, adjustable shelves, backs, sides and fronts are of enameled pressed steel sheets. Doors are usually hollow steel. One sign of the better grades are drawers that move on roller slides rather than friction slides. Since it is impossible to trim metal cabinets and the variety of sizes available

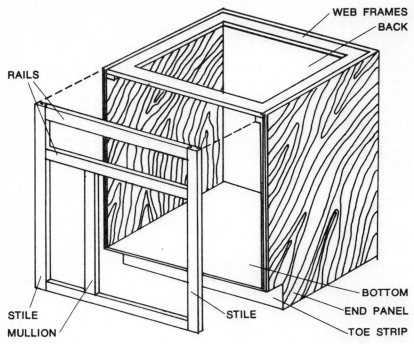

Typical base cabinet construction, showing component parts.

CABINETS

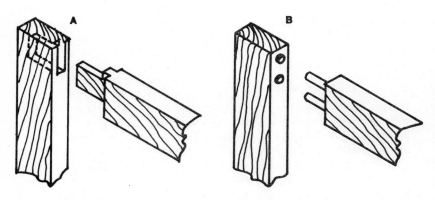

Rails and stiles of frames can be joined by: (A) mortise and tenon, or (B) dowel joint.

is smaller than the variety of wooden cabinets, it is necessary to carefully plan the space to be filled with a cabinet arrangement.

Installation

Cabinets must be installed level, plumb and true to ensure proper operation. They should be screwed, not nailed, to the wall studs. The screws should go through the framing members.

Cabinets are usually installed by first removing any floor or wall irregularities that will interfere with their plumb and level installation. A base corner unit is installed first and then each additional unit is aligned to it and held in place with a clamp until all the screws are installed and tightened. Wall cabinets are installed similarly, starting with the end or corner unit.

Moulding

The ancient Greek designers first used moulding to divide surfaces into small parts, to create interest and variety and to produce highlights and shadows. Later the Romans simplified moulding designs and substituted various compass and mechanical curves for the freehand design of the Greeks. Moulding details of both the Greeks and the Romans have passed down through the centuries and are today the basis of most moulding designs.

Moulding is made from a variety of hardwoods and softwoods for interior and exterior use. It is made at a mill by special machines that cut, plane and sand the lumber surfaces into the desired shapes. A general rule is that the thicker the piece of moulding and the more intricate the design, the more expensive it is.

Most moulding used today is "stock" moulding of one of the standard sizes and shapes. In the past, architects would design a special moulding for a custom house and the mill would make it to his specifications. Although the practice is rare today, there are still a few mills capable of making custom moulding.

Interior Applications

The use of interior moulding in a modern house may be limited to simple casing around the doors and windows, baseboards and ceiling although all of these may not be present. In more elaborate houses and in houses that are in the architectural style of a certain period, extensive or elaborate moulding may be used. The elaboration may be increased by using two or three pieces of moulding together.

Cornices

Interior cornices are strips of wood moulding nailed to the wall-ceiling joint in such a way that they may project away from the wall with a space between the back of the moulding and the wall-ceiling junction. This space may be just left unfilled or blocking may be installed in it for better nailing. It may be made large by attaching a frieze board to the wall and ceiling and then attaching the moulding. Still more ornamenta-

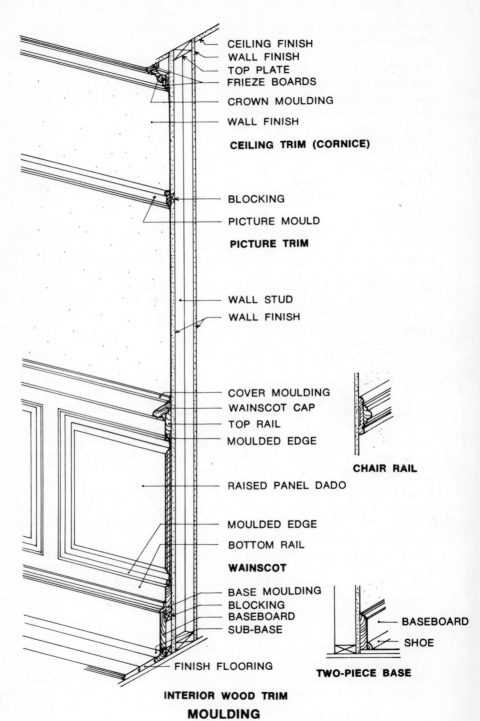

CEILING FINISH
WALL FINISH
TOP PLATE
FRIEZE BOARDS
CROWN MOULDING
WALL FINISH

CEILING TRIM (CORNICE)

BLOCKING
PICTURE MOULD

PICTURE TRIM

WALL STUD
WALL FINISH

COVER MOULDING
WAINSCOT CAP
TOP RAIL
MOULDED EDGE

CHAIR RAIL

RAISED PANEL DADO

MOULDED EDGE
BOTTOM RAIL

WAINSCOT

BASE MOULDING
BLOCKING
BASEBOARD
SUB-BASE

BASEBOARD
SHOE

FINISH FLOORING

TWO-PIECE BASE

INTERIOR WOOD TRIM

MOULDING

tion may be obtained by adding egg and dart moulding, dentils, meander or other details.

Mantels

The mantel is the ornamental facing around a fireplace. (It is also known as a mantelpiece, although the mantelpiece is more specifically the protruding shelf portion of the mantel.) Mantels vary from a simple shelf at the top of the masonry to elaborate facing of the fireplace front with wood, marble, limestone or other ornamental material. However, it is important that the opening directly around the fireplace itself be edged with masonry material and not with wood in order to prevent charring.

Wainscots

When the upper portion of a wall is finished with a different material than the lower portion, the lower portion is called the wainscot and whatever material it is finished with is called wainscotting. The main part, the lower section, is the dado; the upper edge is the surface or cap. Simple wainscotting is solid wood, plywood, hard plaster, cloth, linoleum or other sheet material. Elaborate wainscotting is made of solid lumber paneling usually made in a mill and then installed in large sections. A paneled look is accomplished by nailing wide mouldings to the surface of the plywood dado.

Chair Rails

Older houses have horizontal strips of wood nailed to the wall at the height of a chair to prevent marring of the walls by chairs. These may double as the upper rails of the wainscot.

Picture Moulding

A narrow wooden moulding strip is so shaped that flat metal hooks may be hung from it. They should be nailed into the stud and not just into the plaster.

Bookshelves

Bookshelves may be constructed similiarly to cabinets. Commonly they are adjustable rather than rigidly affixed to the walls. A simple adjustable shelf may be made with a series of holes drilled or punched at regular intervals to the case ends or to the back of the case. Pegs or brackets inserted into these holes support these shelves.

Exterior Applications

Windows may be effectively accented with outside casing. When added to metal windows, mouldings soften the harsh contrast of metal against siding. Eaves, soffits and rakes look more elaborate with moulding and they are used here in both colonial and contemporary homes.

Quality Standards and Grades

The Western Wood Moulding and Millwork Producers, in cooperation with the Western Wood Products Association and the Southern Pine Association, publish *Moulding Patterns,* which has become the standard reference of moulding patterns. Most of the stock moulding made today is from one of the designs in this book. This book also contains the industry grading and other standards.

Doors

Door Types

There are seven basic types of doors: batten doors, sliding glass doors, folding doors, flush solid doors, flush solid core doors, flush hollow core doors and stile and rail doors. Each of these classifications has several subdivisions.

Batten doors consist of boards that are nailed together in various ways. One common way is to have two layers of boards nailed at 45 degree angles to each other. Another way is to nail the vertical boards to ones running horizontally (called ledge boards) or brace boards that run at a 45-degree angle. Batten doors are used where appearance is not important, such as for cellar and shed doors.

Flush solid doors are perfectly flat on both sides and made of planking. This type of door is rare in houses.

Flush solid core doors are made with smooth face panels glued to a core that is made of either a composition material, glued-together wood blocks or glued-together wood pieces. Added strength is obtained by using two layers of face board. Usually strips of the core material are glued around the edge. Solid doors are often used as exterior doors.

Flush hollow core doors are also perfectly flat. They have a core that consists mainly of a grid of crossed wooden slats or some other type of grid construction. The face wood must be three or more plies to give the door strength. The edges of the core are solid wood and wide enough so that hardware may be attached by screwing into the solid edge. These doors are light and are used for interior doors.

Sliding glass doors contain at least one fixed pane plus one or more panes that slide in a frame of wood or metal. The most common type has two panes which provide a maximum of 50 percent ventilation. The three-pane type has fixed end panes and a sliding center pane and provides only 33 percent ventilation. The four-pane type also has fixed end panes with two sliding center panes. It provides 50 percent ventilation.

Accordion folding doors are usually made of wood slats connected with tape or cord forming flexible, drape-like doors. They can be made as long as needed. They are usually so designed that when folded they are about the width of the jamb. Usually they are hung from ceiling tracks and operate on nylon rollers or glides. They are used to provide

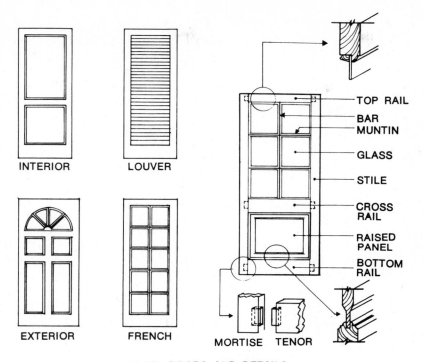

INTERIOR

LOUVER

EXTERIOR

FRENCH

TOP RAIL

BAR
MUNTIN

GLASS

STILE

CROSS
RAIL

RAISED
PANEL

BOTTOM
RAIL

MORTISE   TENOR

PANEL DOORS AND DETAILS

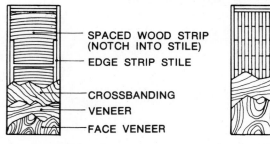

SPACED WOOD STRIP
(NOTCH INTO STILE)

EDGE STRIP STILE

CROSSBANDING

VENEER

FACE VENEER

HOLLOW CORE DOOR

EDGE STRIP
STILE

SOLID CORE
(STAGGERED
BLOCK)

CROSSBANDING

FACE VENEER

SOLID CORE DOOR

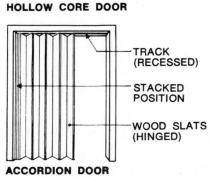

TRACK
(RECESSED)

STACKED
POSITION

WOOD SLATS
(HINGED)

ACCORDION DOOR

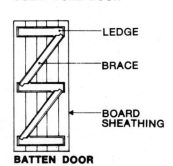

LEDGE

BRACE

BOARD
SHEATHING

BATTEN DOOR

DOORS

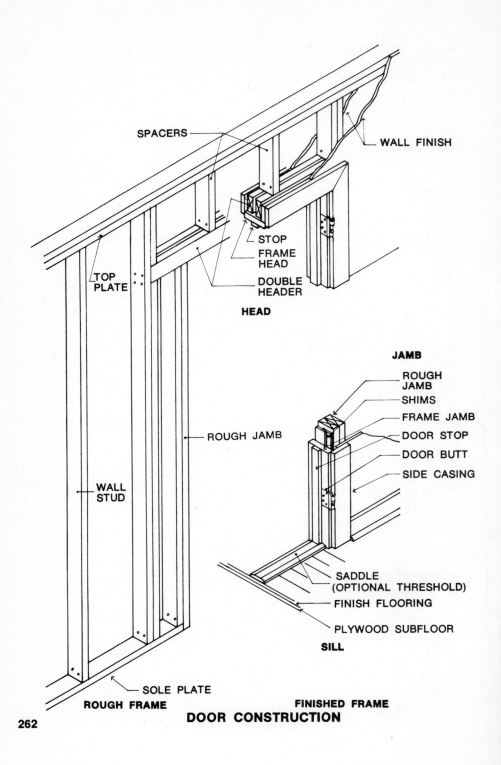

SPACERS

WALL FINISH

STOP

FRAME HEAD

DOUBLE HEADER

TOP PLATE

HEAD

ROUGH JAMB

WALL STUD

JAMB

ROUGH JAMB

SHIMS

FRAME JAMB

DOOR STOP

DOOR BUTT

SIDE CASING

SADDLE (OPTIONAL THRESHOLD)

FINISH FLOORING

PLYWOOD SUBFLOOR

SILL

SOLE PLATE

ROUGH FRAME

FINISHED FRAME

DOOR CONSTRUCTION

262

a visual screen rather than a sound barrier in such places as closets or laundry enclosures.

The solid wood type of accordion door consists of thin wood slats connected to each other with continuous metal, vinyl or nylon fabric hinges. The woven type is made of vertical wood strips ⅜ to 1 inch wide, interwoven basketweave style with tape. The corded type is made with wood slats similar to the woven type, but the slats are connected with cotton cord.

Stile and rail doors consist of a framework of vertical boards (stiles) and horizontal boards (rails). These stiles and rails are usually made of softwood and give the door its strength. They are connected by use of dowels or mortise and tendon joints and are glued together.

What fills the space between the stiles and rails determines the name of the stile and rail door style. The spaces in many of these doors are filled with from one to eight solid wood or plywood panels and are known as panel doors. They may be horizontal, vertical, square or some combination of these. When some of the spaces are filled with glass, the door is a sash door. When all of the spaces are glass-filled, it is a French door. Rim type French doors have one piece of glass and divided light French doors have multiple, smaller lights.

The spaces may also be filled with wood slats attached to the stiles in such a way that air may circulate between the slats. Standard slats are flat and chevron slats are curved. The slatted doors are known as louvered doors. They are used for laundry areas, furnace rooms, closets and other areas where air circulation is needed and privacy and sound insulation is not required.

Storm doors, screen doors and combination storm and screen doors are often hung outside the regular door to provide additional insulation in the winter and ventilation in the summer. The spaces are filled with glass or screens that are removable so they can be interchanged from season to season. The frame is usually made of lightweight wood or aluminum. On the self-storing type, the glass and screens slide up and down so that either may be used without removing the other. The disadvantages of this are high expense and a maximum of only 50 percent ventilation or light available.

Door Operation

Doors can be installed so that they either swing, slide or fold open and shut. The location and use must be considered in selecting which method should be employed.

A swinging door is attached either to the side jamb with hinges or to the ceiling and floor with pivots. Most doors used in houses are the hinge type, with pivots being used mainly for the door between the kitchen and the dining room when it swings in both directions. Swinging doors are called right-handed when the knob of the door is on the right side as it is viewed from the side it opens into. Hinged swinging doors provide maximum security, since the hinge cannot be removed from the outside when the door is closed. (The pins may be removed from the inside in the closed position.) The door may fit tightly when closed and can be weathertight and soundproof. Weatherstripping on the door will increase these qualities. The main disadvantage of a swinging door is that it requires a space to swing into and in some places this clear space is not available.

Sliding doors slide back and forth parallel to the wall. They may be attached to an overhead slide and operate on nylon rollers or be mounted on rollers that roll in a floor track. Sliding doors are popular for use in closets where they don't take up any valuable closet or room space. They may be designed so that two doors are used and each slides past the other. The advantage is that the wall does not have to be cut into to make space for the opened door. The disadvantage is that only 50 percent of the opening is usable at any one time.

The major disadvantages of doors that slide along the wall surface are appearance and the need to keep the wall space clear of interfering objects. The door may also slide into a pocket in the wall. The advantage of this is that the door becomes invisible and out of the way. This is a costly installation to make, however, and usually some wall studs must be eliminated to make room in the wall for the door.

Folding doors are hung on overhead tracks and move on nylon rollers or slides. They may also have a floor track to guide the folding movement.

Installation

For a door to function properly and appear unblemished it must be handled and stored correctly before installation. It should be handled only with clean gloves and stored flat on a level surface in a clean, dry, well-ventilated space.

To ensure proper hanging, the door frame should be square and plumb and doors should fit with a total clearance of about $\frac{3}{16}$ inch in both directions. Too little clearance will cause the door to stick in humid weather and excessive clearance permits the weather and sound to pass through. Inside doors usually require two hinges and outside doors three hinges.

Windows

History

When primitive society moved out of caves and trees, it was into the single unit, one-story hut. The huts were square, rectangular or round (particularly in parts of Africa), but each one had a single opening and no internal divisions. There was no such thing as a window.

In fact, the simple concept of the window evolved slowly and was preceded by the advent of the two-story house and the floor. At first one hut would be joined to another if the family needed more room, but there was still no room division. The next development was placing one hut on top of the other. It is at this point that, for all practical purposes, windows and doors became part of architecture. Some archeologists feel that the Mayan society pioneered this growth, while others feel that a more northern aboriginal society was responsible since windows made more progress in cold climates than in warm. The latter would appear to be the more logical premise because there is less light in northern climates and more need to have the fire indoors. thus making it necessary to let smoke out through windows.

In any event, the window was here to stay and its three basic functions of lighting, insulation and ventilation have not changed since the first one was whittled out of a wall, be it clay, wood or ice. When the window came into its own it seemed to "shed light on itself" and gradually became a thing of beauty as well as a functional item.

The sash window came into popularity about the last quarter of the 17th century when it replaced the casement window. It is still popular today.

Window Types

Double-hung and casement windows are the two most common styles found in houses. In new construction the horizontal sliding window and the clerestory (also clearstory) window, a window placed high in the wall or in the ceiling, is also gaining popularity. There are also a variety of other window styles, each with its own advantages and disadvantages.

The double-hung window consists of two sashes that move up and down in a pair of channels. The old colonial type was held up by metal or wood fittings that projected through the rail or stile into holes in the sash. Many houses built prior to World War II (and some built after as well) have sashes that are connected by a cord or chain to a center balance weight that is hung inside the wall. Modern double-hung windows are held open by tension springs.

The advantages of a double-hung window is that it offers two perfectly flat surfaces at all times, permitting storm windows and screens on the outside and any type of blind, curtain or drape on the inside. Its disadvantage is that only half of the window can be opened at any one time. Another disadvantage is the difficulty of cleaning double-hung windows from the inside. Newest models have spring release devices that permit the window to be removed from the inside for cleaning.

The casement window consists of one or more sashes hinged at the side like a door, swinging horizontally out or in, usually with the aid of a crank. Its advantages are that it is very easy to wash and provides 100 percent ventilation. The disadvantages are that it usually does not fit tightly, drafts result and ugly screens and storm windows must be put on the inside. Newest models have improved on tightness and made appearance of screens and storm windows less objectionable.

The horizontal sliding (traverse) window is like a double-hung window laid on its side. It consists of two sashes that slide back and forth. The outside is somewhat easier to clean than a double-hung window and it has the further advantage of not interfering with storm

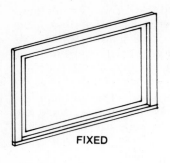

FIXED

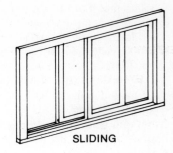

SLIDING

DOUBLE OR SINGLE-HUNG

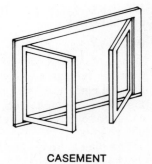

CASEMENT

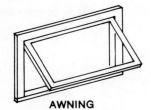

AWNING

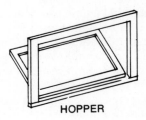

HOPPER

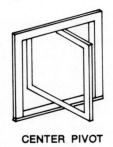

CENTER PIVOT

JALOUSIE

WINDOW OPERATION

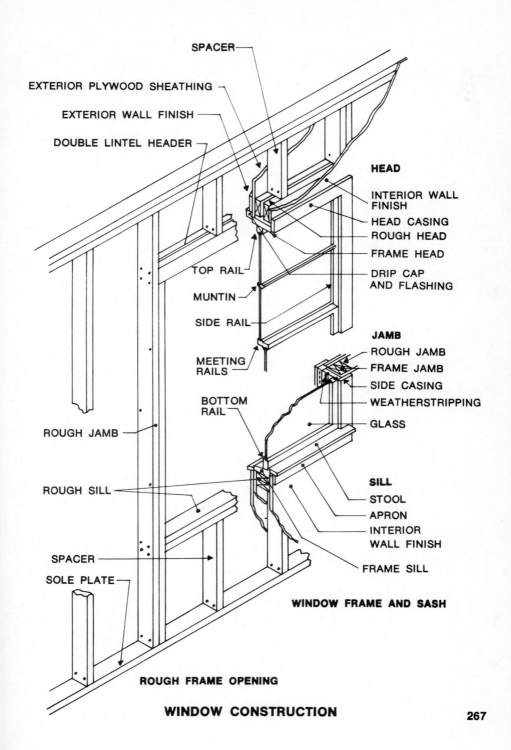

SPACER

EXTERIOR PLYWOOD SHEATHING

EXTERIOR WALL FINISH

DOUBLE LINTEL HEADER

HEAD

INTERIOR WALL FINISH

HEAD CASING

ROUGH HEAD

FRAME HEAD

DRIP CAP AND FLASHING

TOP RAIL

MUNTIN

SIDE RAIL

MEETING RAILS

JAMB

ROUGH JAMB

FRAME JAMB

SIDE CASING

WEATHERSTRIPPING

GLASS

BOTTOM RAIL

ROUGH JAMB

ROUGH SILL

SILL

STOOL

APRON

INTERIOR WALL FINISH

FRAME SILL

SPACER

SOLE PLATE

WINDOW FRAME AND SASH

ROUGH FRAME OPENING

WINDOW CONSTRUCTION

windows and screens on the outside and blinds, curtains or drapes hung on the inside. Additional advantages are extra privacy and safety for children when the window is set high in the wall. So set, it also allows for additional furniture to be placed along the wall. The principal disadvantages are its appearance, which does not blend into many traditional designs, its tendency to stick if there is any settling of the wall or damage to the window and the fact that it permits only 50 percent ventilation.

The jalousie window is popular in warm climates, where it has the advantage of permitting up to 100 percent or controlled ventilation of a large wall area. It is designed something like a venetian blind with slat-like glass panes that open and close with a cranking device. The disadvantage is that the many small panes are difficult to clean and tend to leak.

Awning windows open up and out, usually by means of a cranking device. They have the advantages of keeping out rain when in an open position and providing 100 percent ventilation. Their disadvantage is that storm windows and screens must be placed on the inside.

A hopper window is like an awning window hung upside down and backwards. It opens into the room and is hinged from the bottom. Its advantage is that it throws the incoming wind upwards. Its disadvantage is that it interferes with curtains, drapes and blinds and often presents a safety hazard as it sticks out into the room.

The basement transom window is like a hopper window except it is hinged at the top. Its advantage is that it throws the air downward from its customary high location on the basement wall. A disadvantage is that it provides little rain protection and it also tends to leak.

Weatherstripping

The purpose of weatherstripping on windows is to provide a seal to prevent air and dust leakage. A common kind of weatherstripping used today is the spring tension type of bronze, aluminum, rigid vinyl, stainless or galvanized steel or rigid plastic steel. Other types are woven felt, compression sash guides and compression bulbs.

Storm Windows and Screens

Most windows transmit heat and cold far in excess of a typical house wall. Heat loss and transmission may be reduced by the use of storm windows which are often made in conjunction with screens. In cold weather they save sufficient fuel to justify their expense and inconvenience, which vary with the house construction, climate and habits of the people in the house. Although the advantage of storm windows is that they reduce drafts, they are difficult to clean and are considered by many to be unattractive.

Hardware

The purpose of window hardware is to provide effective closure, to operate the sash and to hold it stationary in an open position. Window hardware may be a simple lock or a complex opening and closing device. Many double-hung windows are still equipped with balancing weights and hardware.

Installation

In order to perform satisfactorily, a window must be properly installed. Since windows are often damaged in shipment, prior to installation they should be checked for squareness and equal spacing to the jambs. Braces are often used to hold the window in shape until after installation. The frame is fitted to the opening and shims are used to support the frame while it is being nailed into place. After the frame is nailed to the rough frame the space between it and the rough frame is filled with insulation, flashing is installed and the entire unit caulked. Hardware and weatherstripping are installed and adjusted for proper operation. Interior trim is then nailed in place to seal the wall cavity between the rough frame and the finish frame. All wood windows should be prime coated as soon as they arrive on the site if they have not been factory-primed.

Defective Windows

Dust streaks or water stains around the window trim can be evidence of leakage.

While checking windows, some with missing locks and window lifts or counter balance weights may be discovered. It may also be difficult to reach over the kitchen sink to open that window if it is the double-hung type. A window in the bathroom over the tub or toilet lets in uncomfortable drafts.

A special look should be taken at the windows in children's rooms. Are they high enough to be safe yet low enough to allow escape in the event of a fire?

Chapter 7
Mechanical
Systems

Without mechanical systems to breathe life into it, a house is only a brick, mortar and wood shell. It can be constructed with the most modern building techniques and using the most modern building equipment, but without properly working systems, the house cannot really perform its functions of providing shelter and comfort. All of these systems—from heating to plumbing to electricity—must be installed properly and constantly kept in efficient working order.

Heating and Cooling

History

Some time, thousands of years ago, long before the wheel, a caveman brought fire inside his cave and thus invented the first home heating system. Most likely the cave filled with smoke, causing his eyes to burn and his lungs to choke. If he was lucky, the cave had a hole in the top and the smoke escaped through this primitive chimney.

In the majority of Greek and Roman houses the hearth was the focus of cooking and heating. The Romans had a system, called a hypocaust, used only by the very rich. Heat from a furnace was collected under the floors and distributed to rooms throughout the house.

In the East the kangi was used for centuries (and still is today in India and Turkey) to provide heat. It is simply an earthenware container filled with hot embers. Often it is placed under the dining table and covered with a heavy cloth.

The principal method of home heating has been, for more than a thousand years, a fireplace in the kitchen and more recently, in other rooms of the house as well (see "Fireplaces" in Chapter 5). Hot water throughout this same period was made in a pot, kettle or other container heated in the fireplace or on the cooking fire. Benjamin Franklin's stove was a major advance. Its warm cast iron sides radiated heat throughout the room.

The first modern central heating systems were built in the 1800's. They consisted of a boiler producing steam which was circulated by pipes into large, ornate cast iron radiators. The boilers were fired by furnaces hand-fed with coal or wood as fuel; ashes and clinkers were also hand-removed at regular intervals.

The central heating and hot water systems of today are generally under 100 years old. In many areas of the world, including England, modern systems were rare until after World War II. It is possible with today's systems to maintain a level temperature throughout a house all year round, regardless of the exterior temperatures.

Central heating and hot water systems can be classified by dividing them into the kinds of fuels or the type of transfer medium used to carry the heat from the ignited fuel to the rooms in the house.

For ease of understanding, this section starts with a short explanation about the principles of heat. Then, heating systems are explained, divided by transfer medium. Heat controls and hot water systems are discussed followed by fuels and methods of calculating their relative costs.

Principles of Heating

In order to understand the various heating systems and how they work, it is first necessary to know a little about heat itself and what makes people feel warm or cold, since the principal purpose of a heating system is to make the people living in a house feel comfortable.

The body itself has its own heating system providing enough heat for human survival in a wide range of temperatures, which explains how the human race survived so long without much heat. Maximum comfort, however, is obtainable in a very narrow range of temperatures and only when the heat loss given off by the body by radiation, convection and evaporation is in balance. When it is too slow a person feels hot and when it is too fast he feels cool.

Radiation is the transfer of heat by direct rays from the body to cooler surrounding objects, such as cold windows, walls and floors. A person putting a hand next to a cold wall will feel a chill as the rays leave the hand.

Convection is the transfer of heat by the circulation of air around the body. The movement of the air by wind or a fan or a draft speeds up the heat loss and chills the body. Evaporation is a minor source of heat loss.

The ideal heating system should supply that amount of heat which will keep the body heat loss in balance. It must also prevent drafts, warm the floors and walls and provide steady temperatures from room to room and floor to ceiling.

Hot Air Gravity Systems

The gravity air system enjoyed wide popularity in spite of its deficiencies and is still being installed in some small houses. Cold air is heated in a furnace that looks like a giant octopus. Because hot air rises, it goes up unaided in a series of ducts and into the rooms being heated through wall, floor or ceiling registers. When it cools, the air descends through other ducts or hallways by the force of gravity. This system is characterized by the distribution of hot, dirty, dry air unevenly through the house. With some rare exceptions, in houses under 1,000 square feet, this furnace should be replaced with a modern forced air system.

Another type of gravity air system is a floor furnace. These are also used in small houses, especially in moderate climates. The furnace is suspended below the floor and the hot air rises from the furnace through a flush grille in the floor. There usually are no ducts. The system is inexpensive to install. However, the larger the area to be heated, the poorer the heat distribution.

Space heaters are rarely satisfactory except in mild climates, cottages or for supplementary heat in areas beyond the main heating system.

Forced Warm Air Systems

The forced air system corrected the deficiencies of the gravity system. With this system, the furnace has a fan or a blower that pushes the warmed air through smaller ducts. These ducts may be run horizontally as well as vertically, allowing considerable flexibility in their placement. Filters can be installed in the system to clean the air and a humidifying system included to add the needed moisture. The air does not have to be as hot as in a gravity system.

A properly maintained and adjusted system does not create dirt or dust. It does, however, stir up already existing dust and redistribute it. This may be turned to an advantage by the installation of throwaway filters which will remove the larger dust particles or an electronic air cleaner that electrically charges the dust particles and removes as much as 90 percent of them by attracting the particles to metal collecting plates. An additional advantage of this system is that often the ducts can be used for air-conditioning.

A common way to lay out the ducts is called the trunk line or extended plenum system. A large trunk line (plenum) extends from the furnace across the length of the house under the first floor joists. From the plenum, ducts run to the various rooms of the house. A return air intake gathers the cooled air through return ducts. The air passes through a filter into the bottom of the furnace, where it is reheated and the cycle begins again. The fan creates a pressure in the plenum which forces the air through the ducts and out the registers as well as creating a negative pressure which draws the air back into the return ducts. In better systems, the warm air ducts are under the windows along the cold outside walls of the house. Money can be saved at the cost of comfort by short duct runs to the inside walls.

An extended plenum system is good because it responds quickly to temperature changes and is economical to install in houses with basements or crawl spaces. It can be adapted by insulating the plenum and altering the registers to force the air upward to use for air-conditioning. It also adapts itself to the installation of air purification and humidification equipment.

Perimeter systems can be used in houses built on a concrete slab or in houses with basements and crawl spaces. They require a downflow furnace that blows the warm air downward and accepts the returning cool air in through the top.

In a perimeter radial system, ducts extend directly from the furnace to the various heated rooms by going under the floor joists or through the walls. The air is discharged through registers, which should be placed along the cold outside walls for best results. Cool air returns to the furnace through return ducts or under doors. A room will not heat properly without a way for the cool air to return to the furnace.

A more sophisticated system is a perimeter loop system. The warm air is fed from the base of the furnace through feeder ducts into a duct that runs around the exterior wall of the house with registers under the windows in each room. This results in a warming of the concrete slab

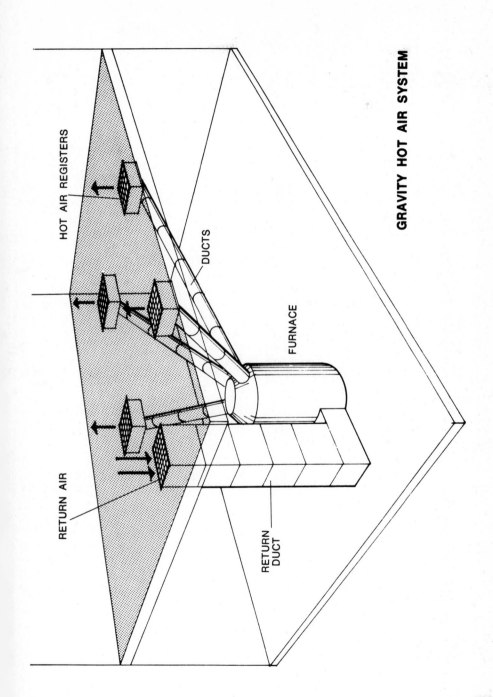

HOT AIR REGISTERS

DUCTS

RETURN AIR

FURNACE

RETURN DUCT

GRAVITY HOT AIR SYSTEM

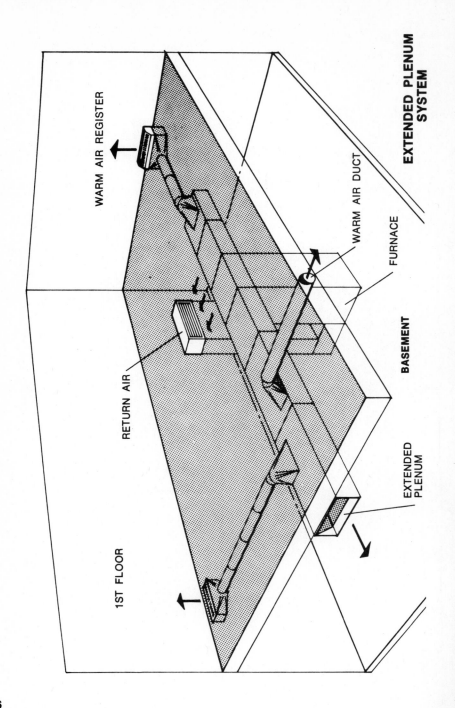

WARM AIR REGISTER

1ST FLOOR

RETURN AIR

WARM AIR DUCT

FURNACE

BASEMENT

EXTENDED PLENUM

EXTENDED PLENUM SYSTEM

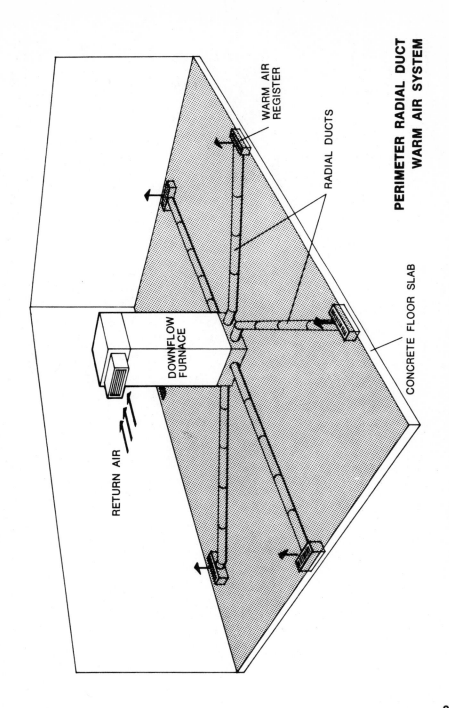

WARM AIR
REGISTER

RADIAL DUCTS

RETURN AIR

DOWNFLOW
FURNACE

CONCRETE FLOOR SLAB

**PERIMETER RADIAL DUCT
WARM AIR SYSTEM**

277

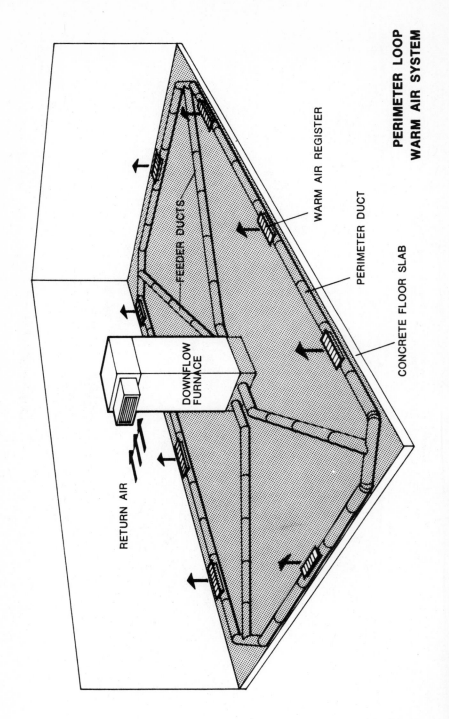

RETURN AIR

FEEDER DUCTS

DOWNFLOW
FURNACE

WARM AIR REGISTER

PERIMETER DUCT

CONCRETE FLOOR SLAB

**PERIMETER LOOP
WARM AIR SYSTEM**

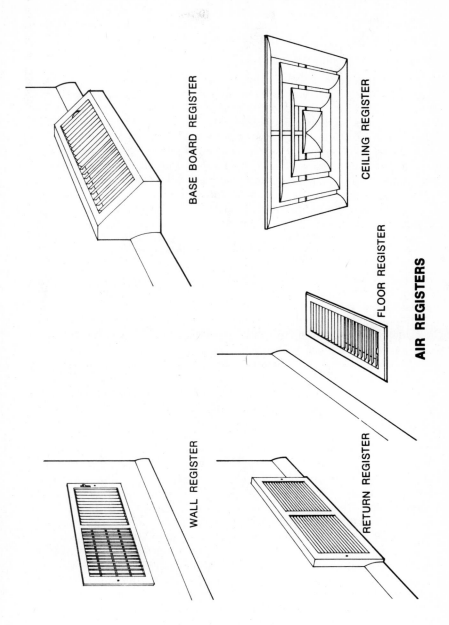

BASE BOARD REGISTER

CEILING REGISTER

FLOOR REGISTER

WALL REGISTER

RETURN REGISTER

AIR REGISTERS

279

which eliminates the cold floor feeling. As with the radial system, provision must be made for cool air return from each room.

Warm air perimeter systems are economical to install. They require little floor area, which is important in houses built on concrete slabs. And they can be adapted for summer cooling systems.

Hot Water Heating Systems

In a hot water (hydronic) system, water is heated in a boiler of cast iron or steel. It is then pumped by one or more circulators through finger-size tubes into baseboard panels, radiators or tubes embedded in the walls, ceiling or concrete slab. Water is an excellent heating medium, retaining heat longer than any other common medium.

Some older systems, like the hot air gravity system, depend upon gravity for circulation. The more modern units have one or more circulating pumps which are controlled by thermostats pumping hot water into the pipes when heat is called for.

In a one-pipe series connected system, the radiators, baseboard panels or convectors are connected so that the pipe runs through each unit. The main advantage of this system is its low cost. The major disadvantage is that because the water must run through each unit there are no individual control valves. Also, the units at the end of the loop are not as hot as the first units on the loops. By dividing the house into more than one loop or zone, each served by a separate circulator connected to a thermostat, some separate control is achieved. An alternate method of control is to have separate loops off a common circulator with a motorized valve connected to a thermostat regulating the flow of water through each loop.

A better single pipe system connects each unit separately to the loop with an individual take-off pipe and shut-off valve. This allows individual regulating of each unit as well as control of the loop, as in the series connected system. Special fittings are required to keep the water running in the proper direction and to induce flow into the radiators.

The best system is a two-pipe reverse return system where one pipe delivers the hot water and another returns it to the furnace. To take maximum advantage of a two-pipe system, it must be set up so that the first radiator to receive the warm water is the last radiator on the return line. This is called a reverse flow system.

A hot water system generally uses radiators, baseboard panels or convectors to distribute the heat into the room. These three units depend upon both convection (air being warmed as it passes over the heated metal and then circulating into the room) and radiation (heat waves being transferred directly from the heated metal to the object being heated by radiant energy). There are also combination systems in which the heat is brought to the radiator by warm water. A fan in the radiator blows air over the radiator fins, heating rooms by convection.

A radiant heating system depends solely upon the direct transfer of heat by radiation from the hot metal to the object being heated. The

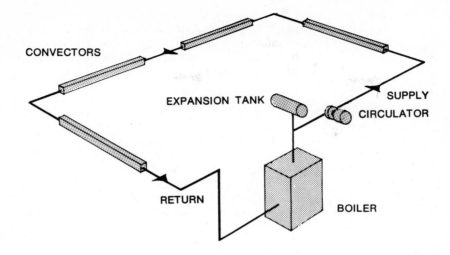

ONE-PIPE SERIES CONNECTED HOT WATER SYSTEM

This system may be identified by looking up at the part of the basement ceiling that is directly under the first floor radiators. In a one-pipe series connected hot water system, one wide-diameter pipe will be seen going in and out of each radiator. There will be no small-diameter feeder pipes.

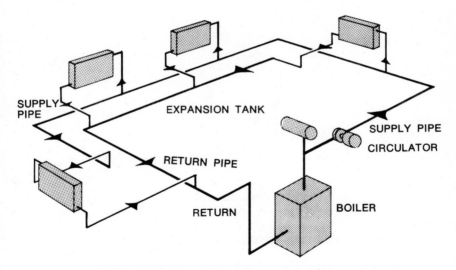

TWO-PIPE REVERSE RETURN HOT WATER SYSTEM

Looking up at the part of the basement ceiling that is directly under the first floor radiators, two wide-diameter pipes will be seen under each radiator. Each will be connected to its radiator by a smaller feeder pipe.

281

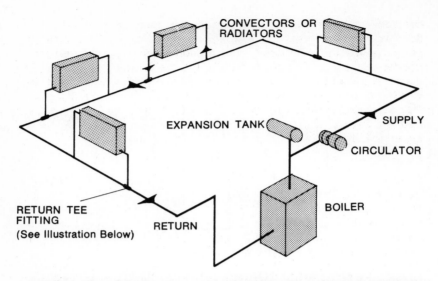

SINGLE PIPE INDIVIDUAL TAKE-OFF HOT WATER SYSTEM

Looking up at the area of the basement ceiling that is directly under the first floor radiators, one wide-diameter pipe will be seen under each radiator. Each of these pipes will be connected to its radiator by two smaller feeder pipes.

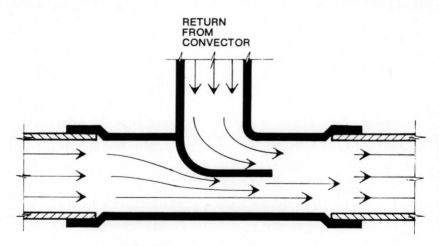

ONE-PIPE RETURN TEE FITTING

heated water is pumped through coils of pipe that are embedded in the floor, walls or ceiling or some combination of these. This system is particularly adaptable to slab construction where the coils are embedded in the slab. A major advantage is the absence of visible radiators or baseboards. A disadvantage is slow response to changing weather or other conditions affecting the temperature of the room.

A hot water system can be expanded to perform other functions, from melting ice and snow on the sidewalks and/or driveway to heating an indoor or outdoor swimming pool or greenhouse.

Steam Heating System

Steam heat is produced by a furnace which is a boiler with a firebox underneath it. The water in the boiler boils, making steam which is forced by its pressure through pipes into the radiators throughout the house. In a single pipe system the steam cools, turns back into water and runs back to the furnace to repeat the cycle. In a two-pipe system it returns via a separate pipe.

A two-pipe system tends to be less noisy than a one-pipe system. Noise is a disadvantage inherent to all steam systems. Another disadvantage is the difficulty in controlling the heat when only small amounts are needed.

Electric Heating System

When electricity is used in the same way as oil, gas or coal to heat the air in a hot air furnace or the water in a hot water furnace, it can be thought of as just another fuel. What makes it unique in house heating is its use with resistance elements which produce heat at the immediate area to be heated.

Electric resistance elements, which convert electricity into heat, are embedded in the floors, walls and ceilings to provide radiant heat. They are embedded at the time of construction or may be factory-installed in the building materials. The advantages claimed for electric radiant heat are the lack of visible radiators or grilles and the comfort obtainable without the necessity to heat the room air hot enough to drive out the moisture. Its acceptance by the public is not universal and substantial experimentation is still being conducted by the manufacturers and others. Electric heat has the potential, however, of being a very satisfactory heating system because baseboard radiators with individual resistance elements are inexpensive to install compared to other systems. Electric heat also provides the advantage of individual temperature control for each room.

Electric heating panels, also with individual resistance elements, are often used for auxiliary heat in bathrooms, additions to the original house and summer homes.

The heat pump electric system provides both heating and cooling from a central system. It is actually a reversible refrigeration unit. In the winter it takes heat from the outside air, ground or well water and distributes it in the house. Its efficiency decreases when it is very cold

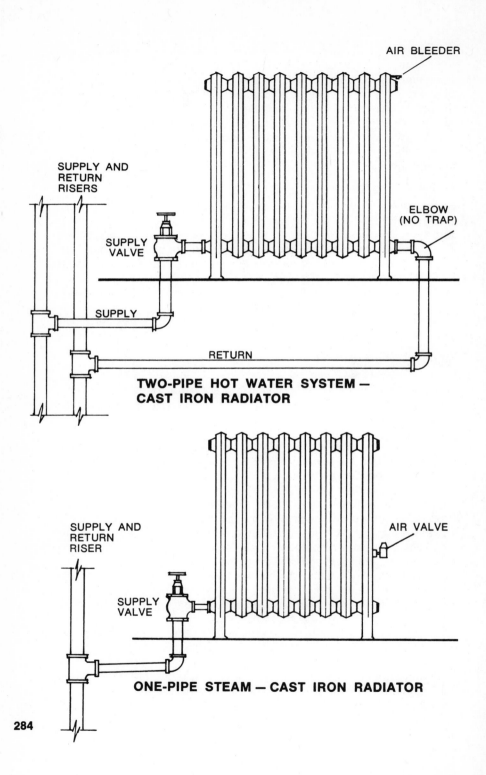

AIR BLEEDER

SUPPLY AND
RETURN
RISERS

ELBOW
(NO TRAP)

SUPPLY
VALVE

SUPPLY

RETURN

**TWO-PIPE HOT WATER SYSTEM —
CAST IRON RADIATOR**

SUPPLY AND
RETURN
RISER

AIR VALVE

SUPPLY
VALVE

ONE-PIPE STEAM — CAST IRON RADIATOR

284

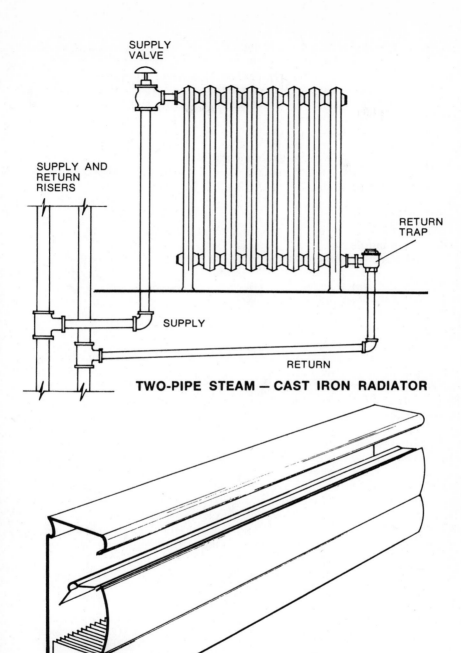

SUPPLY
VALVE

SUPPLY AND
RETURN
RISERS

RETURN
TRAP

SUPPLY

RETURN

TWO-PIPE STEAM — CAST IRON RADIATOR

BASEBOARD CONVECTOR

outside and it must be supplemented with resistance heating. In the summer the system cools by extracting heat from the inside of the house like a typical air-conditioning unit. This modern heating and cooling system still constitutes a small percentage of systems being installed. However, it is increasing in popularity and may turn out to be the system of the future.

Solar Heat

Solar heat is still in the experimental stages. With it, the sun's heat is collected and used to heat water, which in turn is pumped throughout the house to provide heat. (See Fuels)

Heat Controls

All of the systems described above can be operated by simple, automatic controls.

The first control one thinks of is the thermostat which is designed to turn the system on and off to produce heat only when it is needed or route heat to the areas where it is needed. A thermostat is a temperature-sensitive switch. While a thermostat often controls the heat in more than one room, it can read the temperature only at the point where it is located. Therefore, it must be located at a spot typical of the area being regulated.

Different heating systems allow various levels of control, ranging from some electric systems that control each room individually to steam, hydronic and air systems with only one control zone for the entire house. It is best to place the thermostat 2 to 4 feet from the floor. Common mistakes are to place it too high or to put a lamp, radio or TV near it.

More sophisticated thermostats have a device in them, called an anticipator, which shuts off the heating shortly before the desired temperature is reached. Because the system does not cool off instantly, the room will be brought to the desired temperature. Thermostats that automatically lower the heat during the sleeping hours are also available. In many houses this will result in fuel savings. However, in some houses it is just as economical to leave the heat at a constant temperature. Many people think that if they turn a thermostat up higher than the desired temperature, a room will warm up faster. However, since a thermostat only turns the heat on or off and does not control the timing of temperature changes, this is not true.

When it is very cold outside, additional comfort may be obtained by setting the temperature a few degrees higher than usual to compensate for the increased body heat loss by radiation to the cold outside walls. It also compensates for a tendency of thermostats with anticipators in them to run a few degrees colder under these conditions.

Humidifiers

Dry air in the winter can be a problem with all types of heating systems, contrary to the popular misconception that it is a warm air heating deficiency. The major controlling factor is the construction of the house. Moisture is generated by the activities of daily living, such as cooking and bathing, and is drawn out of the house in the winter to the dry air outside through leaks and cracks. A tight, very well-insulated house will retain sufficient moisture for comfortable living, which is about 25 to 30 percent relative humidity.

A dry air problem can be corrected by the addition of a humidifying device to a hot air heating system or with portable humidifying devices. Often a small vaporizer in the bedroom at night will solve the problem.

Occasionally, in a super-insulated, electrically heated house the reverse problem will develop, resulting in wet walls and windows. In this case, an exhaust fan connected to a humidistat control will effectively control the humidity.

Poor Heating

The major causes of poor heating are insufficient insulation and an inadequate or poorly functioning heating system. Insulation may often be added, as may storm windows and weatherstripping.

The condition of the furnace is often reflected in its appearance. An old furnace encased in asbestos probably is getting ready to give trouble. An adequate-sized, clean furnace without rust may require minor repairs but usually has plenty of good life left in it.

A free or nominally priced inspection of the heating system is often available from the fuel suppliers in the area. Fuel costs may be checked by asking for a year's bills from the homeowner or fuel supplier. Fuel costs can vary considerably according to the habits of the residents.

The performance of many furnaces can be improved by a good cleaning, adjustment and replacement of clogged air filters.

Domestic Hot Water

Hot Water from the Furnace

Hot water may be made in a hot water or steam furnace by the installation of a coil in the boiler. A storage tank may, optionally, be added to the system. This type of system, called a "summer-winter hook-up" in some parts of the country, provides a steady, small supply of hot water which may be exhausted by too much use at one time. Another disadvantage is the need to run the furnace year-round. The recovery rate is fast, however, and the hot water will shortly replenish itself when exhausted.

Electric Hot Water Heaters

There are two basic types of electric hot water heaters. The "quick recovery" type has one or two heating elements which turn on as the water is used and bring the water temperature in the tank up, starting immediately after it is lowered. In some areas, an "off peak" system is used. The tank produces most of the hot water during the night when the electric rates are low and makes hot water during the day only when the night-produced supply has been exhausted.

One and two-element "quick recovery" tanks range in size from 30 gallons, suitable only for one or two people who use a limited amount of hot water, to the more popular 66 and 82-gallon sizes. Most of the "off peak" tanks are the 82-gallon size.

Gas Hot Water Heaters

Gas hot water tanks range in size from 30 to 82 gallons. Gas produces a faster recovery rate than electricity, so a slightly smaller tank than would be required for electric heat can be used. However, the cost per gallon of gas hot water tanks of similar quality is higher, so the difference becomes academic. A 30 or 40-gallon tank is suitable only for one or two people whose use of hot water is limited. The more popular sizes are the 52 and 82-gallon sizes.

Oil Hot Water Heaters

The recovery rate of an oil hot water heater is very fast, compared to gas or electricity. Because of this, a 30-gallon tank will provide enough hot water for the needs of most families. The operating cost in most areas is less than gas or electricity. What limits the popularity of oil hot water heaters is their initial high cost and high installation expense. This is especially true if there is not a flue and oil storage tank already available.

The length of the pipe from the source of hot water to the fixture should be less than 15 feet. Longer pipes cause a great deal of heat loss and the inconvenience of having to let the water run for awhile before it becomes warm. This can be partially corrected by insulating the hot water pipe. Also, it is possible to install a return system from the fixture to the hot water source, which provides a continuous circulation of water from the heater to the fixture and back to the heater. Sometimes, in larger houses, the best way to keep the pipes short is to install two tanks.

The final selection of a hot water heating process is usually based on the relative fuel cost in the area. Using the same fuel that the heating system uses often results in a lower overall rate (especially if the cooking appliances also use the same fuel).

As nothing is more annoying than the lack of hot water when needed, care should be exercised in selecting a system adequate for the family's needs. Utility companies have recognized the problem and offer a practical solution to inadequate hot water by making available tanks that are rented from the utility company and paid for along with the utility bill.

Fuels

War has broken out between the purveyors of heating fuels, each proclaiming the merits of his own products and the disadvantages of the competitors'. Each fuel has its own significant advantages and disadvantages which must be considered against individual needs and preferences and then weighed against the relative costs in each area.

The ideal time to select a heating system, of course, is before the house is built. It rarely pays to tear out an existing, working system and replace it. Often, however, it does pay to rebuild or repair a system to bring it up to peak efficiency. If the furnace needs to be replaced, consideration might be given to changing fuels.

In spite of advertising claims to the contrary, natural gas, liquid petroleum (L.P.), gas, fuel oil and electricity are all about equally safe and clean. Coke and coal are definitely dirtier. Except when these are used, the extent of safety and cleanliness is governed by the condition of the heating system itself. With the exception of a few isolated areas of the country, coal and coke are rarely used in new systems for residential heating. In general, systems using them as fuel are fast becoming obsolete.

Fuel Oil

Oil is still the least expensive fuel in the northeast and northwest sections of the country and is competitive in many other sections. It may be stored in the basement of the house in one or two free-standing tanks not more than 275 gallons in size (larger tanks and more than two tanks in a basement are considered unsafe). Outside tanks buried in the ground commonly have 550 or 1,000-gallon capacities. Many oil companies offer automatic delivery, regulated by a system of measuring the degree days of heat required, that virtually ensures a continuous supply with a reserve in the event of interrupted supply. The larger the storage tank the less the per gallon cost of the oil. New kinds of burners have increased the potential efficiency, but they are more complex than gas burners, require more maintenance and have a shorter life expectancy. Oil tanks require maintenance and periodic replacement and take up living space in the basement.

The price of oil tends to fluctuate substantially throughout the season. The new laws requiring the use of only low-sulphur fuel in many areas have raised the price. However, the construction of more super-tankers and the changing world political situation may eventually produce price reductions.

Natural Gas

Natural gas offers the convenience of continuous delivery via pipeline without the necessity of storage tanks. Many suppliers offer lower rates if the hot water heater and cooking ranges are also gas. A gas furnace and burner is less complex than an oil furnace and burner. It requires less maintenance and has a longer life expectancy. It needs no on-site storage, but in the event of a supply interruption there is no reserve supply. In most areas of the country (the major exceptions being the Northeast and Northwest), gas is the most economical fuel.

Liquid Petroleum Gas

L.P. (liquid petroleum) gas is used in rural areas. It requires on-premises storage tanks and is usually more expensive than natural pipeline gas. In other respects it is similar to natural gas.

Electricity

Electricity appears to be the fuel of the future. It requires no on-premises fuel storage and can be used like coal, gas or oil to heat air in a hot air furnace or water in a hot water furnace or water heater. When used with resistance units at the area to be heated, it is unique. This system is the least expensive to install, since it requires no furnace, no furnace room, no ducts, no flue and no plumbing.

It does require, however, a much larger electric service into the house and wiring to each unit. When used in colder climates, special care in construction is required to eliminate air leaks and provide sufficient insulation. It has the disadvantage of no emergency on-site supply. Electric resistance heating systems and heat pump systems are unquestionably the most versatile, convenient and controllable systems. To date, in spite of advertising to the contrary, electric heat costs remain high except in lower-cost power areas.

Solar Energy

Hopefully the 1970s will go down in history as the decade that solar energy became a practical fuel for heating homes. Almost daily the newspapers carry another release about another experimental solar heating system. At the start of 1976 there were probably less than 500 houses that depended primarily upon solar energy for their heating fuel. It is estimated, however, that almost one million houses worldwide have some type of rooftop solar heater used either to heat water or supplement the heating system.

The basic principles of most solar systems developed so far are similar. The sun's energy is collected with an array of pipes or flat metal sheets which are painted, coated, plated or otherwise treated to increase their ability to absorb heat. Often they are encased in glass or plastic and positioned to catch the maximum amount of sun.

Air, water or other chemical mixtures in the piping collect the heat and distribute it either through a standard heating system or one that has been modified for solar energy use. Sometimes the heated water is pumped in a storage area for use when no sunlight is available.

As of February 1976, when this was written, solar energy was still experimental. The cost is not competitive with that of conventional fuels. One problem is that the efficiency of the system tends to fall off rapidly as the system ages.

The government predicts that by 2000, 20 percent of our energy will be solar. They also predicted 30 years ago that nuclear energy was the fuel of the future; yet today less than 2 percent of our power comes from this source.

Comparison of Fuel Costs

One does not have to be a mathematical genius to do the calculations required to figure the relative cost of fuel. Since there is so much confusion on this subject, it seems worthwhile to provide instructions on how to make the comparison for those who wish to know.

Since it is impossible to compare directly the cost of one fuel against the cost of another, it becomes necessary to first find some common denominator to use as a unit of comparison. Heat is measured in units called calories or therms. These units are so small, however, they are hard to use. A bigger unit was developed called a British Thermal Unit, the B.T.U. This is the amount of heat it takes to raise the temperature of one pound of water one degree Fahrenheit. This is still a very small amount of heat. In home heating the cost per million B.T.U.'s is the standard measurement used.

To compare the cost of fuel, figure out how much it will cost to make a million B.T.U.'s with each kind of fuel in a specific area. Once the relative costs of the fuels are known, this can be weighed against the other advantages and disadvantages of each type of system and a final selection can be made.

Fuel Oil Cost Calculation

Step 1: Estimate the efficiency with which a residential fuel oil furnace will convert a Number 2 oil (the most common type used in house heating) into heat. The efficiency ranges from about 50 percent for an older converted furnace to 80 percent for a new, properly-adjusted furnace. Most major oil retailers will perform an efficiency test on a furnace at no or nominal cost. However, if no such test is made, just make an estimate for an existing furnace or use 80 percent.

Step 2: Obtain the price of fuel oil in the area. Use an average price, taking into consideration seasonal fluctuations. Don't be afraid to ask for a discount; they are a common practice in many areas. Also, check the saving offered by installing a larger tank. The conclusion may be that its installation will soon pay for itself.

Step 3: Calculate the cost to produce a million B.T.U.'s of heat. This is done by multiplying the oil cost per gallon by a factor of 7.15 and then dividing by the system's efficiency percentage. Example: if the oil cost is 18 cents and the furnace tests at 60 percent efficiency:

$$\frac{.18 \times 7.15}{.60} = \$2.14 \text{ per million B.T.U.'s}$$

Natural Gas Cost Calculation

Step 1: Estimate the efficiency of the gas furnace to convert gas into heat. The efficiency ranges from about 60 percent for an older, converted furnace to 80 percent for a new, properly adjusted furnace. **291**

The gas company will usually make an efficiency test at no charge and at the same time offer recommendations to increase the system's efficiency.

Step 2: Obtain the price of gas in the area. If it is not published in therms, ask the gas company for the price per therm. If the rate is on a sliding scale, ask them for the typical average rate for a one-family house.

Step 3: Calculate the cost to produce a million B.T.U.'s of heat by multiplying the gas cost per therm by a factor of 10 and then dividing by the system's efficiency percentage. Example: if the gas cost is 9.5 cents per therm and the furnace tests at 60 percent efficiency:

$$\frac{0.95 \times 10}{.60} = \text{\$1.58 per million B.T.U.'s}$$

The same formula can be used for bottled liquid petroleum (L.P.) gas.

Electric Heat Cost Calculation

Step 1: The efficiency of an electric baseboard system is close to 100 percent since it converts all the electricity into heat and there is no flue through which the heat can escape.

Step 2: Obtain the price of electricity in the area (be certain this is the residential heating rate). Also check for a possible rate reduction by converting the hot water heater and stoves to electricity.

Step 3: Calculate the cost to produce a million B.T.U.'s of heat by multiplying the electricity cost per kilowatt hour by a factor of 293 and then dividing by 100. Example: electricity costs $1.50 per kilowatt hour:

$$\frac{1.50 \times 293}{100} = \text{\$4.39 per million B.T.U.'s}$$

Air Cooling

A few attempts were made in the late 1800's to cool houses, but little was effectively done other than moving the air around with fans until after World War II.

Fans

Fans are still an accepted method of cooling. The air inside the house may be cooler than the outside air yet may feel warmer because of the lack of air movement. A simple room fan (especially when the temperature and humidity are not too high) will often make the room seem much cooler. The fan in a warm air heating system may be used to move the air and to bring cooler basement air up into the house. (Again, this works best when the temperature and humidity are not too high.)

An attic ventilation fan cools in several ways. During the day it removes the hot air from the attic, which when left in place soon seeps down into the house, substantially raising the overall interior temperature. For best results the attic fan should be large enough to remove air equal to the volume of the house every minute. For example, a house that is 10,000 cubic feet should have an attic fan rated to remove 10,000 cubic feet of air per minute.

In many areas of the country the outside air temperature drops sharply during the night. The attic fan is used to replace the hot daytime air with cool night air. In some areas of the country a timing device is needed to turn off the fan to prevent the house from over-cooling. If the house is well-insulated and kept closed during the following day, the inside air may remain substantially lower than the hot outside air.

Window exhaust fans work on the same principle as attic fans. They are especially effective in changing cool night air for hot day-time air in bedrooms and living areas.

Evaporation Cooling

In some areas of the West the humidity is low most of the time, even in periods of high heat. In these places, a simple system of blowing air across wet excelsior or some other water-absorbing material will cool the air substantially. What happens is that the water evaporates and, in the process, cools the air. Package units are manufactured for home installation in windows. They are simple machines and are less expensive than conventional air-conditioning.

Air-Conditioning Principles

With the exception of fans and evaporation cooling, most house air-conditioning uses either electric power compressor-cycle equipment or gas absorption-cycle equipment. These are the same two types of systems used in household refrigerators. The basic law of physics that applies to both types is that when a liquid is changed to a vapor or gas, heat is absorbed and when the vapor is compressed back into a liquid, the heat it previously absorbed is given off.

In an electric system the refrigerant that is changed back and forth from gas to liquid is usually Freon. The equipment consists of a device that compresses the Freon into liquid inside an arrangement of finned tubes that are located outside the area to be cooled (usually outside the house). As the Freon is compressed in these tubes, they become hot. The heat is disbursed by either running water over the tubes in a water-cooled unit or more often by blowing outside air over them with a fan.

The liquid is then run through a pipe to another set of thin tubes inside the room to be cooled. The liquid is allowed to expand back into gas inside these tubes which become cold and absorb heat from the room air which is blown over them with another fan. Moisture in the air will condense on the fins and tubes, drip off and must be drained away. This also reduces the humidity in the air which, for

293

human comfort, is just as important as temperature reduction. The gas is then pumped into the compressor and the cycle repeats itself. The condensation tubes and evaporation tubes can be close together as they are in a room window and sleeve units or separated and connected by two insulated tubes as they are in most central air-conditioning systems.

In a gas powered absorption cycle system, ammonia is often used as the refrigerant with water as the absorbent or water is used as the refrigerant and a lithium bromide solution as the absorbent. Instead of a compressor there is a generator where heat is applied in the form of a gas flame and the refrigerant is boiled out of the absorbent; it passes in gas form into the condenser under high pressure caused by the boiling in the generator. Here, it is cooled by passing outside air or water over it and it condenses back into liquid while still remaining under pressure. From the condenser it goes through a metering valve into the evaporator tubes located in the room to be cooled. It expands in the evaporator, absorbs heat and condenses moisture from the air in the room. As an alternate, the expanding refrigerant can be used to cool water that has been pumped into coils or convectors in the room. The refrigerant, now in vapor form, passes on to the absorber where it forms the solution that is pumped back into the generator.

Electric Room Air-Conditioning Units

Millions of room compressor-cycle air-conditioning units are being manufactured and sold in this country each year. They now provide an economical solution to summer heat. Their main disadvantages are their ugliness, their high noise level and the fact that they are a haphazard solution to complete year-round home temperature control. On the other hand, if the family budget does not permit a central air-conditioning system, air-conditioning units provide a good deal of comfort for their relatively low cost.

Widely available are 4,000 and 5,000 B.T.U. units that can be carried home from the store and installed by many only moderately handy homeowners without the help of an electrician. These units run on 110 volts and use so little electricity that they can be plugged into a regular wall outlet in most homes with modern wiring. They will nicely cool a small room in many climates.

Larger room air-conditioners range upward in size from the popular 8,000 to 12,000 B.T.U.'s all the way to as high as 35,000 B.T.U.'s. These units generally require 220 volts and should be installed on their own separate electric line with a built-in ground wire.

The most common place to install a room air-conditioning unit is in the bottom half of a double-hung window. The window is secured tightly shut against the unit and the cracks plugged up with special felt and boards usually supplied with the unit. Specially shaped units are also made for installation in casement windows and through wall sleeves. The advantages of sleeves compared to window instal-

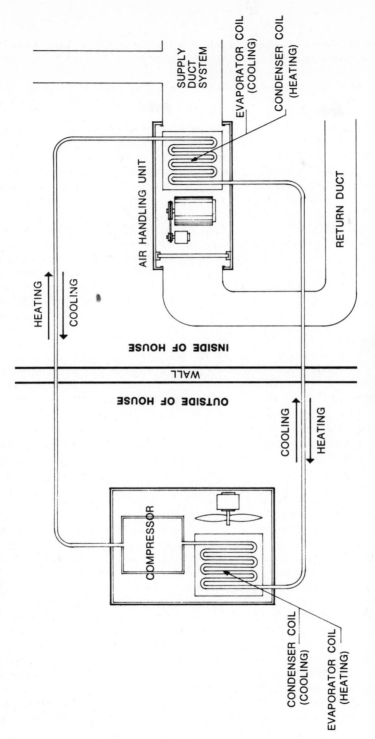

SUPPLY DUCT SYSTEM

EVAPORATOR COIL (COOLING)

CONDENSER COIL (HEATING)

AIR HANDLING UNIT

RETURN DUCT

HEATING

COOLING

INSIDE OF HOUSE

WALL

OUTSIDE OF HOUSE

COOLING

HEATING

COMPRESSOR

CONDENSER COIL (COOLING)

EVAPORATOR COIL (HEATING)

HEAT PUMP

lations are that sleeve units do not obstruct window vision, tend to be more permanent and may be placed high on the wall for more efficient air distribution. If there is a choice of windows or walls, the units should be installed on the north or east side of the house.

Ducted Central Air-Conditioning Systems

These systems can be custom made or can be prewired, precharged factory-assembled packages that are connected at the home site. Their condensor portion is set outside the house on the ground or on the roof. It is connected by pipe to the evaporator air-handling unit that is inside the house. The air-handling unit, consisting of the evaporator and a fan, is located inside the house and is connected to a system of ducts that distributes the cool air throughout the areas of the house to be cooled. If the house has a warm forced air heating system the air-conditioning system can use the same fan, filter and duct system. However, the ducts of the warm air system may not be suitable for air-conditioning because cooling generally requires double the duct size as heating and the cooling system works much better if the registers are high on the wall or are the type that directs the air steeply upward.

A hydronic heating system can also be combined with a cooling system. A water chiller is coupled to the boiler to supply cold water to a special type of convector that has been placed in each room as a replacement for the old style radiator. Conversion of old style hydronic systems to combination systems is often so costly that it is less expensive to install a separate new set of air-conditioning ducts.

Heat Pump Systems

The heat pump is a reversible heating and cooling system. It both heats and cools the house but works best in climates where the winters are not very cold. This system has been described previously in the electric heating section above.

Plumbing

Plumbing is like an iceberg—only a small part of it is visible. There has been plumbing of a sort in houses for more than 2,000 years. Emphasis gradually shifted from running water used for washing and cooking to all kinds of bathing devices and finally to the inside toilet. Probably the biggest difference between houses being built today and those built prior to World War I is the plumbing system. However, the plumbing systems in many of these pre-World War I houses have now been completely modernized. The old style plumbing fixtures are becoming so rare, many are selling as antiques.

The best way to cut through the mystery that exists in the minds of so many when it comes to plumbing is to discuss it in its various parts rather than as an incomprehensible whole system.

First, municipal water supplies and wells, water pumps, water softeners and water pipes will be discussed. Then the bathroom,

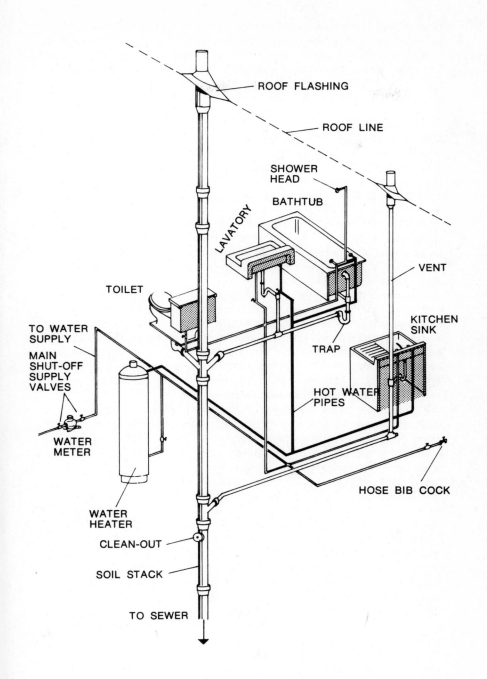

ROOF FLASHING

ROOF LINE

SHOWER HEAD

BATHTUB

LAVATORY

TOILET

VENT

TO WATER SUPPLY

MAIN SHUT-OFF SUPPLY VALVES

TRAP

KITCHEN SINK

WATER METER

HOT WATER PIPES

WATER HEATER

HOSE BIB COCK

CLEAN-OUT

SOIL STACK

TO SEWER

PLUMBING SYSTEM

kitchen and other fixtures in which water is used will be covered, as well as methods of removing the water from the fixtures through waste pipes to the sewer, septic tank or cesspool.

Technically, parts of the heating, hot water and air-conditioning systems are also plumbing. However, for ease of understanding they are discussed separately in this book.

Water Supply

Common sense and MPS both require that when a public water supply is available it should be used. An attempt to save money by using a well when this is not necessary is an economy made at the possible expense of the family health.

In the absence of available public water the next best supply is a well which is dug on the lot and water from it pumped into the house. There are some houses that obtain water from rivers, streams, lakes and even rain water collected from the roof and stored in tanks. None of these latter systems are considered to be satisfactory water supplies, however, since they do not provide a sufficient quantity of dependably pure water.

Water supplied by public water system must meet purity standards of local health officials. It usually is piped in the streets in water mains. A copper pipe of at least ¾-inch diameter taps into this line and runs underground into the house. Instead of copper, 1 to 1¼-inch galvanized iron pipe may also be used. There should be a cut-off valve at the street edge to shut off the water in an emergency. It is usually found in a hole covered with a steel or iron plate. Another cut-off valve for the same purpose should be located where the pipe enters the house. The pipe should not be run where it is likely to be disturbed, such as under the driveway. When the pressure in the water main is more than 80 pounds per square inch (p.s.i.), it is necessary to install a special valve to reduce the pressure in the house to below 80 pounds to prevent damage to the plumbing system. A water meter is generally installed right after the in-house emergency cut-off valve and often another cut-off valve is installed on the other side of the meter as well. In some older communities the water company sells water on a flat rate or so much per fixture rather than by metering it.

As MPS states, "A well shall be capable of delivering a sustained flow of 5 gallons per minute. The water quality shall meet the chemical and bacteriological requirements of the health authority having jurisdiction." Generally, except in arctic conditions, the well should be located outside the building foundation at a minimum depth of 20 feet. It should be located at least 100 feet from an absorption field or seepage pit and 50 feet from a septic tank. Bored wells should be lined.

An artesian well is one drilled through impermeable strata, deep enough to reach water that is capable of rising to the surface by internal hydrostatic pressure. It still requires a pump to develop sufficient pressure for household use. The only way to be sure a well meets these important standards is to have it tested professionally.

There are two basic types of well pumps, the submergible pump located inside the well and the basement pump. The pump capacity should not exceed the flow rate of the well or it will drain it and bring up dirt. At least a 42-gallon storage tank is needed to provide a smooth flow of water.

Water Pipes

The water pipes must carry water throughout the house to the various fixtures without leaking, making noise, reducing the pressure or imparting any color or taste to the water.

Brass has been used for many years, but it is now expensive. Older brass pipes tend to crystallize and become coated on the inside in areas where the water is corrosive. It is, however, easily worked and is generally installed by threading the ends and screwing them into joints.

Galvanized steel is used in some areas. However, it is easily attacked by corrosive water. Like brass, it is easily worked and is connected with threaded joints and fittings. Galvanized wrought iron is similar to steel but is more resistant to corrosion.

Copper comes as rigid pipe and flexible tubing. In many areas it is the only acceptable material. Joints are made by soldering copper joints to both pipe ends. Some areas permit the use of mechanical joiners. Lead used to be a popular material. It is still used for the pipe from the water main to the house in some areas but is rarely used inside the house.

Plastics are the newest material used for pipes. The manufacturers claim and many builders agree that plastic is as good as any other material and may indeed be better. It is gaining acceptance although it still is not permitted in many cities.

Water Softeners

In many areas of the country, especially where water is obtained from deep wells, large amounts of calcium, magnesium, sulphates, bicarbonates, iron or sulphur are often found in the water. These minerals react unfavorably with soap, forming a curd-like substance difficult to rinse from clothes, hair and skin. The bicarbonates, when heated, form a crust inside pipes and cooking utensils and a ring in the bathtub. Iron will stain clothing and sulphur makes the water taste and smell bad.

The simplest water softener is a manual single tank that is connected to the water line. Inside the tank is a mineral called zeolite which will exchange the offensive minerals in the water for sodium chloride (common salt). The zeolite must be poured into the tank. In a typical area with normal water use, salt must be added and the zeolite flushed out about twice a week. Those not willing to perform that task can obtain a two-tank automatic unit that will add the salt

299

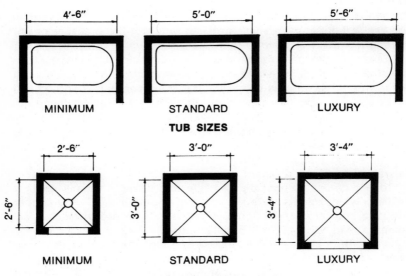

TUB SIZES

SHOWER STALL SIZES

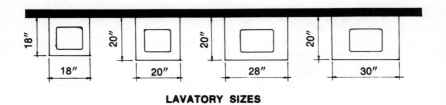

LAVATORY SIZES

BATHROOM FIXTURE SIZES

and flush the zeolite automatically. Of course, salt still has to be fed into the second tank every month or so. The zeolite is best at removing the calcium and magnesium. If the amount of iron, sulphur and other minerals is still very high, a third tank with special filters and chemicals may be needed for their complete removal.

Plumbing Fixtures

The parts of the plumbing system that compare to the above-water portion of an iceberg are the fixtures. Bathroom fixtures consist of lavatories, wash basins, bathtubs, showers, toilets (also known as water closets in the plumbing trade) and, occasionally, bidets.

Kitchen fixtures consist of sinks, laundry sinks, dishwashers and garbage disposals. In other areas of the house additional laundry tubs, sinks, bar sinks, outside sill cocks and other specialized fixtures are found.

Technically, hot water heaters and hot water heating systems may be considered part of the plumbing system and fixtures, but they are handled separately in this book.

Wash Basins (Lavatories)

The most satisfactory way to install a wash basin is to set it into a vanity counter. This provides the much needed space alongside the basin on which to stand beauty aids, shaving equipment and the baby's things.

Counter-mounted lavatories come in four basic types. The flush mount requires a metal ring or frame to hold it in place. Although it is very popular and inexpensive, dirt tends to collect around the edges of the rim. The self-rimming style eliminates this dirt collection problem on the basin side of the rim but not on the counter side. A handsome effect can be obtained with an under-the-counter lavatory when used in conjunction with a counter top that is or looks like marble. The seam where the lavatory meets the underside of the counter is likely to collect dirt and is difficult to clean, however. The newest style is a lavatory that is an integral part of the counter top, making cleaning very easy. Wall-hung lavatories come in several models. Generally they are installed as an economy measure or where space is at a premium.

Basins are made of the same materials as tubs and these materials have the same advantages and disadvantages as when they are used for tubs. The most common satisfactory material is cast iron covered with acid resistant vitreous enamel. The product of the future is a fiberglass basin molded together with the vanity counter top as one piece.

The size of the basin is important. For a bathroom, an 18-inch wide basin is minimum standard. Larger sizes range from 20 to 28 to 30 inches in luxury models. Fifteen inches is a minimum satisfactory depth, but 18 to 20 inches is better. One should look under the basin for the imprint of a national manufacturer because if there is no name the fixture may be a second or the lowest grade.

Bathtubs

The bathtub is the most expensive fixture in the bathroom and builders are often tempted to economize with it. Four different materials are generally used for tub construction.

Ceramic tile tubs are built in place usually at the same time as the ceramic walls and floors are laid. These tubs are expensive and uncommon. They require skilled craftsmanship for proper installation and are prone to leaking unless perfectly constructed.

By far the most common type of tub is made of cast iron coated with enamel. The cheaper grades have regular enamel and the better grades acid-resisting enamel. Unfortunately it is very difficult to tell by looking what type of enamel has been used, except that most colored tubs are the acid-resistant type.

Steel tubs covered with vitreous enamel are less expensive than cast iron tubs. The problem with them is that they are less rigid and more likely to crack. They can be identified by tapping and listening for the less solid sound or by pressing the bottom which often will be springy.

Fiberglass is gaining in popularity and well may be the product of the future for bathtubs. It is light and often is cast in one piece in conjunction with the wall, thus making the tub easy to clean and waterproof at the edge joining the wall. The finish, however, is not quite as hard as vitreous enamel and is more susceptible to scratches.

The most common tub size is 5 feet long and 14 to 15 inches deep. Even a few inches difference in depth, however, makes a big difference in comfort, as does 6 inches extra length. Therefore there is a big comfort difference between a standard 5-foot, 14-inch deep tub and a 5½-foot, 16-inch deep tub. The cost difference is about $50.

Sunken tubs are status symbols which appeal to some. Their prestige must be weighed against their hazardous nature and the back-breaking chore of bathing children in them. Square tubs are useful in certain layouts and will provide more tub area than a regular tub in the same spot.

The imprint of the name of the national manufacturer stamped on each fixture should be checked because a fixture without a name stamped on it often is the lowest grade or a second. In general, a cast iron tub is better than steel.

The best grade of tub fixtures is solid brass coated with chrome, nickel or brushed or polished brass. Most good grade fixtures have solid handles with grooves for fingers. Poor grade faucets are zinc or aluminum castings and often have cross-shaped handles.

Tub-Shower Combinations

By far the most popular shower arrangement is to have the shower located over the end of the tub. This arrangement costs little more than the tub alone to install, because the cost of a separate shower enclosure is eliminated. This saving is mostly lost, however, if there is a sliding glass (or translucent plastic) shower enclosure installed on the edge of the tub. A shower rod and curtain is much less expensive and preferred by many who do not like the closed-in feeling of the glass enclosure or the difficulty of keeping it clean. A shower rod with a double curtain works quite satisfactorily for many families.

Showers and Stall Showers

A separate shower stall is a troublesome, expensive luxury which many people feel is still worthwhile. They are usually located in the master bedroom. A popular type of construction includes a floor pan of concrete covered with ceramic tile, walls covered with ceramic tiles and a glass door. Anything but a one-piece floor pan will eventually leak and other wall coverings rarely work.

A less expensive prefabricated steel model is also commonly used. Here the floor pan is also a one-piece concrete pan and the walls are painted steel or galvanized iron. The door is usually open with a rod for a curtain. These units are quite satisfactory although not very luxurious in appearance.

The newest product is a one-piece fiberglass stall. These are attractive, come in many shapes, often eliminate the need for a shower curtain and are reported to be quite satisfactory. Cleaning the ceramic tile and steel models is difficult and time-consuming, but the fiberglass model cleans easily.

Leaking usually takes place through the walls, at the joint between the wall and floor pan and around the seam at the edge of the floor pan and drain. In new houses it is impossible to tell by inspection if the shower will leak, so a good builder's guarantee is important. In older houses inspection of the ceiling under the shower usually will show telltale marks of any leaking. New painting or papering in this area may be suspect as it may have been done to hide leak marks. Ceramic tile walls are most susceptible to leaking, but this problem is less likely with steel and is unlikely with fiberglass. Ceramic tile showers and steel showers both may leak at the wall pan joint. This is unlikely with fiberglass. All types of stall showers may leak at the pan-drain joint.

Toilets

The design and quality of the toilet is more important than the other fixtures because the tub and lavatory primarily hold water while the toilet is a much more sophisticated mechanism. Most residential toilets consist of a bowl and tank which stores sufficient water to create a proper flushing action. There is another type of toilet, often found in commercial buildings, which does not require a tank. Few houses have sufficiently large pipes or water pressure for this tankless type, however.

Briefly, what happens when the flush lever on a toilet is depressed is that it lifts a plug off an opening in the bottom of the tank and the **303**

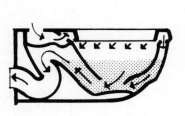

WALL-HUNG SIPHON JET

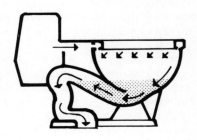

LOW PROFILE SIPHON ACTION

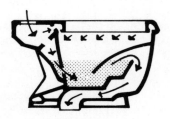

WASHDOWN

REVERSE TRAP

SIPHON JET

TOILET TYPES

water flows into the flushing unit. As the water flows out of the tank, a large air-filled metal ball, which is called a float and is attached to a valve by a metal arm, drops down, opening the valve to let fresh water into the tank. As the tank refills, the float rises and when the water reaches the correct level (as indicated by a line inside the tank) the valve is turned off.

There is also a stand pipe (overflow) next to the valve with a small pipe emptying into the top of it. After the toilet has been flushed, water continues to flow from the small tube into the bowl, filling it again to the correct level. The stand pipe also serves as an overflow protection in case the valve fails to shut off.

The quality of a toilet is judged by its performance in the following areas: self-cleaning properties, free rapid flushing action, quiet during flushing action and ease of cleaning around exterior.

There are four basic types of toilets, practically all of which are now made of vitreous china.

The wash-down bowl type toilet is an inferior product that is used only to save money. It is flushed by a simple washout action. The interior water level is very low, making the bowl subject to fouling, staining and contamination. Its flushing action is noisy and its self-cleaning properties poor. It can be recognized by its almost round shape and its straight line profile in front.

The majority of toilets currently made are the reverse trap bowl type. The inside water level covers about two-thirds of the interior surface. It is flushed by creating a siphon action in the trapway, assisted by a water jet located at the inlet to the trapway. This siphon action pulls the waste from the bowl. Its self-cleaning characteristics and flushing action are quite satisfactory but moderately noisy. The reverse trap bowl toilet is only slightly more expensive than a wash-down and a much better buy.

An improved version of the reverse trap toilet is a siphon jet. It covers a larger surface of the bowl with water. The trapway is larger and thus less subject to clogging and noise during the flushing action.

A still better siphon jet is the wall-hung model which makes for easy cleaning around the toilet area. Because it is suspended from the wall in the back and does not touch the floor, the wall studs to which it is fastened must be specially braced. Unfortunately, this toilet costs substantially more to install and is usually found only in custom built homes.

The most luxurious toilets are the one piece, low profile siphon action toilets. They provide almost silent flushing action, almost no dry interior bowl surface and therefore excellent self-cleaning action.

One may easily test each toilet in a house. Just throw in a piece of crumpled facial tissue, toilet paper or cigarette butt and flush the toilet. The water should flow swiftly over all the interior wall, thereby performing a self-cleaning action and the noise should not be excessive. Watch how quickly and surely the paper goes down. Watch and listen to the refilling action to determine if it is quick and quiet.

Bidets

Bidets are a standard fixture in continental baths. In America they are rare but slowly gaining popularity. They are usually installed by people of European background, those who travel to Europe often or more likely as a status symbol or conversation piece.

The bidet is perhaps the most misunderstood bathroom fixture. Many people incorrectly believe that it is used for internal feminine hygiene. Actually it is used by the entire family to wash the perineal area after using the toilet. A pop-up stopper holds water in the basin while washing. A spray is provided for rinsing and a flushing rim serves to rinse the entire inner surface of the bowl after using.

Bathroom Fittings

Fittings include faucets, spigots, shower heads, pop-up drains—the working parts of the plumbing system. These will require repair and replacement many times during the life of a house.

A faucet controls the flow of water into the fixture and often is also designed to mix hot and cold water. The most common type of arrangement is two separate valves, one each for the hot and cold water. In older or very cheap installations there is a separate spout for each valve, too. Most faucets now being installed feed into a single spout, but each handle is turned separately to control the water temperature.

A better arrangement is a single control valve feeding through one spout. The temperature of the water is controlled by moving a knob or lever to the right or left. Water volume is controlled by moving the same knob or lever in and out or backward and forward.

A pressure balancing valve is most commonly found controlling a shower head. The water volume is usually preset and the user selects only the temperature by turning the valve handle. A pressure sensing device adjusts the flow of hot or cold water to compensate for pressure changes, thus keeping the temperature even and safeguarding against scalding.

The ultimate is a thermostatic control valve that automatically senses the water temperature and adjusts the flow to maintain the preselected desired temperatures. This type of valve gives more precise temperature control than the pressure balancing valve and it usually also permits control of the water volume as well.

Most manufacturers make three grades of fittings. One way to spot the low grade is by the cross shape or inexpensive looking handles. The best buy is the middle grade which usually has attractive rounded handles with contours for the fingers. Luxury grade fixtures have elaborate handles and platings of such exotic materials as gold or brass.

A quality shower head swivels in any direction, has an adjustable
spray control handle and is self-cleaning. A shower should have an

"automatic diverter control" that switches the flow of water back to the tub after each shower so that the next user will not accidentally get wet or scalded.

Kitchen Fixtures

Most kitchen sinks are installed in some type of counter top. It may be either a single or a double bowl type set so as to have a drain board on one or both sides. Single bowl sinks range in size from 12 by 12 to 20 by 24 inches and double bowl sinks from 20 by 32 to 20 by 40 inches. The drain board type has a drain board made of the same material as the sink rather than the counter top. A type seen more often in apartments is the combination sink and laundry tub which has one deep and one regular depth sink. The deep sink has a removable cover that serves as a drain board when closed.

Kitchen sinks may be made of acid resistant enamel cast iron, enameled steel, stainless steel or Monel metal. Most modern kitchen sinks have a combination faucet with a swing spout. A separate spray on a flexible tube is also very common.

The drain should have a removable crumb cup or better, a combination crumb cup and stopper. There is now also available an attachment for a kitchen sink that provides boiling water instantly.

Though technically part of the plumbing system, a dishwasher is really a household appliance like the stove, refrigerator and washing machine.

The best place for a dishwasher is under a kitchen counter near the sink. Normally it has to be connected by a plumber to the hot and cold water pipe lines and to the drain pipe. It also must be wired into the electric system. These two installation expenses plus cutting into the kitchen cabinets if installed after the house is built make installation costs quite high. Standard models come in front-loading and top-loading types. The front-loading type has the advantage of making the counter top available for other uses. Mobile dishwashers are available where permanent installation is not possible or desired. Generally, however, their convenience is not as great as those installed in the counter.

Waste disposal devices are installed under the kitchen sink, connected to the drain. When filled with garbage and flooded with running water, their fast rotary action breaks up the garbage into particles small enough to go down the drain to the sewer or septic tank.

They are unquestionably a convenience, but they also present some problems. First, some cities will not permit their use because the local sewer plant cannot handle the additional waste produced by them. Likewise, many septic systems cannot handle the additional waste. The safety hazard of putting a hand into a running garbage disposal has been eliminated on some models that will not run without a top on them. Since this solution may cut down on the convenience of the disposal, many are wired to go on with a switch.

There is still much to be said for a laundry tub in the basement or laundry room even if there is an automatic washing machine and clothes dryer. It can be used for soaking clothes without tying up the washing machine. It also provides a good place to wash household articles, plants, paint brushes and other supplies.

Laundry tubs come in a wide range of sizes and materials. The most common are the one and two-tub models made of enamel covered cast iron on steel legs. Some styles permit the installation of a counter top which provides handy extra work space.

Outside Sill Cock

An outside sill cock (or bib cock) is a water faucet on the outside of the house with a screw nose to which a hose can be connected. MPS requires a minimum of two per house. They should be located on opposite ends of the house. In areas subject to freezing they should be the frostproof type and have individual shut-off valves inside the house so they can be drained and turned off for the winter months.

Waste Pipes

The biggest difference between waste pipes and water pipes is the lack of pressure in the waste pipe system. The water pipe must be strong enough to contain the water pressure and does not have to depend upon gravity to make the water flow to the fixture. Since there is no pressure in a waste drain line, the pipes must be slanted so that the waste will flow from each fixture through the main lines into the sewer or sewerage disposal system. Generally, drain lines are much larger than water pipes. Pipes from the toilets must be 3 to 4 inches in diameter, from showers, 2 inches in diameter and 1½ inches in diameter from all other fixtures, according to MPS.

The drainage system starts at each fixture with a curved pipe called a trap. A popular misconception is that the principal purpose of the trap is to catch objects that fall down the drain. The real purpose is to provide a water seal to prevent the seepage of sewer gas into the house. Some municipalities require a special trap to catch grease before it enters the sewer line.

Drainage lines that run horizontally are called branches and those that run vertically, stacks. The pipes that receive the discharge from the toilets are called soil lines and those that receive the rest of the discharge, waste lines. Vent pipes from each stack to the roof prevent the sewer gas from building up pressure in the system.

Pipes for the drainage system are often made of cast iron, copper, plastic, tile, brass, lead or fiber. Special fittings are often used, especially on the cast iron pipes, that aid the flow of the sewage.

Defective Plumbing and Noises in the Plumbing System

Plumbing suffers from the two major problems of leaking and clogging with rust and mineral deposits. Leaking can be detected by visual inspection. Old style iron or steel pipes are much more likely to develop leaks than corrosion-resistant copper and bronze. Iron and steel pipes can be detected with a small magnet which will be attracted to an iron or steel pipe but not to a copper or bronze pipe.

Insufficient water pressure can be caused either by clogged pipes, an undersized water main from the street, low water pressure in the street main or problems in the well or plumbing system. Water pressure can be tested by turning on full all faucets in the bathroom on the highest floor and then flushing the toilet. A substantial reduction of flow in the faucets is a sign of trouble and the system should be checked by a plumber to determine the cause and the cost to correct it.

Stains in the bathtub and lavatories are a sign of rusting pipes or unsoftened hard water. If hard water is suspected a sample can be professionally tested by a firm selling water softeners. Such a firm can also recommend what equipment will be needed and tell how much it will cost to provide soft water.

Leaks under sinks may be only from a loose washer but may also be caused by a cracked fixture.

A high-pitched whistling sound when the toilet is flushed is caused by the valve in the toilet closing too slowly. A simple adjustment by a plumber will eliminate the noise. A sucking sound when the water runs out of a fixture is often made by a siphoning action in the trap caused by improper venting of the waste stack. If unclogging the vent doesn't work, only a major change in the vent system will eliminate the noise.

A hammering noise in the water pipes when the water is turned off is caused by a build-up of pressure in the pipe. In high pressure areas, air chambers, which are pipes filled with air, are installed at the fixtures connected to the water line. They provide a cushion of air which lets the pressure build up more gently. Pressure build-up is a serious problem which, if gone uncorrected, will result in broken or leaking pipes. It may be possible for a plumber to install one or two large air chambers in the system or a variety of other mechanical devices designed to correct the trouble.

The sound of running water is caused by undersized pipes and pipes that run in walls that are not sound-insulated. Wrapping the pipe with a noise insulation material may help. If the noise is very objectionable the pipe may have to be replaced with a larger one.

Sewers, Septic Tanks and Cesspools

Few will argue the substantial advantage of being connected to a municipal sewer system by a single outlet or separately into the sanitary and storm water disposal system. MPS requires that, when available, the municipal sewer system be used. With increasing awareness of the damaging effects of pollution on our environment, rapid improvement and expansion of these systems can be expected in the near future. Health experts estimate that 50 percent of the septic systems now in use are not working properly. Still, it is estimated that almost 50 million people in 15 million homes, especially in the

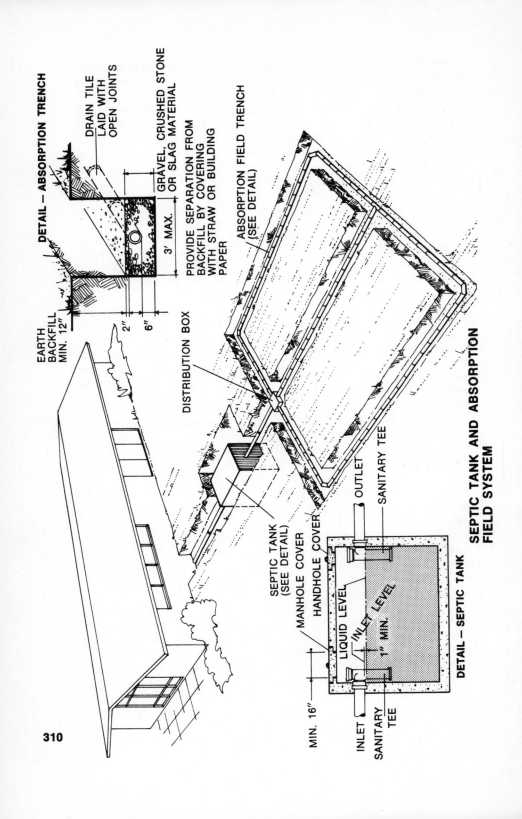

DETAIL — ABSORPTION TRENCH

DRAIN TILE
LAID WITH
OPEN JOINTS

GRAVEL, CRUSHED STONE
OR SLAG MATERIAL

PROVIDE SEPARATION FROM
BACKFILL BY COVERING
WITH STRAW OR BUILDING
PAPER

3' MAX.

EARTH
BACKFILL
MIN. 12"

2"

6"

ABSORPTION FIELD TRENCH
(SEE DETAIL)

DISTRIBUTION BOX

SEPTIC TANK
(SEE DETAIL)

MANHOLE COVER

HANDHOLE COVER

OUTLET

SANITARY TEE

LIQUID LEVEL

INLET LEVEL

1" MIN.

MIN. 16"

INLET

SANITARY
TEE

DETAIL — SEPTIC TANK

SEPTIC TANK AND ABSORPTION
FIELD SYSTEM

310

suburbs and rural areas, depend upon a septic system for their waste disposal and that 25 percent of the new houses being constructed do not connect to municipal systems.

A typical septic system consists of a large concrete tank with a capacity of 900 gallons (about 8 by 4 by 4 feet) buried in the ground. One end accepts the waste material from the house drain line. Once inside the tank, the waste tends to separate into three parts. The solid waste materials (only about 1 percent of the total volume) sink to the bottom. The grease (also less than 1 percent of the total volume) rises to the top. The rest is liquid. Bacteria in the tank decompose the solid wastes and grease and a relatively clear liquid flows from the opposite end through the drain line either into a distribution box that directs the liquid into a network of buried perforated pipes called a leaching field or into a seepage pit. From here the liquid runs off into the ground to be absorbed.

The required capacity of the tank depends upon the size of the house and usage. The size of the leaching field depends on the soil's capacity to absorb water. The rate at which the soil will absorb water can be measured by making a percolation test. A hole at least 12 inches deep is dug in the ground and filled with water. Each hour the depth of the water is measured. Anything less than an inch decrease in depth each 30 minutes is substandard. This test should be carried out in the wettest season of the year and preferably by an expert. Usually the local health department will make the test at no cost or for a nominal charge. Also, it is likely that the local health authorities will have previous knowledge of the individual system.

Septic tanks must be checked frequently to make sure they are not clogged and that the bacterial action is working properly. Chemicals must be used with care as they can kill the bacteria. Often the tank must be pumped out and the cycle started anew.

A cesspool is similar to a septic system except that instead of a tank there is a covered cistern of stone, brick or concrete. The liquid seeps out through the walls directly into the ground rather than into a leaching field or seepage pit. It is important to learn about a house's particular system, including the location of the clean-out main, which is often buried in an unmarked spot, so that inspections and repairs can be made as required. A properly working system should produce no odor which is one of the first signs of trouble.

Anyone wishing to gain information on septic systems should find out how often the system has to be pumped out. In many towns the local health officer is very knowledgeable about many systems in his jurisdiction and of problems in general in a particular neighborhood. Learning the location of the clean-out main from the current owner saves a lot of digging and searching if it is buried.

Septic system problems may sometimes be corrected by simply pumping out the tank. Sometimes new leaching fields are required. Unfortunately, there are situations when the soil absorption rate is poor or the water table is close to the surface and little can be done to make the system function properly.

Electric System

History

There is evidence in prehistoric caves showing that man was using inside lighting before the beginning of recorded history. Homer's *Illiad,* written sometime before 700 B.C., tells of torches used for light. Metal torch holders are found in medieval houses and inside castles, indicating that torches were a popular light source. Abraham Lincoln, according to legend, read by the light of the hearth, which had been a light source long before his time.

Braziers filled with pitch and chips of resinous wood will make a surprisingly bright, but smoky light. The candle was probably invented by the Etruscans and passed on to the Romans many hundred years before Christ. Candles are still very popular today for decorative, romantic, religious and emergency lighting.

Another basic light source was the liquid-filled lamp, such as rudimentary stone dishes burning melted fat; an advanced form was the whale oil lamp widely used in this country. The discovery of oil led to the kerosene lamp and the gas light fixture.

Thomas Edison created the first commercially practical light bulb in 1879. In addition, he developed a complete distribution system for electric light and power which included everything from the generator to the light socket and bulb, junction boxes, fuses, underground conductors and the other devices needed to make modern electric living a reality. In 1882 he built the world's first central electric power plant.

It is possible to judge the total impact of Edison's genius on modern life by comparing two Sears, Roebuck and Company catalogs. The 1898 Sears catalog had more than 700 pages and claimed to have "about everything the customer uses." It had books on electricity, an electric motor for no stated purpose and some telephones. It contained no electric light fixtures or bulbs, no toasters, a waffle iron that sat on top of the stove burner, no radios or TV sets.

The latest Sears catalog contains around 1,600 pages and now modestly stresses "guaranteed satisfaction" rather than completeness, although it certainly is a very complete representation of things now in use. It has books on electricity, an assortment of electric motors, telephones, numerous kinds of light bulbs, electric radios and televisions, plus ten pages of light fixtures. For the kitchen alone there are electric stoves, refrigerators, freezers, ventilation fans, mixers, blenders, can openers, choppers, coffee pots, frying pans, waffle irons, oven broilers, hot plates, juicers, steam irons, ice cream makers, knives and garbage disposals. It is apparent that we have become addicted to electric appliances and have no intention of being cured.

Principles of Electricity

Electricity is a form of energy. A few simple universal and unchangeable physical laws apply to it. One of these laws, stated in very simple terms, is that energy can be changed from one form to another. This is what Thomas Edison did when he built the first commercial generator plant on Pearl Street in New York City in 1882. In this plant, energy stored by nature thousands of years ago in coal was changed into electricity. Today, that is what is being done in generating plants all over the country by our utility companies. These plants change energy stored in coal, oil, water and atoms into electricity.

Another law, again simply stated, is that energy may be converted into heat or light by running it through some resistance. This is what happens in a stove, light bulb, resistance heating unit and any other electric appliance that produces light or heat.

Electricity will travel from its source to the ground at the speed of light (which is instantaneous for all practical purposes) unless stopped by resistance. A substance that has high resistance is called an insulator (glass, wood, rubber and porcelain are a few common items that have high resistance). A substance that has low resistance is called a conductor (copper, aluminum, iron, steel and many other metals have low resistance).

The way to get electricity to go to a desired location is to provide a path of low resistance material. We commonly do this with wire made out of low resistance metal. To keep the electricity from going into the ground before it gets to its intended destination the wire is surrounded with insulation. Air is an excellent form of insulation. Therefore, bare wire running through the air is really surrounded with insulation.

The electricity in the wire flowing from its source to the lights and appliances in the house can be compared to water flowing through the pipes from the well or water company. The rate of flow of water is measured in terms of gallons per second. The rate of flow of electricity is measured by amperes (coulombs per second). The pressure in the water main and pipe is measured by pounds per square inch (p.s.i.).

The pressure forcing the electricity through the wire is volts. The water company may have very high pressure in the street main and reduce it to below 80 p.s.i. with a special valve for home use. The electric company may have high voltage in the street lines and reduce it with transformers (devices for changing the voltage of electricity) to 110 to 120 volts or 220 to 240 volts, the two standard pressures used for house lighting and appliances. The actual amount of water used may be measured in terms of gallons. The amount of electricity used may be measured in terms of watt hours.

When a bulb is stamped on its end with "60 watts," this means that it will consume 60 watts of power for each hour it is lit. A kilowatt hour (kwh) is 1,000 watt hours. For example, an electric light bulb that is rated 100 watts will use 1 kwh of power for every ten hours it is lit (100 watts x 10 hours). If a kwh costs 3 cents it would cost 3 cents to leave that light on overnight.

It takes a great deal of electricity to make heat. A typical range top may take 5,000 watts. To keep it running for just one hour will use

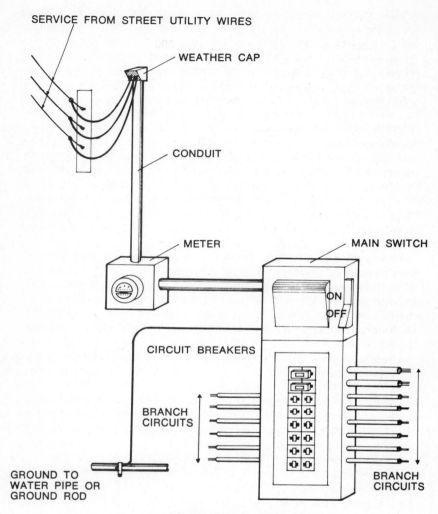

SERVICE FROM STREET UTILITY WIRES

WEATHER CAP

CONDUIT

METER

MAIN SWITCH

ON
OFF

CIRCUIT BREAKERS

BRANCH
CIRCUITS

GROUND TO
WATER PIPE OR
GROUND ROD

BRANCH
CIRCUITS

DISTRIBUTION PANEL

Here is the usual relationship between ampere service, number of circuit breakers and maximum number of watts:

Size of Service	No. of Branch Circuits (fuses or circuit breakers)	Maximum No. of Watts
30 ampere	4	6,900
60 ampere	6 to 8	13,800
100 ampere	12 to 16	23,000
150 or more amperes	20 or more	30,000 or more

ELECTRIC SERVICE ENTRANCE

5 kwh (5,000 watts x 1 hour). At the same 3-cent rate it would cost 15 cents to cook dinner. Some typical watt ratings are as follows: light bulbs, 25 to 300 watts; refrigerator, 300 to 400 watts; garbage disposal, 350 to 500 watts; television set, 300 to 500 watts; toaster, 1,000 to 1,500 watts; washer-dryer, 4,000 to 5,000 watts; water heater, 3,000 to 5,000 watts; range top, 3,000 to 12,000 watts and air-conditioners, 3,000 to 12,000 watts.

Electric Service Entrance

The home wiring system starts with the service entrance which brings the power from the street through an electric meter to a distribution panel. The service entrance may be designed to bring in 30, 60, 100, 150, 200, 300 or 400 amperes of electricity into the house.

A 30-ampere service will provide electricity for a total usage at one time of 6,900 watts. This amount of service is still found in many older one-family homes. The panel box (usually black) most often has four fuses in it. It will provide enough electricity for lighting and very limited appliance use. However, it is below FHA standards, is obsolete and should be replaced.

A 60-ampere service was the standard for many years for small to medium houses. It is still acceptable, according to MPS, when it can be demonstrated that the demand will be no more than 13,800 watts. A builder installing a 60-ampere service today, however, is probably trying to save money and should be judged accordingly. Many homeowners with 60-ampere services are converting to larger services. A 60-ampere service panel box usually has only 6 to 8 fuses or circuit breakers.

A 100-ampere service is the standard today for most small and medium sized houses without electric heat or central air-conditioning. It provides 23,000 watts of power. A typical panel box will have 12 to 16 fuses or circuit breakers. There may also be some small separate distribution boxes for the major appliances. A house with less than a 100-ampere service is or soon will be obsolete.

In larger houses and where electric heat, central air-conditioning or a large number of appliances are used, 150 to 400-ampere services are needed. The 150-ampere service is being used for houses without electric heat where pennies are not being pinched by the builder.

Most homes today are served by a three-wire, 220 to 240-volt service. The electricity is brought from the transformer at the street into the house through three wires. Two wires each carry 110 to 120 volts and the third is a ground wire. The wires are strung through the air overhead and connected to a service head, then run through a piece of conduit pipe down the wall to the electric meter (which may be inside or outside the house) into a distribution panel box. An alternate method is to bring the wires in from the street through underground conduit pipe. The use of underground wires is increasing as public resistance to the unsightly overhead wires increases.

Distribution Panel and Protective Devices

The distribution box must first of all provide a switch that will cut off all electric service to the house when the switch is manually pulled. This is needed in the event of an emergency, when work is to be done on the system or for any other reason the system may be required to be disconnected. It also must contain either a fuse or a circuit breaker that will disconnect the entire system automatically if the system is overloaded. This occurs when the system is called upon to provide more watts than it is capable of providing or if a short circuit takes place because of a break in a wire, a faulty appliance or two wires touching each other.

Fuses and circuit breakers are the two types of devices used to cut off the electricity automatically. A fuse is nothing more than a piece of wire that will melt when more than the prescribed amount of electricity flows through it, thus making a gap in the wire system across which the electricity cannot flow. The type of fuse used to protect the entire service is usually cartridge-shaped. The fuse holder should be so designed that it will not hold fuses that are too large for the size of service being protected.

Circuit breakers are a special type of automatic switch that will turn themselves off when excess electricity passes through them. They must be turned back on manually and will again turn themselves off if the problem that originally tripped them off has not been corrected. Circuit breakers are designed to wait a moment before they shut the electricity off because of an overload. This is helpful because electric motors draw a large amount of electricity for a brief moment when they start. This overload is harmless if it lasts just the normal short time.

Another advantage of a circuit breaker is that, unlike a fuse that must be replaced when used only once, a circuit breaker only needs to be turned back on to restore the electric service once the problem has been corrected. The only disadvantage of circuit breakers is that they are more expensive than fuses, which accounts for the lack of their widespread use.

The distribution box also divides the incoming electric service into separate branch circuits that lead to the various areas throughout the house. Special branch circuits should be provided for those appliances that use a large amount of electricity such as heaters, air-conditioners (some small room air-conditioners do not require a separate circuit), ranges and ovens, dishwashers, washing machines and dryers or any other high-use electric appliance.

Each individual circuit must also be protected by a fuse or a circuit breaker. If an overload or short circuit occurs on the circuit, it will automatically shut off without tripping the main fuse or circuit breaker and shutting off the whole house service.

The type of fuse used to protect an individual circuit is different from the cartridge types used to protect the whole service. Here a

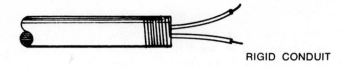

RIGID CONDUIT

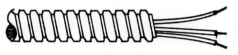

FLEXIBLE CONDUIT

ARMORED (BX) CABLE

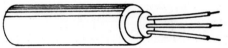

NON-METALLIC CABLE

SURFACE RACEWAY

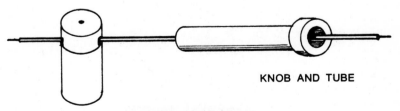

KNOB AND TUBE

TYPES OF WIRE

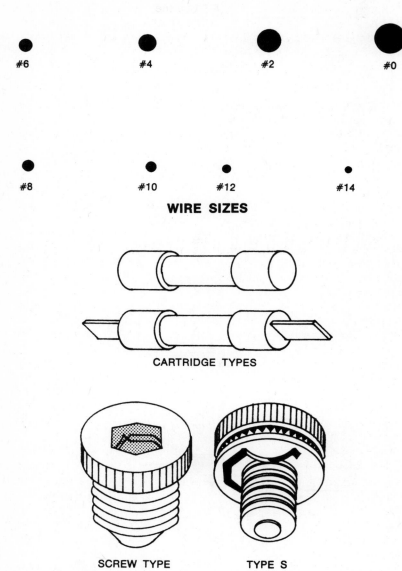

WIRE SIZES

#6   #4   #2   #0

#8   #10   #12   #14

CARTRIDGE TYPES

SCREW TYPE   TYPE S

TYPES OF FUSES

screw-in fuse is used. Its size may be 15, 20 or 30 amperes depending upon the capacity of the circuit being protected. A special type of fuse known as Type S is now required by the National Electric Code and MPS. It has the same momentary delay feature as a circuit breaker. It also requires a special socket that holds only the correct size fuses. Once these sockets are installed in the distribution panel they cannot be removed except with a special tool.

Circuits and Wiring

In the panel box the electric power is divided into separate branches known as circuits. Each circuit will serve a separate area of the house or an individual appliance. General circuits run to each area of the house. Connected to them are the permanently installed lighting fixtures and receptacle outlets, into which are plugged lamps and appliances. These general circuits are wired with two No. 12 or larger wires and protected with a 15-ampere fuse or circuit breaker. They provide 1,800 watts of 110 to 120 volt power. The National Electric Code and MPS require three watts of power per square foot of floor space for general circuits. A better standard is five watts per square foot. Therefore, each general circuit should serve 360 to 500 square feet of floor area. A 1,500-square foot house could meet the minimum standards with three general circuits but should have five for convenience and expansion.

Special circuits for small appliances are wired with No. 12 or larger wire and protected with 20-ampere fuses. They provide 2,400 watts of 110 to 120 volt power and cannot be used for large appliances. MPS requires at least two special circuits in each house.

Large appliances such as clothes dryers, water heaters, ranges, dishwashers, freezers and large window air-conditioners require large amounts of watts and often 220 to 240 volts and special three-wire circuits using wire from No. 12 to No. 6 size (the lower the number the larger the wire). These circuits are protected with 30 to 60-ampere fuses or circuit breakers.

If the size of the wire used in a circuit is too small the lights and appliances do not work at peak efficiency. The bulbs shine dimmer than they should and the appliances do not work as well or get as hot as they are designed to do. In this case, the amount of electricity comes through the electric meter and is paid for even though the electricity is not providing maximum performance. This is because the undersize wire itself is creating extra resistance and turning the electricity into heat.

The old system of wiring a house was to run two insulated wires parallel to each other from the panel box to the outlets and fixtures. A separate pair of wires was run for each circuit. The wires were run a few inches apart and attached to the house with white porcelain insulators called knobs. When the wire passed through a wall or joist, it went through white porcelain tubes, hence the name, knob and tube wiring. This system is obsolete and often must be replaced, which can be a major expense.

Today, non-metallic cable is the next cheapest system available. Each wire, in addition to its own insulation, is wrapped with a paper tape and then encased in a heavy fabric and treated to be water and fire resistant. A similar cable has a thermoplastic insulation and jacket. The cables are attached to the joists and studs with staples. They must be protected from damage and should not normally be used outside or underground unless they are plastic coated. Both of these cables are prohibited in many major cities.

Armored cable (or B.X. cable) consists of insulated wires wrapped in heavy paper and encased in a flexible, galvanized steel covering wound in a spiral fashion. This system has wider approval but not universal acceptance even though it is less susceptible to physical damage. Surface raceways made of metal or plastic are sometimes used in houses, mostly for repairs and in solid-core walls and partitions.

Flexible steel conduit is constructed similarly to B.X. cable except that it is installed without the wire. Wires are drawn through the conduit after installation.

Rigid steel pipe, which looks like water pipe, is still the most preferred and expensive method and meets the most rigid codes. Like the flexible conduit, the wires are pulled through after it is installed. The larger the size of the conduit the more and larger wires can be pulled through it. Most outdoor installations are rigid conduit, although a vinyl insulated wire has been developed that can be buried in the ground.

Telephones and doorbells use low voltage wiring which does not present a safety hazard and therefore can be run loose throughout the walls and along the joists. Many houses now are prewired for telephones, which eliminates the necessity of the telephone company cutting into the walls and floors after the house is complete.

Inter-communication and music systems are also becoming very popular. These systems also use low voltage, hazard-free wiring. Music at will throughout the house, answering the door without opening it, talking from room to room without screaming are all possible with this equipment.

The increase in crime has brought an increase in houses wired with burglar alarm systems. Some houses also have fire warning systems. And for the rich and the nostalgic there is the button under the dining room table that, when stepped on, summons the maid from the kitchen and a more complicated set of buttons throughout the house that summons the butler from the pantry.

Inadequate Wiring

The problem of inadequate wiring starts with inadequate voltage and amperage coming into the house. A minimum 220 to 240 volts and 100 amperes should be supplied and more if the house is large, has many major electric appliances (especially ranges and clothes dryers) or has electric heat and air-conditioning.

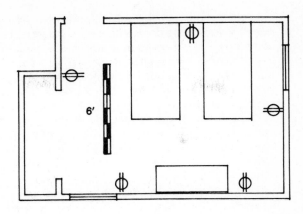

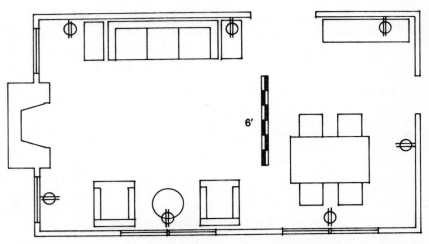

Notice that no point on usable wall space is further than six feet from an outlet.

MINIMUM FHA-MPS CONVENIENCE OUTLETS

Lack of sufficient branch circuits to the various appliances and rooms of the house can be corrected by installing a bigger distribution panel and additional wiring. Fuses being used that have higher ratings than called for is a sign that the wiring is under the needed capacity. Insufficient wall outlets in rooms leads to the use of dangerous extension cords and monkey plugs. Lack of outside outlets is an inconvenience.

Knob and tube wiring must be viewed very suspiciously. It is often old and its insulation has a tendency to crack with age, leaving exposed wires which are very dangerous.

Outlets

The duplex receptacle was, until 1960, the most common type of household outlet used. It accepts a two-prong plug, the type most often found on lamps and small appliances. In 1960 the National Electric Code and MPS required that all receptacles be the grounding type designed to also accept a three-prong plug. Many small appliances are wired with a third ground wire that is attached to the frame or metal housing of the appliance. The third slot in a grounded outlet is connected to a water pipe or other grounding metal. Grounding of an appliance by using a three-prong plug and receptacle reduces its shock hazard.

There are special waterproof receptacles with caps for outside use, clock outlets, TV outlets, locking outlets and a variety of other special purpose outlets. The receptacle for a 220 to 240-volt line is designed to accept only special plugs. A standard two or three-prong plug cannot be plugged into it. It is also so designed to accept only plugs for appliances using only the exact number of amperes it will supply.

Outlets should be conveniently located throughout the house. MPS requires that three-prong duplex grounding outlets be installed in all habitable rooms so that no point along the floor line is more than 6 feet from an outlet. It also requires that an additional outlet between all doors and between doors and a fireplace be supplied (unless the wall is too small for a piece of furniture); that in rooms without permanent light fixtures at least three outlets be provided regardless of the room size; that two outlets be installed in the kitchen over the counters and that an outlet be installed next to the mirror in the bathroom.

Switches

Wall switches are used to control permanently installed light fixtures and may also be used to control wall outlets. Rooms without permanent light fixtures are becoming very common as ceiling fixtures in some rooms continue to go out of style and lamps gain in popularity. However, it would be inconvenient to have to walk into a dark room to turn on a lamp. A wall switch near the door that controls the wall outlet into which the lamp is plugged eliminates this problem. It is also convenient to control fans, garbage disposals and some other

322

small appliances by a wall switch rather than by the switch on the appliance itself.

The simplest and most common switch is a two-way snap switch which has two copper contact points inside. When the switch is up in the "on" position a spring snaps the copper contact points together and lets the electricity pass to the fixture or outlet. When the switch is flipped down to the "off" position the spring pulls apart the copper contacts and the electricity is turned off.

Three-way switches are used to control a fixture or outlet from two different places. With this wiring and switching arrangement, the fixture or outlet is turned from "off" to "on" or "on" to "off" when the position of either switch is changed; therefore the up position is not necessarily the "on" position or the down position the "off" position. This arrangement is very useful for a stair light, so that it can be controlled from the top or bottom of the stairs. Other good places are in rooms with two doors, for garage lights, outside lights and bedroom lights (a switch at the door and another at bedside).

Two and three-way switches are also made with silverplated contacts. Because silver is an excellent conductor, less pressure is required at the contact points and the switch is much quieter. A completely silent type switch uses mercury in a tube to open and close the circuit. The only disadvantage of a silver contact or mercury switches is their higher cost. They are often a sign that someone was willing to pay for quality.

Some other, more expensive switches are those with lighted handles (some glow all the time and others only when the switch is in the off position). There are also switches that turn on and off when a door is opened or closed (often used in closets), key controlled switches, pull chain switches, outside weatherproof switches, touch switches, time delay switches and a variety of other specialized switches.

Some houses are controlled by a low voltage switching system. Instead of the switch directly opening and closing the circuit, it controls a relay which in turn operates the switch. The advantage of this system is that control panels, located thoroughout the house, can control many lights and outlets from one place. These control panels are often located at the main entrance or in the master bedroom.

Dimmer switches are used to vary the intensity of the light while leaving the same number of fixtures on rather than by the old way of just turning some fixtures off. This more sophisticated method allows much better light distribution and better decorative effects.

A good indication of an adequate switching arrangement is being able to walk anywhere in the house and turn on a path of light and then turn off the lights without having to retrace steps or walk in the dark. This includes getting in and out of bed and entering and leaving the house by both entrances and through the garage. Some houses have additional switches to control light at remote locations, such as a switch in the master bedroom controlling the outside lights.

Artificial Lighting

Studies continue to show that only a small percentage of houses have adequate lighting or the wiring necessary to add the necessary lamps and fixtures to correct this deficiency. However, the public is becoming more and more aware of what adequate lighting is and all of its advantages. As more and more people demand this, houses without it will become obsolete.

Most home lighting is provided by either incandescent bulbs or fluorescent tubes. Each has very different light characteristics.

The amount of light a bulb emits is measured in lumens. In general, the more watts of power a bulb uses, the more lumen or light it gives out. To get an idea of how much light a lumen is, think of a candle, which gives off between 10 to 15 lumens. A frosted 100 watt bulb emits about 1,600 lumens (16 lumens per watt) when it is new and getting sufficient electricity. One hundred watts of fluorescent tubing will emit about 7,000 lumens (70 lumens per watt). More than four times as much light from fluorescent tubes is available than from incandescent bulbs using the same amount of electricity.

The amount of lumens required varies according to activities carried on in the room. The following is a guide to the amount of lumens needed in each room according to a recent study: kitchen, 80 lumens per square foot; bathroom, 50 lumens; bedrooms, 35 lumens; living room, 40 lumens; family room, 50 lumens; dining room, 40 lumens and hallways, 30 lumens.

It is easy to figure the amount of bulb watts required to adequately light a room. For example, a bedroom 10 by 12 feet has 120 square feet which when multiplied by 35 lumens per square foot indicates a need for 4,200 lumens of light. When divided by 16 lumens a watt, the answer indicates that about 260 watts of bulbs are needed. There is a difference in the number of lumens given per watt depending upon the size of the bulb but not enough to make a significant difference. The same room can be adequately lit with about 60 watts of fluorescent tubing, as indicated by dividing the needed 4,200 lumens by 70 lumens per watt produced by the fluorescent tubing. It is apparent that a fluorescent light gives a good deal more light for the same amount of watts used. However, fluorescent fixtures are much more complex and expensive.

The trend is to build more lighting into the house in the form of recessed fixtures using fluorescent tubes, spotlights, floodlights and other specialized fixtures that each produce a distinct pattern of light best suited to a specific area.

Bathrooms require special lighting. When the mirror is used for shaving or makeup application, both sides of the face and the area under the chin should be well-lit without shadows. The light should be bright but not so bright that pupils contract or squinting results from the glare. The color balance should make skin look lifelike.

There should also be additional general lighting in the bathroom and in the shower stall. Pull chains and switches located in areas where they can be reached from the tub should be avoided since they present a shock hazard.

It is beyond the scope of this book to go into greater detail about house lighting. Suffice to say that the entire interior appearance of the house is affected by the use of artificial light and that adequate light is important to a family's health and happiness.

Chapter 8
Materials
Manufacture

In addition to knowing how a house is constructed and how its mechanical systems operate, it is also valuable to know the methods of manufacture of the raw materials and components that go into house construction. Every component part of a house has a complex manufacturing process. In addition, most building materials and supplies are graded for quality according to industry regulations. These standards are outlined in this chapter along with brief descriptions of manufacturing processes.

Wood

A tree grows by adding a new layer of cells directly under the bark each year. Each year's growth is distinguishable from the previous year's, which is what gives wood its grain. When a branch develops in the tree and grows through the grain, a knot composed of rings of grain running in small circles is formed. Sometimes a separation, called a shake, develops between the layers of grain. When the grain separation becomes filled with pitch (liquid or solid resin), it is called a pitch pocket. Knots, shakes and pitch pockets are considered imperfections in the wood and when wood is graded, their presence results in a lower grading for the wood.

The chemical composition of wood is about 70 percent cellulose. The material that binds the cellulose together is called lignin and makes up most of the balance of the wood except for the small remainder of materials distinctive to each species that give individual color, odor and decay resistance.

An important property of wood is that it is hydroscopic (absorbs moisture). It expands when it absorbs moisture and shrinks when it dries out, a characteristic that must be considered whenever wood is used. In the tree, the wood is in a wet (green) condition, but when wood is used as a building material, it must be dry. When the tree is felled and cut into lumber its moisture content begins to drop. Eventually the lumber will reach a point of equilibrium at which the moisture in the wood will equal the moisture in the air. This ensures that the wood will shrink or expand only a small, tolerable amount after it is fabricated or installed.

Lumber

The manufacturing process of lumber begins with felling the tree in the forest and transporting it to one of more than 30,000 sawmills. There the logs are first debarked by equipment that peels, scrapes or blasts the bark off the log with high pressure water jets. Next, the log is fed into the head saw, which saws it into rough pieces called cants, timbers, planks or boards. Then these pieces are fed into a series of smaller saws called edgers which remove the rounded edge and rip wide pieces into narrower widths. The pieces go into the trimmer saws which trim off the round ends and end defects and cut the wood into the desired lengths. The next step is the green chain where the pieces are sorted (usually manually) according to grade, species and size as they move along a conveyor belt. Once the lumber is sorted some of it may be dried and the rest may be sent without drying directly to manufacturing plants for resawing, dressing, ripping, planing or other treatment.

The best way to reduce the moisture content of lumber in the shortest time is to dry it in a kiln, a large building in which the heat, humidity and air circulation are controlled. Kiln drying can reduce the moisture to any desired percentage, usually 6 to 12 percent for softwoods and 15 to 19 percent for hardwoods.

Lumber may also be dried by letting it stand outdoors or indoors in an unheated building or shipped undried to the lumber yard where it is air dried. The use of unseasoned lumber or improperly dried lumber results in poor construction caused by the shrinking and bending of the wood after it has been installed.

After the lumber is seasoned it is sent to a planing mill where the surface is planed to make it smooth and the lumber is further sawed. The amount of planing and sawing varies. Little is needed to produce rough boards, dimension stock and timbers, but more planing is necessary for the finished grades of these items.

Some of the lumber is further processed or "worked." It may be matched for use where tight joints are needed between boards as in sheathing and subflooring. A tongue is cut into one side of the board and a groove into the other. When the board is installed the tongues are fitted into the grooves.

Shiplapping is another cut that will produce a tight joint but not as good as the one produced by matched lumber. Shiplapping is also used for subflooring, wall and roof sheathing or siding. Each edge is rabbeted and, when installed, the edges overlap.

Patterned lumber is a type of lumber that is cut into various shapes and is used for siding and mouldings.

Lumber is graded and sized according to American Lumber Association standards. The standards vary from species to species. Each individual board is stamped with its size, grade and manufacturer. Therefore, a builder may easily obtain the proper grade of lumber for each part of the house.

Plywood

Plywood is manufactured from several thin sheets of hardwood or softwood (veneer) bonded together or to a core of lumber or particle board.

Manufacturing starts with the selection of large peeler logs, 8 feet long and 1 to 4 feet or more in diameter, cut by a giant lathe into continuous sheets from $\frac{1}{10}$ to $\frac{3}{16}$ inch thick. An alternate process is to slice the sheets from large blocks of wood. The sheets are cut to the desired widths by a clipping machine and then are oven-dried to reduce their moisture content so that the glue will bond properly.

The sheets are then sorted in various grades depending upon the number of knots and other imperfections.

The final panel is made by spreading glue on both sides of the veneer and core, laying them on top of each other and putting them into a hydraulic press where, under pressure, they are bonded into plywood panels.

Shingles and Shakes

Manufacture of shakes starts with the selection of cedar logs cut to the length of the finished product, 18, 24 or 32 inches. The logs are then quartered into blocks that will fit the splitting machines or, if they are to be handsplit, cut into convenient, workable sizes. Most shakes are split by machines, although the trade of splitting shakes by hand with a hardwood mallet and a steel froe still thrives.

The three basic types of shakes made are tapersplit, straight split and handsplit and resawn. Taper split shakes are produced by turning the block over after each shake is split off, while straight split shakes are cut from the block without turning it. Handsplit and resawn shakes are produced by splitting blocks into boards of desired thickness and then passing them through a thin bandsaw to form two shakes, each with a handsplit face and a sawn back. When they are guided through the saw diagonally thin tips and thick butts are formed. Packers then bundle the finished shakes in standard size frames, compressing the bundles slightly and binding them with wooden "bandsticks" and steel strappings.

The manufacture of a shingle begins like that of a shake, with the selection of logs, cutting them to shingle length and then quartering them into blocks that will fit into the sawing machine. The sawing machine has two blades, one that cuts the shingle from the block and the other that trims the edges. After being sawed, the shingles are graded and bundled like shakes. They are held flat in the bundles by two bandsticks and metal straps while they are stored and seasoned.

Standard shingles may be from 3 to 14 inches wide. Dimension shingles have one equal width. There are also complex standards for thickness which vary with the length. Shingles with extreme cross grain are not permitted in any grade. There are four grades of shingles. The best is No. 1 (blue label), most often used for roofs and sidewalls. They must be 100 percent heartwood, 100 percent clear and 100 percent edge grain. No. 2 (red label) is a good grade with some imperfections but usable in most places No. 1 can be used. No. 3 (black label) is a utility grade for economy applications. No. 4 is suitable only for undercoursing.

Concrete

Concrete contains portland cement, water, fine and coarse aggregates and admixtures.

Cement Manufacture

Portland cement is a compound of lime, silica, alumina and iron. The materials are carefully selected and the manufacturing process closely controlled. The materials these compounds are obtained from are limestone dug from a quarry, oyster shells, clay, marl (a rock that contains limestone), iron ore, silica sand and blast furnace slag.

These ingredients are mixed in proper proportions. They are then ground up by either a wet grinding or dry grinding process. The ground up mixture is fed into a rotary kiln that burns it at a temperature of approximately 2,700 degrees. At this heat the ingredients combine into clinkers. The clinkers are cooled, combined with a small amount of gypsum and pulverized into a very fine powder which is then packed in paper bags or shipped in bulk in special railcars and trucks.

Cement Quality Standards and Grades

Manufacturers of portland cement market it under their own trade and brand names. Specifications have been established by the American Society for Testing and Materials (ASTM). Practically all cement is made to these specifications.

Normal portland cement Types I and IA are suitable for almost all residential construction purposes. In areas where there is a large amount of sulfates present, Types II and IIA should be used since they are more resistant to sulfate attack. Types III and IIIA are used when fast setting is required, such as in cold weather construction. Type IV is not used for residential construction. Type V is used mostly in the West where the soil and water subject the cement to very high sulfate attacks. The "A" after the type number indicates that it is resistant to frost action and to the effect of salt application, an advantage in the North where salt is used for snow and ice removal.

Aggregates

Fine and coarse aggregates make up about 75 percent of the final concrete mixture. Fine aggregate consists principally of sand. The type of sand best suited for concrete manufacture ranges from pieces that are very fine to pieces up to ¼ inch in size. It is important that the range of sizes be even. Coarse aggregate consists principally of crushed stone and gravel larger than ¼ to 1½ inches. The maximum size should depend on the final use of the concrete.

Aggregates should be sound, hard and durable. If soft and flaky aggregates are used, the concrete is weakened. They must also be free of loam, clay and other vegetable matter as these too will weaken the concrete.

Water

The water used to make concrete must be free from oil, alkali or acid. In general, water that is suitable for drinking is suitable for use in concrete. Water that contains high quantities of sulfates should not be used since sulfates attack cement.

Admixtures

Admixtures are materials that are added to the concrete mixture to produce special qualities such as improved workability, reduction of the separation of coarse and fine aggregates, entraining of air or acceleration or retardation of setting or hardening. They must be used carefully as they may, in addition to producing the desired effect, also produce undesirable effects. There are three frequently used admixtures: accelerators, retarders and air-entraining agents.

Accelerators, usually calcium chloride, increase the rate of early strength development. These may be needed to reduce the waiting time for the finishing operations to start, allow earlier removal of the forms, reduce curing time, lengthen the time that the structure can be used, offset effects of cold weather in slowing up setting time and perform emergency repairs when delay of use causes problems.

Retarders, a wide variety of chemicals, delay the early stiffening action. They are often used in hot weather and in difficult installations where more than normal installation time is required.

Air-entraining agents are mixtures that improve the workability and durability of concrete and increase the concrete's resistance to damage from frost action and the damaging effects of salt. Their use is so common and their cost so low that they are being added during the manufacturing process. There is little excuse not to use them for all residential installations.

Mixtures

The goal of a proper concrete mixture is to surround completely each particle of aggregate with the cement water paste. This will produce the best workability, strength, durability, watertightness and wear-resistance.

The proportion of water to cement, stated in gallons of water per 94-pound sack of cement, is the water-cement ratio. The more water that is added, the less the strength, durability and watertightness of the cement. Therefore, just enough water should be used to produce the required workability. The longer the mixture remains moist after it is poured, the stronger it becomes. It may be kept moist by covering it and spraying it with water.

It is beyond the scope of this book to go into the details of how a proper mix is made for each job. The type of job must be considered along with the size and moisture content of the aggregate, temperature and humidity. Usually a small trial batch is made to verify that the proposed mixture is correct.

A slump test may be used as a rough measure of the consistency of concrete. This is done by filling a specially designed cone with the mixture. The cone is filled in stages and "puddled" with a metal rod a prescribed number of times between each stage. The top is struck off (smoothed) with a trowel and the cone is gently removed immediately after being filled. The amount that the top of the pile drops from its original height is then measured.

Another test is used to measure the compression strength of the concrete. At regular intervals throughout the discharge of the concrete mixture, samples are taken, poured into a cone, puddled and left to set. The strength of the mixture is then measured.

All concrete ingredients are mixed until they are uniformly distributed. This may be done by hand or in mechanical mixers on the site. However, the use of ready-mixed concrete is increasing. It is mixed in a special plant and hauled to the site in agitator trucks or transit mixer trucks operated at agitator speed. In any case the cement must be discharged from the truck within 1½ hours after the water has been added to the mixture.

Clay Brick and Tiles

Manufacture

Winning (mining) is the first step in the brick and tile making process. Clays are dug from the open pits or mines with power shovels, crushed and put in storage bins. To minimize variation in the properties of the clays, clays from different areas of the pit or mine and from different mines and pits are blended together.

After the stones are removed, the mixture is pulverized on huge grinding wheels weighing up to 8 tons each. Usually the mixture is then passed through vibrating screens to further control the particle sizes. In a pug mill, water is added to the clay and huge blades on a revolving shaft mix the clay and water together into a homogeneous mixture.

The mixture is then formed into bricks or tiles by one of three processes. The stiff mud process is the most common. During this process, just enough water is added in the pug mill to make a mixture of plastic consistency. The mixture goes into a de-airing machine which has a vacuum inside it. This vacuum removes many of the air holes and bubbles in the clay. The mixture is then forced through an extrusion die which forms it into bars the final width and shape of the brick or tile. Automatic cutters cut the bars into bricks or tiles which are larger than the final product in order to allow for shrinking during drying and burning. Textures, if any, are also stamped or scratched on at this point.

The soft mud process is used mostly when the clay is too moist for the stiff mud process. The clay is poured into molds, the sides of which have been lubricated with sand (bricks called sandstruck) or water (bricks called waterstruck). This process is not used for tiles.

The dry-press process is used for dry clays. The bricks or tiles are formed by putting the clay into steel pressure molds which press the clay into the desired size and shape.

After the bricks or tiles are shaped they are dried in kilns at temperatures from 100 to 300 degrees for one to two days. An optional step either at this point or after the next step is glazing, which produces a glasslike colored coating on the exterior surface. High-fired glazes are sprayed on the units before and after the preceding drying process. The bricks are then completed in the normal way: low-fired glazes are applied and then the bricks are refired at lower temperatures.

Burning, the next process, hardens the clay and water mixture. The units are put into a tunnel kiln or a periodic kiln. In a tunnel kiln the bricks move along from one stage of the process to the next in special cars, whereas in a period kiln they remain in place.

Burning takes place in stages. In the first water-smoking stage, any free water is evaporated at a temperature of about 400 degrees. The temperature is then raised to 1,800 degrees during the dehydration and oxidation stages and finally to about 2,400 degrees in the vitrification stage. Near the end of the burn, the units may be flashed to produce different colors and shades by reducing the oxygen in the kiln.

After the temperature has reached its final height, it is slowly reduced and the cooling process begins and lasts from two to four days. The kiln is then unloaded (called drawing) and the bricks or tiles are sorted, graded and either stored or shipped by rail or truck.

Concrete Brick and Block

Manufacture

Raw materials are received and elevated to storage bins as the first step in manufacture of concrete bricks and blocks. Then they are released into a weight batcher under the storage bins which regulates the amount of each ingredient by weight. The weight batcher is moved over the mixer and the ingredients are released into the mixer together with the correct amount of water. After mixing, the ingredients are discharged into a hopper above the block machine and controlled quantities are fed into the machine. The ingredients are compacted into the mold with pressure and vibration.

From the block machine emerge green concrete bricks or blocks molded onto steel pallets which are placed on a steel curing rack. The curing rack is moved into an autoclave or steam curing kiln.

Kiln curing begins with a holding period of one to three hours inside the kiln at normal temperatures. The heating starts when saturated steam or moist air is injected into the kiln and the temperature is gradually raised over several hours to about 180 degrees. When the desired heat is reached, the steam is turned off and the units soak for 12 to 18 hours while their strength develops. Next they are dried as the temperature continues to drop. The drying process may be speeded up by again raising the temperature, this time with dry heat. The total time for curing and drying is about 24 hours.

High pressure steam curing is performed in autoclaves that are steel cylinders from 6 to 10 feet in diameter and 50 to 100 feet long. The green molded units are put into the autoclave and held at normal temperatures for two to five hours. Saturated steam is forced into the autoclave at about 150 pounds per square inch of pressure. The only purpose the pressure serves is to hold the steam at the high temperature. The temperature is raised gradually over about a three-hour period until 350 degrees is reached. Temperature and pressure are maintained from five to ten hours while the units soak. The pressure is then released over about a half hour period, which facilitates rapid loss of moisture from the block without setting up shrinking stresses.

It is also possible to cure the green units by keeping the moisture at normal temperatures for about a month without the use of an autoclave or kiln.

After the curing is completed in the kiln or autoclave the units are taken to the cubing station where they are assembled into cubes and then stored until they are shipped. Blocks are usually made locally and therefore are almost always shipped by truck.

Mortar

Straight lime mortar, made of lime, sand and water, hardens at a slow, variable rate. It develops low compressive strength and has poor durability to the freeze-thaw cycle. However, it does have high workability and high water retention. It tends to self-mend small cracks that appear, which helps keep water penetration to a minimum.

Portland cement mortar, made of portland cement, sand and water, has the opposite characteristics of straight lime mortar. It hardens at a quick, constant rate, develops high compressive strength, has good resistance to the freeze-thaw cycle but has poor workability and low water retention and does not self-mend small cracks.

Portlant cement-lime mortar, made of portland cement, lime, sand and water, combines the characteristics of the above two mortars and makes a much more satisfactory mortar for general use.

Masonry cement mortar is made of masonry cement, sand and water. Masonry cement consists of portland cement, natural cement, finely ground limestone, air-entraining agents and gypsum. This is a convenience item, since these ingredients are premixed by the manufacturer rather than at the site.

Glass

Most house glass is a mixture of silicone, usually in the form of silica obtained from beds of fine sand or from pulverized sandstone, soda used as an alkali to lower the melting point, lime as a stabilizer and cullet (waste glass) to assist in melting the mixture. These are the same materials that have been used from ancient times. Small amounts of other materials are added to produce special types of glass.

The ingredients are mixed together and then melted in tank furnaces with capacities as high as 1,500 tons or in pot furnaces that hold up to 2 tons. The glass is released from the furnaces and rolled into sheets, cooled and cut into the desired sizes. Plate glass is polished on one or both sides.

Glass used to be made by blowing large bubbles, cutting them and allowing the half spheres to flatten out on a hot bed. Making large panes (called "lights" in the building trades) was almost impossible, which explains why most windows in colonial houses consisted of many small panes of glass.

Resilient Flooring Products

Linoleum

True linoleum is made of oxidized linseed oil, resin binders, wood floor pigments, mineral fillers and ground cork. The ingredients are mixed together and bonded to an organic backing of burlap or asphalt-saturated rag felt. Hot linseed oil is placed in closed cylinders where it solidifies into a jelly-like mass. Next, the oxidized oil is fused with the resins into a tough cement that is blended with the other ingredients and onto the backing material. It is seasoned in ovens for several weeks to cure it and then waxed or lacquered to improve its stain resistance.

Asphalt Tile

Asphalt tile is made of asbestos fibers, finely ground limestone fillers, mineral pigments and asphaltic or resinous binders. Asphaltic binders are made from a blackish, high-melt asphalt, mined principally in Utah and Colorado. Resinous binders are made from asphalt which is a product of petroleum cracking or from coal tars. The ingredients are mixed at a high temperature and rolled into sheets that are sprinkled with colored chips. The sheets are cooled and cut or stamped into tiles. Sometimes they are prewaxed.

Rubber Tile

Rubber tile is made of natural or synthetic rubber, clay and fibrous talc or asbestos fillers, oils or resins and nonfading organic or chemical color pigments. The ingredients are mixed, rolled into sheets and then vulcanized in hydraulic presses under heat and pressure. The sheets are sanded on the back to uniform thickness and cut into tile.

Cork Tile

Cork tile is composed of the bark of the cork tree, found in Spain, Portugal and North Africa, and synthetic resins. The cork is granulated, mixed with the resin and pressed into sheets or blocks and baked. It is then cut into tiles. Often wax, lacquer or resin is applied under heat to the surface to provide a protective coating.

Vinyl Asbestos Tile

Vinyl asbestos tile is made of asbestos fibers, ground limestone, plasticizers, color pigments and polyvinyl chloride as a binder. The ingredients are mixed together under heat and pressure and rolled into blankets to which decorative chips are added. Additional rolling produces the desired thickness. The sheets are waxed and cut into tiles.

Solid Vinyl Tiles

Solid vinyl tile is made of polyvinyl chloride resin, mineral fillers, pigments, plasticizers and stabilizers. The ingredients are mixed at high temperatures, hydraulically pressed or pressure-rolled into sheets of the required thickness and cut into tiles.

Asbestos and Rag-Backed Vinyl Tiles and Sheets

These products are made with a vinyl wear surface bonded to a backing of vinyl, polymer-impregnated asbestos fibers or asphalt-saturated or resin-saturated felt. New vinyl products have a layer of vinyl foam bonded between the backing and the wear surface. The process expands the decorative possibilities substantially, making possible tiles that look like brick and many other materials. These tiles are very attractive but tend to be quite expensive.

Carpeting

Woven carpet is made on a loom by one of four weaving processes known as velvet, Wilton, Axminster and loomed. Velvets are the most formal weave and also the simplest. The Wilton loom is a complex form of the velvet loom fitted with a special Jacquard mechanism which uses punched cards to select various colors of yarn and to make the complex pattern. The Axminster loom simulates hand weaving because each tuft of yarn is individually inserted and theoretically each tuft could be a different color. This flexibility offers unlimited design possibilities. Loomed carpet has recently been developed for use with bonded rubber cushioning.

Tufted carpet is made by inserting tufts of pile into a premade backing material that is then coated with a layer of latex to hold the tufts permanently in place. The tufting machine has thousands of sewing machine-like needles that operate simultaneously.

Asphalt Roofing and Siding Products

Manufacture

Usually asphalt roofing and siding products are made by a continuous process. The felt rolls are put on a dry looper which provides enough slack so that one roll can be attached to the next without stopping the machine. The felt goes through a saturator tank filled with asphalt that soaks into the felt. Next, a wet looper holds the saturated asphalt and allows it to shrink. For products that require it, the next step is a coating machine. The machine applies a thin coating of asphalt over both sides of the saturated felt. The thickness of the coat is controlled by rollers and automatic scales.

When smooth roll roofing and siding is being made, the surface is coated with talc or mica and rolled into the surface by pressure rollers. When mineral surfaced products are made, the mineral surfacing materials, which have been stored in hoppers, are spread over the hot asphalt-coated surface and run through a series of pressure and cooling rollers. To make shingles, the coated sheets are cut, stacked and packaged. When rolled materials are made, the coated sheet is rolled on rollers, cut and packaged.

Gypsum Products

Gypsum rock is mined and crushed in a hammer mill to a ½-inch particle size. It is calcinated in a rotary steel, 150-foot cylinder lined with fire brick or a kettle calciner by heating it to 350 degrees. When aggregates are used, they are washed, crushed and graded. Vermiculite and Perlite aggregates are heated to 2,000 degrees to expand them. When lime is used it requires special processing. It is calcinated and slaked (combined with water).

Plaster is manufactured by taking the gypsum from the kettle calciner and treating it in a tube mill, a rotary cylinder containing thousands of steel balls which grind up the gypsum. Retarders are added to control the rate of set which should be up to four hours. For fibered plasters, organic fibers are added and for ready-mixed plasters, lightweight aggregates are added. The plaster is then packaged and shipped.

Plasterboard manufacture is started by taking the gypsum from the calciner and mixing it with water to form a slurry, which is fed onto a conveyor belt between two layers of paper. It sets in a few minutes to a hardness that permits it to be cut into pieces and conveyed to kilns for drying. Special backings and finishes, such as aluminum foil, decorative vinyl or paper finishes are affixed and the boards are stacked and shipped.

Asbestos Cement Products

Manufacture

The first steps of the three common manufacturing processes (wet mechanical, dry and wet formed) are the same. In most plants it is a continuous process starting with the mixing of asbestos fiber, portland cement and water and forming the mixture into a plastic sheet. The sheet is hardened, trimmed to final size, punched with nail holes and cured. Colors are either mixed in during the mixing stage or coated onto the hardened sheet. Textures are embossed on the sheet while it is still plastic.

The most commonly used wet mechanical process is similar to paper making. The ingredients are dry mixed, after which water is added to form a thin slurry. The slurry is picked up by wire mesh rollers, deposited on a continous carrier blanket and then transferred to an accumulator roll where the thickness is built up to the required amount. The sheets, now in a plastic form, are removed and processed as described above. This process produces a built-up, laminated sheet with asbestos fibers aligned in one direction, giving the sheet significantly greater strength in the direction of fiber alignment.

The dry process is used by some manufacturers to produce roofing and siding sheets that are homogenous in nature rather than laminated. These sheets have uniform strength in all directions since the asbestos fibers are not aligned in any one direction. The ingredients are dry-mixed and distributed in an even layer on a continuous moving belt. The layer of material is sprayed with water and compressed with steel rollers. The rest of the process is the same as described above.

Large flat sheets of asbestos cement are made by a molding process called the wet form process. The ingredients, including water, are mixed into a slurry and poured into the mold of a hydraulic press that forces out most of the water and compresses the mixture to the required density. The fibers are non-aligned and the sheet produced has uniform strength in all directions.

After the product has been formed by any of the previously described three processes, it must be cured. Like any product made with portland cement it will become harder and stronger with age and proper moisture conditions.

Normal curing is accomplished by stacking the finished products when they leave the production line in a curing room for approximately a month. The temperature and humidity in the curing room are carefully controlled. Steam not under pressure is sometimes used to provide the required warm, moist air.

Curing time may be reduced substantially by the autoclave curing (high pressure steam) method: after a few days of normal curing the material is put into an autoclave and subjected to high pressure steam.

Cabinets

Manufacture

A standard cabinet is made up of various components: front frames, end panels, door, backs, bottoms, shelves, drawers and hardware. The front frame usually is made of ½ to ¾-inch thick hardwood. The pieces, known as rails, stiles and mullions, are doweled or mortise-and-tenoned and glued together.

The end panels usually are made of ¾₆ to ¼-inch plywood. When thin pieces are used they are glued to a frame to give them stiffness, but when thick pieces are used no frame is necessary.

Doors are made a variety of ways. Some are solid wood or plywood, others are hollow or filled with particle board, wood, high density fiber or plastic. Some doors are faced with plastic; others are solid plastic. Backs are made of hardboard or plywood. The bottoms and tops are ¼ to ½-inch hardwood plywood. Shelves are usually boards, plywood or particle board from ½ to ¾ inch thick. Drawers often are made of hardwood lumber sides and backs and plywood bottoms.

The large cabinet-making factories are almost completely integrated operations. The raw material comes in and is completely processed into a finished cabinet. Only a few minor components are manufactured elsewhere. These plants include complete milling operations as well as assembly facilities. Other smaller cabinet-making factories only assemble parts which have been milled elsewhere. In the large plants, cabinets are made on a production line like automobiles. Each part is made in its own department and then fed simultaneously into the production line for final assembly.

The front frame is made from stiles, rails and mullions that have been cut previously and shaped in the mill section. They are assembled in an air press which holds them in place until they are glued and stapled. After being left to dry for a day, the frame is sanded and holes are drilled in it for hardware.

Doors

Manufacture of Wood Doors

Most doors used in houses are factory-made in standard sizes and with standard detailing. However, in very expensive houses, doors may be specially made to an architect's specifications. Builders today rarely make doors themselves, although in the past this was more common.

Wood doors may be made out of many hardwoods and softwoods. Most flush doors are made with hardwood veneered faces. Most stile and rail doors are made from ponderosa pine and other softwoods. Accordion folding doors are made from hardwoods and softwoods.

The stiles, rails and lock blocks for flush doors are cut from kiln-dried, surfaced lumber. The stiles and rails are joined by metal fasteners or dovetailed joints. The frame is then glued to the face panel and core. Hollow flush doors are pressed in a cold press to assure a good bond. Solid core doors are pressed in a hot press to assure proper setting of the glue. Additional optional operations on some doors include routing for louvers and glass openings and installing muntins and bars prior to glazing.

The kiln-dried lumber for stile and rail doors is surfaced and cut to exact width and length. Holes are bored in the stiles and rails to accept the dowels. A molding machine is used to form the sticking slot (the slot for the glass or panel) on the edge. The dowels are inserted into place and glued. In a separate operation the panels are cut, sanded and shaped to the desired size. The panels are then set into the sticking slot, glued with a water resistant adhesive and clamped in a jig until the glue has set. The door is then sanded to a smooth finish on both sides.

If the door has glass in it the glass will have been previously cut to the correct size. It is set in place with a glazing component and the stops which hold the glass are nailed in place.

Additional processes that may take place at the factory to reduce on-site preparation are coating the door with a seal coat or complete surface finishing. Usually, prefinished doors are prefit to exact opening measurements so that they will not have to be trimmed and their finishes marred. Sometimes the hardware is factory-installed.

Prehung doors, where the door is attached to the frame, are also factory made. The complete frame is then set into the wall frame or partition, thereby eliminating the difficult step of hanging the door at the site.

Manufacture of Aluminum Doors

The use of aluminum for sliding doors and combination storm doors and screens continues to increase. The principal materials used are wrought aluminum and aluminum screening. Cast pieces are used for the locks, hinges, handles and corners.

Doors are produced with natural mill finish, protective finish and decorative finish. The natural mill finish is a silver sheen when new which turns to a soft gray with age. In sea coast and industrial areas a protective finish known as anodizing is often applied. This electrolytic process provides a heavy oxide film on the surface which can be clear or contain permanent color tones. Decorative finishes of colored or opaque synthetic resin enamel are also used. They last up to 20 years but are not as permanent as anodizing.

Aluminum doors are fabricated by taking the extruded and cast stock and sawing, drilling and punching it to obtain the desired size and shape and then assembling it into the finished product including the glass, screens and hardware.

Windows

Manufacture of Wood Windows

In the past the vast majority of windows installed in houses were wood stock windows, which are still very popular. Wood has good insulating qualities and takes either natural or painted finishes. In addition, wood windows are easy to repair with simple tools.

Techniques of kiln drying the wood have reduced the shrinking and warping problem. Water repellent preservatives and chemical treatments further reduce swelling and warping, improve paint retention and increase the wood's resistance to decay and insect attack under all climate conditions.

The best wood for windows seems to be ponderosa (western) pine because of its excellent workability, gluability, nail-holding capacity and uniform light color suitable for natural or painted finishes. The wood must be kiln dried, sound and free from defects such as loose knots and checking.

The lumber is cut into the proper lengths and widths and then milled to the desired profile. A cross milling machine cuts the grooves and channels and routing machines cut notches for the hardware. The pieces are then treated to make them water repellent and decay resistant. Sometimes they are also factory paint-primed or even paint-finished. If they are to be weatherstripped, it is done at this stage of fabrication. The pieces are then assembled into the desired shape and nailed together on an automatic nailing machine. The completed sash is then sanded to a smooth finish by a three-step (coarse, medium, fine) sanding machine. The final steps consist of glazing, fitting the sash to the frames and installing the hardware.

Manufacture of Aluminum Windows

The use of aluminum windows continues to increase and they now account for about half of the stock house windows being produced today. Their popularity can be attributed to their ease of production, attractive appearance and long maintenance-free life.

The principal materials used in manufacturing these windows are wrought aluminum alloys which are formed by the extrusion process and then artificially aged by heat treatment (see section on aluminum products). The piece is extruded into the complex shape required for glazing, weatherstripping, condensation control and assembly. Cast pieces are used for the locks, hinges, handles and corners.

Most windows are produced with a mill finish which is a natural, silvery sheen. As the window ages, an aluminum oxide coating forms which dulls the surface to a soft gray and protects it from further corrosion.

Protective finishes are necessary only in industrial and sea-coast areas where the air carries excess corrosive materials. The most common protective finish used is an electrolytic process that provides a heavy oxide film on the surface known as anodizing. This process can leave a clear coating or it can contain permanent color tones. Other protective coatings occasionally used are paint and methacrylate type lacquer.

Decorative finishes of colored or opaque synthetic resin enamel are not as permanent as anodizing but will last up to 20 years when baked on at the factory.

Aluminum windows are fabricated by taking the extruded and cast stock and sawing, drilling and punching it to obtain the desired size and shape and then assembling it into the finished window, including the glass and hardware.

Appendix

THE CTS SYSTEM:
A UNIFORM METHOD FOR DESCRIBING HOUSES

Introduction

The CTS System was developed to fill a need that has long plagued the real estate industry. An examination of the forms used by the government, both local and federal, lending institutions and REALTORS® throughout the country confirms that there is no uniform system to describe and classify houses. The lack of such a system has created a series of problems. For example it is difficult for multiple listing systems to classify houses for easy retrieval by their members.

A salesperson may have a buyer who seeks a "colonial style" home which by the buyer's definition is a modern house that looks like a house built in the early part of our history. Unfortunately in many areas the word "colonial" is used to describe almost any two-story house. Therefore, the salesperson must screen many more listings than would be necessary if the word "colonial" were applied only to a colonial American style house. If the salesperson is using a mechanical or computer sorting system, there is no simple way to know if the "colonials" selected are the style the buyer is seeking.

Tests conducted among REALTORS® revealed that when they were given a group of pictures from a multiple listing book and asked to describe the house, over 50 percent of the answers given differed substantially from the description supplied by the listing broker.

The use of a uniform system will also aid in appraising homes because appraisers like to compare houses by the number of dollars per square foot. When homogeneous houses are compared the value prediction will be more accurate.

When mass data is collected and analyzed by assessors or others using the new computerized regression analysis, identification of similar houses will improve the results of the analysis.

The CTS System

The CTS System was developed in response to these needs. This system provides a precise method of describing houses. The description is divided into three parts: Class, Type and Style.

A uniform narrative description, standard abbreviation and computer code are provided for each part of the system. Which of these will be used will depend on how the CTS System is being utilized. If the materials will be read by the general public, then narrative descriptions should be used. For multiple listing systems and internal records the standard abbreviations will be appropriate. When the data is to be stored in a computer then the computer codes should be used.

How the System Works
Step 1: Identify the Class

344 Class is used to denote the number of families the unit being described will accommodate and whether the units are detached or have a party wall.

For example, a one-family town house would be described "one-family, party wall." This same description would also describe a single unit in a 20-story condominium highrise.

Another example is a two-family house on its own lot which would be described as a "two-family, detached."

If the house being described was a fourplex or quadro and the title included all four units, it would be described as "four-family, party wall." On the other hand, if the title was to only one of the four units, it would be described as "one-family, party wall."

All houses can be classified into one of nine basic classes.

Code	Description	Abbreviation
1	One-family, detached	1 FAM D
2	Two-family, detached	2 FAM D
3	Three-family, detached	3 FAM D
4	Four-family, detached	4 FAM D
5	One-family, party wall	1 FAM PW
6	Two-family, party wall	2 FAM PW
7	Three-family, party wall	3 FAM PW
8	Four-family, party wall	4 FAM PW
9	Other	OTHER

The system in its present form does not include apartment houses except single units in apartments that are to be sold individually in condominium or cooperative forms of ownership.

Step 2: Identify the Type

Type refers to the structural nature of the house. Chapter 2 explains and provides illustrations of the nine basic types of houses. All houses can be classified into one of these nine types.

Code	Description	Abbreviation
1	One-story	1 STORY
2	One-and-a-half story	1½ STORY
3	Two-story	2 STORY
4	Two-and-a-half story	2½ STORY
5	Three-or-more stories	3 STORY
6	Bi-level	BI-LEVEL
6	Raised ranch	R RANCH
6	Split entry	SPLT ENT
7	Split-level	SPLT LEV
8	Mansion	MANSION
9	Other	OTHER

Step 3: Identify the Style

Style refers to the decorative design of a house based on historical or contemporary fashions. Chapter 4 explains and provides illustrations and information about 58 different style houses found in America.

All houses can be identified by one of nine style groups:

Colonial American
English
French
Swiss
Latin
Oriental
19th Century American
Early 20th Century American
Post World War II American

However, some houses are fine examples of specific historical styles. The owner often knows about the style and the REALTOR® demonstrates professionalism by recognizing and using the correct style name.

For example a REALTOR® in New Mexico will sooner or later list and sell an Adobe style home; a California REALTOR® a Monterey style; a Georgia REALTOR® a Southern Colonial and a New Jersey REALTOR® a Garrison Colonial. The only professional way to describe these houses is by their correct name. Here is a list that classifies all the house styles found in America into their nine broad categories and their specific style descriptions.

Code	Description	Abbreviation
100	COLONIAL AMERICAN	COL AMER
101	Federal	FEDERAL
102	New England Farm House	N E FARM
103	Adams	ADAMS CO
104	Cape Cod	CAPE COD
105	Cape Ann	CAPE ANN
106	Garrison Colonial	GARR CO
107	New England	N E COL
108	Dutch	DUTCH CO
109	Salt Box	SALT BOX
109	Catslide	CATSLIDE
110	Pennsylvania Dutch	PENN DUT
	Pennsylvania German Farm House	GER FARM
111	Classic	CLASSIC
112	Greek Revival	GREEK
113	Southern Colonial	SOUTH CO
114	Front Gable New England	F GAB NE
114	Charleston	CHARLES
114	English Colonial	ENG COL
115	Log Cabin	LOG CAB

200	ENGLISH	ENGLISH
201	Cotswold Cottage	COTSCOT
202	Elizabethan	ELIZ
202	Half Timber	HALFTIM
203	Tudor	TUDOR
204	Williamsburg	WILLIAMS
204	Early Georgian	E GEORG
205	Regency	REGENCY
206	Georgian	GEORGE
300	FRENCH	FRENCH
301	French Farm House	FR FARM
302	French Provincial	FR PROV
303	French Normandy	FR NORM
304	Creole	CREOLE
304	Louisiana	LOUISIA
304	New Orleans	NEW OR
400	SWISS	SWISS
401	Swiss Chalet	SWISS CH
500	LATIN	LATIN
501	Spanish Villa	SP VILLA
502	Italian Villa	IT VILLA
600	ORIENTAL	ORIENT
601	Japanese	JAPAN
700	19th CENTURY AMERICAN	19th CTY
701	Early Gothic Revival	E GOTH
702	Egyptian Revival	EGYPT
703	Roman Tuscan Mode	RO TUSC
704	Octagon House	OCTAGON
705	High Victorian Gothic	HI GOTH
706	High Victorian Italianate	VIC ITAL
707	American Mansard	MANSARD
707	Second Empire	2nd EMP
708	Stick Style	STICK
708	Carpenter Gothic	C GOTH
709	Eastlake	EAST L
710	Shingle Style	SHINGLE
711	Romanesque	ROMAN
712	Queen Anne	Q ANNE
713	Brownstone	BROWN S
713	Brick Row House	BR ROW
713	Eastern Townhouse	E TOWN
714	Western Row House	WEST ROW

714	Western Townhouse	W TOWN
715	Monterey	MONTEREY
716	Western Stick	W STICK
717	Mission Style	MISSION
800	EARLY 20th CENTURY AMERICAN	EARLY 20C
801	Prairie House	PRAIRIE
802	Bungalow	BUNGALOW
803	Pueblo	PUEBLO
	Adobe	ADOBE
804	International Style	INTERNAT
805	California Bungalow	CAL BUNG
900	POST WORLD WAR II AMERICAN	POST WW2
901	California Ranch	C RANCH
902	Northwestern	NORTH W
902	Puget Sound	P SOUND
903	Functional Modern	FUN MOD
903	Contemporary	CONTEMP
904	Solar House	SOLAR
905	"A" Frame	A FRAME
906	Mobile Home	MOBILE
907	Plastic House	PLASTIC

Here is how these four common houses would be described using the CTS System.

	Class	Type	Style
NARRATIVE:	One-family, detached	one-story	Adobe
ABBREVIATION:	1 FAM D	1 STORY	ADOBE
COMPUTER:	1	1	803
NARRATIVE:	One-family, detached	two-story	Southern Colonial
ABBREVIATION:	1 FAM D	1 STORY	SOUTH CO
COMPUTER:	1	1	113
NARRATIVE:	Two-family, detached	two-story	Monterey
ABBREVIATION:	2 FAM D	2 STORY	MONTEREY
COMPUTER:	2	2	715
NARRATIVE:	One-family, detached	two-story	Garrison Colonial
ABBREVIATION:	1 FAM D	2 STORY	GARR CO

Many houses have little or any style. One of the many development houses constructed in the 1950s could be described as follows.

	Class	Type	Style
NARRATIVE:	One-family, detached	two-story	Post World War II
ABBREVIATION:	1 FAM D	2 STORY	POST WW 2
COMPUTER:	1	2	900

In our cities we find many multiple-story, multiple-family houses built around 1910. They don't seem to be of any clearly recognizable style. They could be described as follows.

	Class	Type	Style
NARRATIVE:	Three-family, detached	three or more stories	Early 20th Century American
ABBREVIATION:	3 FAM D	3 STORY	EARLY 20 C
COMPUTER:	3	5	800

Every house can be described by the CTS System with almost no training by the user. However, the system also provides a vehicle to more precisely describe many houses by those who know how and want to use their knowledge.

It is optional with the user whether the style is described as one of the nine broad style categories or one of the 58 specific style descriptions.

With the CTS System it is possible to accurately classify and describe any house with the use of just five numbers, a simple set of abbreviations or a brief narrative description. The particular method chosen will depend upon who will read or use it.

HOW TO READ A HOUSE PLAN

The real estate professional often needs to know how to read house plans. They may be plans of an existing house, an addition to an existing house, one under construction or a contemplated structure.

It certainly would be easier for the inexperienced user if architects and builders used pictures or photographs or at least line drawings in perspective to illustrate their planned work.

Unfortunately, it is almost impossible to show complex spatial relationships in perspective. Therefore, the orthographic projection, a picture or drawing without perspective, evolved.

In a perspective drawing the lines are drawn to appear as the eye sees them in real life. Lines that are actually parallel in construction are not parallel in a perspective drawing. Below is a perspective drawing of a house looking at it from the outside. The dotted line runs parallel to the floor about halfway between the floor and the ceiling. It is horizontally sectioned at this level to make a floor plan. The house is sectioned at this level and not the floor level because if it were sectioned at the latter, the doors and windows, which are important parts of the house, would not show. (Fig. 1)

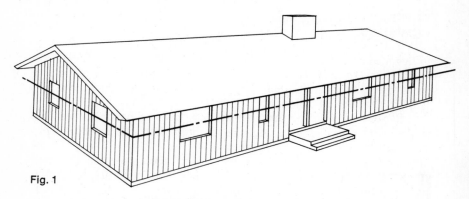

Fig. 1

PERSPECTIVE DRAWING OF A HOUSE

The next illustration is a perspective drawing of the house sectioned at the same plane indicated by the dotted line in the first illustration. It cannot be used for house plans because it is difficult to draw and when it is reduced in size the lines do not reduce proportionally. (Fig. 2)

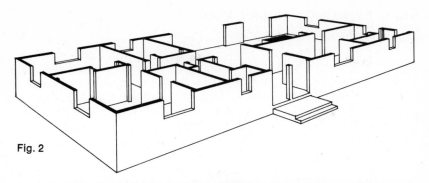

Fig. 2

PERSPECTIVE VIEW OF A HOUSE SECTIONED AT THE PLANE AT WHICH THE FIRST FLOOR PLAN IS DRAWN

To solve this problem an orthographic projection is used. Unlike a perspective drawing, an orthographic projection can easily be reduced in size because all the lines are reduced exactly the same amount.

Following is the same house, sectioned at the same plane, shown in an orthographic projection. Unlike the perspective view, the walls that are parallel in the actual house are also drawn parallel in the orthographic projection. (Fig. 3)

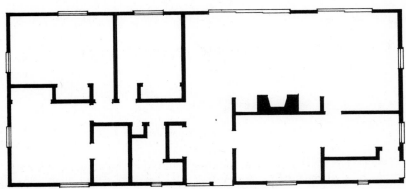

Fig. 3

FIRST FLOOR PLAN (ORTHOGRAPHIC PROJECTION) SHOWING MAJOR STRUCTURAL FEATURES ONLY

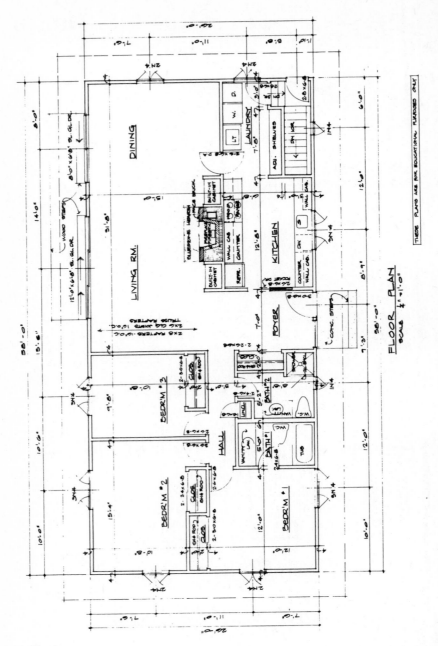

FLOOR PLAN
SCALE ¼" = 1'-0"

THESE PLANS ARE FOR EDUCATIONAL PURPOSES ONLY

Fig. 4

352

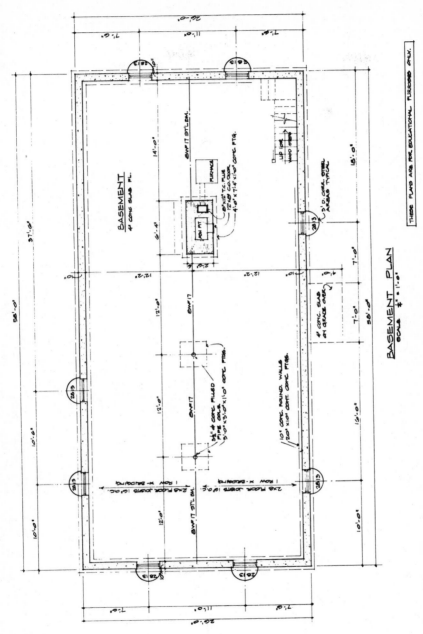

BASEMENT PLAN

SCALE $\frac{1}{4}$" = 1'-0"

Fig. 5

353

A complete set of house plans consists of the following:
1. Orthographic projections of each floor and the basement
2. Electrical plans
3. Plumbing plans
4. Wall sections
5. Elevations of all sides
6. Plot plan
7. Door and window schedules
8. Specifications

Next, look at the actual first floor plan and basement plan of our house. At first glance it looks very complex. It can be even more complex than shown here. Many architects combine the floor plans with the plumbing and electrical plans. This type of combined drawing is very difficult for the layman to understand. Complexity can be still further increased if the architect writes the specifications and the door and window schedules on the same sheets.

The advantage of combining all these things on one or two sheets is that everything the builder needs is included on a few sheets of paper.

Our plan does not combine the plumbing and heating plans or show the door and window schedules and specifications on the same sheets. They are each shown on separate sheets. (Figs. 4 and 5)

To interpret these floor plans it is necessary to know how the architect indicates dimensions on the plan. Also, architects make extensive use of symbols. At first they seem to complicate the plans but in reality they simplify them. To understand the plans it is necessary to recognize the many symbols that are used.

On the plans there are many dimension lines. They indicate the distance between two points on the plan. Three popular styles of dimension lines are shown. The selection of arrows, slashes or dots is a matter of personal choice. Often more than one type is used on the same plan. (Fig. 6)

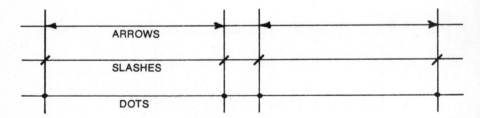

Fig. 6   **ACCEPTED STYLES OF DIMENSION LINES**

354

When the novice checks the dimensions of an existing house against those indicated on the plans, he is often surprised to find they do not appear to agree. This is because one cannot tell just by looking at the plans what the actual points of measurement are. For example, the dimension lines that indicate the size of a room will not agree with the measurements that would be read on a tape run from the wall surface on one side of the room to the wall surface on the opposite side of the room. The measurement read on the tape may be from about 1″ to more than 5″ less than the dimension indicated on the plan.

Architects rarely indicate dimensions from one wall surface to another.

On drawings of frame houses they prefer to indicate the dimensions between either the distance from the surface or the studs on one wall to the surface of the studs on the opposite wall or from the center of the studs on one wall to the center of the studs on the opposite wall. When stud surface to stud surface is used the actual tape measurement will be about one inch less than the indicated dimension line measurement (the thickness of the sheetrock being the difference). When the center of the stud on one wall to the center of the stud on the opposite wall is used the tape measurement will be about 5″ less than the indicated dimension line measurement (half the thickness of the stud which is usually 2″, plus the thickness of the sheetrock).

Following are illustrations of a stud partition and a stud wall. The drawing on the left of the stud partition shows both the stud surface to stud surface (bottom arrows) and the center of stud to center of stud (top arrows) dimensioning techniques. Note that on stud wall and stud partitions the sheetrock and individual studs are not indicated. (Fig. 7)

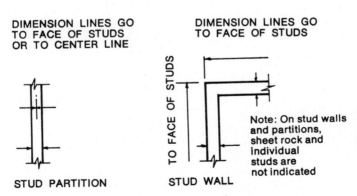

DIMENSION LINES GO TO FACE OF STUDS OR TO CENTER LINE

DIMENSION LINES GO TO FACE OF STUDS

TO FACE OF STUDS

STUD PARTITION

STUD WALL

Note: On stud walls and partitions, sheet rock and individual studs are not indicated

Fig. 7 **ACCEPTED METHODS FOR DIMENSIONING FRAME CONSTRUCTION**

Other dimensioning techniques used for frame construction are shown in the below wall elevation and wall section. The dimension lines for door openings run from the surface of the finished floor to the bottom of the top of the door opening. The dimensions for window openings also run from the top of the finished floor to the finished surface of the window opening.

The dimension lines shown on the wall section show the position of the door and window in the wall and run from the center line of the door or window. Door and window dimensions are not indicated with dimension lines. They are either noted on the plan or shown in the door and window schedule. (Fig. 8)

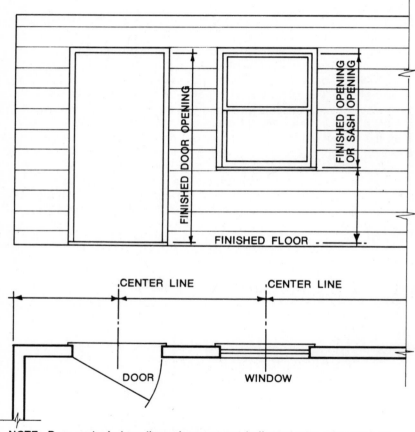

NOTE: Door and window dimensions are not indicated with dimension lines. They are either noted on plan or shown in door and window schedule.

Fig. 8 **ACCEPTED METHODS FOR DIMENSIONING FRAME CONSTRUCTION**

The techniques for dimensioning masonry construction are different from the techniques for dimensioning frame construction.

Dimension lines on plans of masonry construction usually run from one masonry surface to another masonry surface rather than to the surface of the sheetrock or other wall coverings.

The below wall elevation and wall section illustrate this technique. The dimension line in the below elevation of a masonry wall runs from the top of the finished floor to the surface of the masonry in the window opening—not to the wood or the surface of the sheetrock.

The dimension lines on the wall section that show the position of the window in the wall run from the surface of masonry in the window opening to the surface of the wall edge. (Fig. 9)

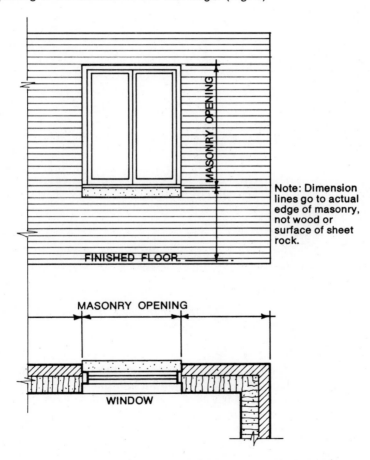

MASONRY OPENING

Note: Dimension lines go to actual edge of masonry, not wood or surface of sheet rock.

FINISHED FLOOR

MASONRY OPENING

WINDOW

Fig. 9 **ACCEPTED METHODS FOR DIMENSIONING MASONRY CONSTRUCTION**

357

Three common types of masonry walls are illustrated below. They are concrete block walls with plaster coating, masonry walls with furring strips and sheetrock or plywood interior wall surfaces and a back wall with brick veneer on the exterior surface and plaster on the inside surface.

The dimension lines on each of these wall drawings runs from the surface of one masonry wall to the surface or edge of another portion of the same wall. (Fig. 10)

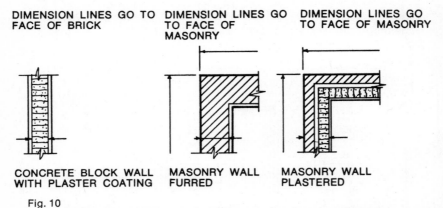

DIMENSION LINES GO TO FACE OF BRICK

DIMENSION LINES GO TO FACE OF MASONRY

DIMENSION LINES GO TO FACE OF MASONRY

CONCRETE BLOCK WALL WITH PLASTER COATING

MASONRY WALL FURRED

MASONRY WALL PLASTERED

Fig. 10

ACCEPTED METHODS FOR DIMENSIONING MASONRY CONSTRUCTION

Shown are a variety of wall types with many common types of doors and windows and how the dimension lines are indicated. (Fig. 11)

For the sake of convenience and ease in handling, plans must be reduced from lifesize. Unfortunately architects, engineers and surveyors all use different reduction scales.

The architect usually reduces the foot to ¼″ or ⅛″. Occasionally he reduces the foot to ½″, ¾″, 1″, 1½″ or 3″. Whatever reduction scale is used, it is indicated on each sheet of the drawing. (Fig. 12)

Engineers and surveyors reduce the foot into units of 10ths, 20ths and up to 60ths. They prefer this system as they feel it reduces the chances for mathematical error. (Fig. 13)

The metric system which will soon be adopted in this country may help architects, engineers and surveyors to finally agree on a standard reduction scale.

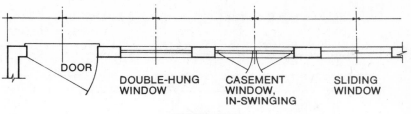

DOOR

DOUBLE-HUNG
WINDOW

CASEMENT
WINDOW,
IN-SWINGING

SLIDING
WINDOW

FRAME WALL

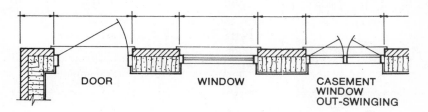

DOOR

WINDOW

CASEMENT
WINDOW
OUT-SWINGING

**BRICK VENEER ON CONCRETE
BLOCK WALL**

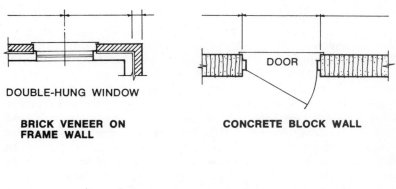

DOUBLE-HUNG WINDOW

**BRICK VENEER ON
FRAME WALL**

DOOR

CONCRETE BLOCK WALL

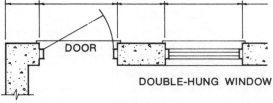

DOOR

DOUBLE-HUNG WINDOW

CONCRETE WALL

Fig. 11   **INDICATIONS OF WINDOWS AND DOORS IN
VARIOUS WALL TYPES (SHOWING DIMENSION LINES)**   359

ARCHITECTS SCALE

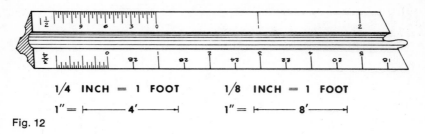

1/4 INCH = 1 FOOT   1/8 INCH = 1 FOOT

1" = |——— 4' ———| 1" = |——— 8' ———|

Fig. 12

ENGINEERS SCALE

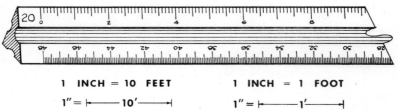

1 INCH = 10 FEET   1 INCH = 1 FOOT

1" = |——— 10' ———| 1" = |——— 1' ———|

Fig. 13

Metric Conversion Imminent

Since the earliest days in this country's history, leaders have been concerned with the need to standardize the U. S. weights and measures system with those of the rest of the world. The first attempts were made by Thomas Jefferson, supported by President Washington. Later, John Quincy Adams strove to bring the U. S. system into agreement with international standards. The first major success came during Abraham Lincoln's administration through the establishment of the National Academy of Sciences, which recommended adoption of the metric system nationally in 1865. It was not acted upon, nor were subsequent attempts successful.

Finally, in December 1975, the U. S. Congress passed legislation which President Ford signed into law, providing for voluntary conversion and establishing a national Metric Board.

The NATIONAL ASSOCIATION OF REALTORS® is participating in this work through its Metric Conversion Committee. Their prominence in the American National Metric Council will also enable them to encourage members to "think metric" and will provide information and conversion tables helpful in the sale of land and properties.

A good beginning point for real estate professionals to start to understand and use the new system is provided in the tables on the following page.

METRIC CONVERSION CHART
Approximate Conversions to Metric Measures

Symbol	When You Know	Multiply by	To Find	Symbol
		LENGTH		
in	inches	*2.5	centimeters	cm
ft	feet	.3	meters	m
yd	yards	0.9	meters	m
mi	miles	1.6	kilometers	km
		AREA		
in²	square inches	6.5	square centimeters	cm²
ft²	square feet	0.09	square meters	m²
yd²	square yards	0.8	square meters	m²
mi²	square miles	2.6	square kilometers	km²
	acres	0.4	hectares	ha
		VOLUME		
ft³	cubic feet	0.03	cubic meters	m³
yd³	cubic yards	0.76	cubic meters	m³

LINEAL
   10 mm = 1 cm
   100 cm = 1 m
   1000 m = 1 km
*1 inch = 2.54 cms (exactly)

AREA
   100 mm² = 1 cm²
   10,000 cm² = 1 m²
   10,000 m² = 1 hectare (ha)

Approximate Conversions from Metric Measures

Symbol	When You Know	Multiply by	To Find	Symbol
		LENGTH		
mm	millimeters	0.04	inches	in
cm	centimeters	0.4	inches	in
m	meters	3.3	feet	ft
m	meters	1.1	yards	yd
km	kilometers	0.6	miles	mi
		AREA		
cm²	square centimeters	0.16	square inches	in²
m²	square meters	10.8	square feet	ft²
km²	square kilometers	0.4	square miles	mi²
ha	hectares (10.000 m²)	2.5	acres	
		VOLUME		
m³	cubic meters	35	cubic feet	ft³
m³	cubic meters	1.3	cubic yards	yd³

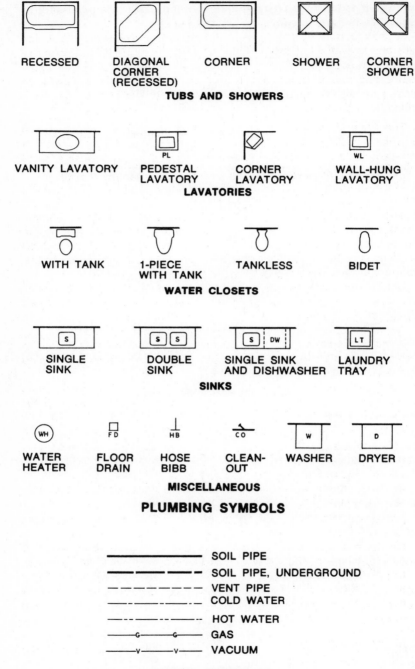

RECESSED   **DIAGONAL CORNER (RECESSED)**   **CORNER**   **SHOWER**   **CORNER SHOWER**

TUBS AND SHOWERS

VANITY LAVATORY   **PEDESTAL LAVATORY**   **CORNER LAVATORY**   **WALL-HUNG LAVATORY**

LAVATORIES

WITH TANK   **1-PIECE WITH TANK**   **TANKLESS**   **BIDET**

WATER CLOSETS

SINGLE SINK   **DOUBLE SINK**   **SINGLE SINK AND DISHWASHER**   **LAUNDRY TRAY**

SINKS

WATER HEATER   **FLOOR DRAIN**   **HOSE BIBB**   **CLEAN-OUT**   **WASHER**   **DRYER**

MISCELLANEOUS

PLUMBING SYMBOLS

————————— SOIL PIPE

— — — — — SOIL PIPE, UNDERGROUND

– – – – – – – VENT PIPE

—·—·—·— COLD WATER

—··—··—·· HOT WATER

—G———G— GAS

—v———v— VACUUM

   Fig. 14   **PIPING SYMBOLS**

In order to read the plumbing and heating plans it is necessary to recognize the symbols for plumbing and heating.

In this illustration, starting from the top, are shown the symbols for tubs and showers, lavatories, water closets (the name for toilets in the building trades), bidets (which are standard in many European bathrooms and gaining popularity in our country, too) and sinks. All of these symbols are easy to work with and remember as they look very similar to the fixtures they represent. (Fig. 14)

The symbols for water heaters, floor drains, hose bibbs (which are also called sill cocks, bib cocks, stop cocks and bib nozzles in different parts of the country) and clean outs are not quite as easy to remember as they look less like the items they represent.

The symbols for clothes washers and dryers are identified by the letters "W" and "D" respectively.

At the bottom of the page are the illustrations for piping symbols. From the top, the thick solid line is the symbol for the soil pipe within the house and below it the thick broken line symbol for the soil pipe underground outside the house. The vent pipes that connect to the soil pipes are symbolized by thinner evenly dashed lines.

Cold water pipes are shown as a series of short and long dashes in a line. The hot water pipe is a line consisting of a long dash followed by two short dashes.

A gas line is broken by "G's" and a vacuum line by "V's."

The plumbing and heating plans of our house shown on the following two pages contain examples of many of these plumbing and piping symbols.

On the Plumbing and Heating Floor Plan illustration, Arrow 1 points to the baseboard radiators under the windows in the bedroom. Under the windows is the best location for radiators. (Fig. 15)

Arrow 2 points to the bathroom and a half, or three-quarter bath as it is known in various parts of the country. In them are symbols for the vanities, water closets, tub and shower. Notice the hose bibb on the outside wall.

Arrow 3 indicates the kitchen area. Notice the symbol for the dishwasher and sink.

Arrow 4 points to the laundry room. Here you can see the symbols for the laundry tray, washer and dryer.

On the Plumbing and Heating Basement Plan illustration, Arrow 5 shows the water service from the street and the connection to sewer.

In this area you can see the solid broken line piping symbols for a soil pipe underground outside the house and unbroken solid lines for soil pipes inside the house. Notice the shut off valve on the cold water line coming in from the street. (Fig. 16)

The furnace is indicated by Arrow 6. It is a cast iron boiler with a tankless hot water heater and a single circulator pump. It is connected to an 8" x 12" flue in the chimney.

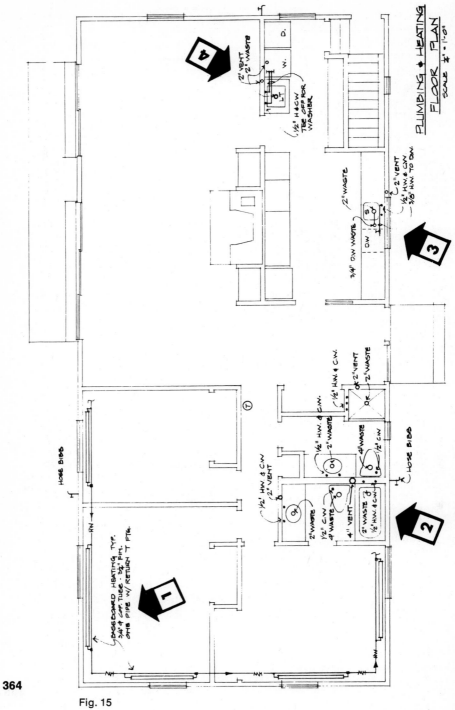

364

Fig. 15

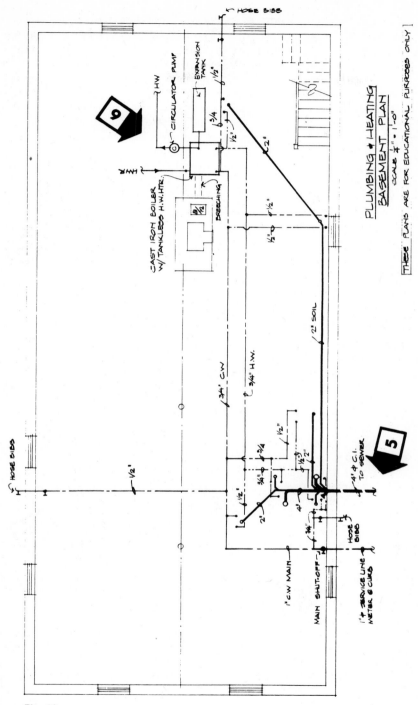

PLUMBING & HEATING
BASEMENT PLAN
SCALE ¼" = 1'-0"

THESE PLANS ARE FOR EDUCATIONAL PURPOSES ONLY

Fig. 16

The electrical plans have special symbols, too. There are many different switches, outlets, panels and fixtures in even a simple, small house. The symbol for all types of outlets is a circle. By adding additional lines and letters the exact type of outlet is indicated. (Fig. 17)

Ceiling and wall outlets provide power for the lighting fixtures that are mounted on them. The small letters "p.s." next to the symbol indicate they are controlled by a pull switch.

The most common convenience outlet used to plug in lamps, appliances and other household gadgets is the duplex convenience outlet. It has two receptacles. Formerly they accepted a common two-prong male plug. Now they are all manufactured to accept a third ground prong, too.

When the convenience outlet has two receptacles the number indicating the number of receptacles is customarily left off. When the outlet has one, three or more receptacles in it the number of receptacles is noted next to the symbol.

Waterproof outlets (Arrow 7) are indicated by the letters "W.P." next to the outlet symbol. There is one on the patio of this house which might be used for an electric barbecue grill, lawnmower or lamp. (Fig. 18)

In the kitchen area (Arrow 8) are shown a special range outlet and fan outlet. Notice that the refrigerator and dishwasher are connected to regular duplex outlets.

In the laundry area (Arrow 9) the clothes dryer requires a special purpose outlet. It is a 230-volt outlet. The voltage and other special features are explained on the plan and in the specifications.

Look carefully at Fig. 18. Can you find symbols that indicate the location of the television, telephone, door chimes, thermostat and push buttons?

In the basement (Arrow 10) are good examples of the symbols for ceiling outlets with switches. (Fig. 19)

Switches are indicated by a capital "S" followed by a small number or letter which shows exactly what kind of switch it is. A capital "S" without a number or letter after it indicates the common single pole switch.

Our plans show many single pole switches. A line runs from each switch to the outlets that it controls. It is not intended that this line show the actual path of the wire. Wire is not shown on the plans. The electrician works out the best wiring layout based on the information in the specifications.

A three-way switch indicated by an "S" followed by a small number 3 allows an outlet to be turned on from one location and turned off at another place. Our plan calls for such an arrangement on the stairs to the basement, in the hallway, in the private-sleeping area and in the kitchen and laundry area.

The distribution panel and power panel have their own symbols, too

Symbol	Description	Symbol	Description
○	CEILING OUTLET	Ⓕ	FAN OUTLET
⊢○	WALL OUTLET	▲	SPECIAL PURPOSE OUTLET (DESCRIBED IN SPECS)
○PS	CEILING OUTLET WITH PULL SWITCH	⊙	FLOOR OUTLET
⊢○PS	WALL OUTLET WITH PULL SWITCH	▬	LIGHTING PANEL
⊖	DUPLEX CONVENIENCE OUTLET	▨	POWER PANEL
⊖WP	WEATHERPROOF DUPLEX CONVENIENCE OUTLET	▭	FLUORESCENT FIXTURE
⊖1, 3,	CONVENIENCE OUTLET (NUMBER INDICATES RECEPTACLES)	CH	CHIMES
⊖R	RANGE OUTLET	Ⓣ	THERMOSTAT
⊖S	SWITCH AND CONVENIENCE OUTLET	▣	PUSH BUTTON
⊏▷	BELL	S₃	3-WAY SWITCH
TV	TELEVISION	S₄	4-WAY SWITCH
◀	TELEPHONE	S_P	SWITCH WITH PILOT LAMP
S	SINGLE-POLE SWITCH	S	WIRE TO OUTLET OR FIXTURE
S₂	DOUBLE-POLE SWITCH		

Fig. 17   **ELECTRICAL SYMBOLS**

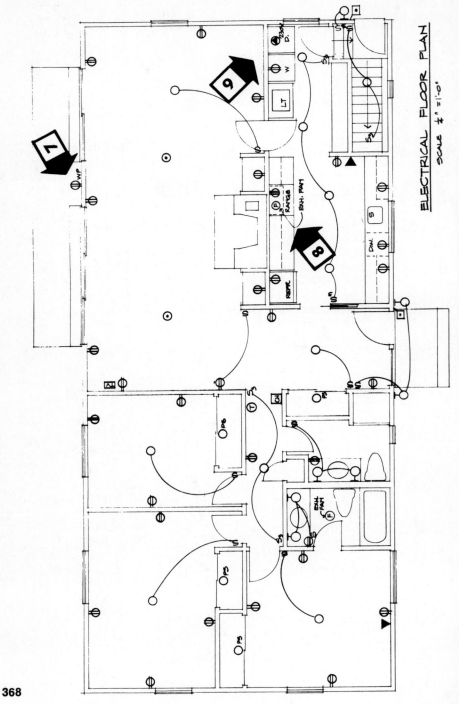

ELECTRICAL FLOOR PLAN

SCALE ¼"=1'-0"

Fig. 18

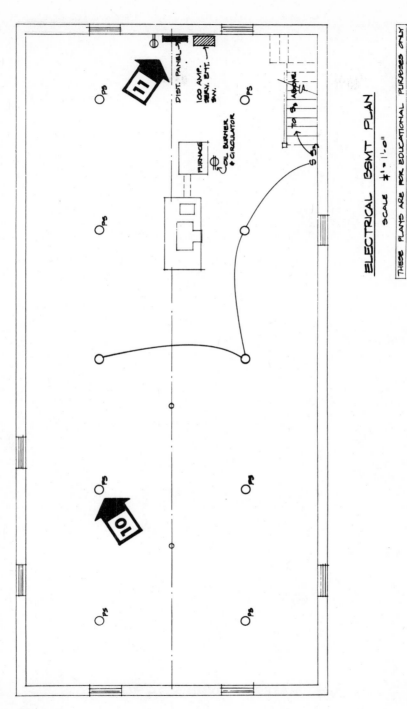

ELECTRICAL BSMT PLAN

SCALE ¼" = 1'-0"

THESE PLANS ARE FOR EDUCATIONAL PURPOSES ONLY

Fig. 19

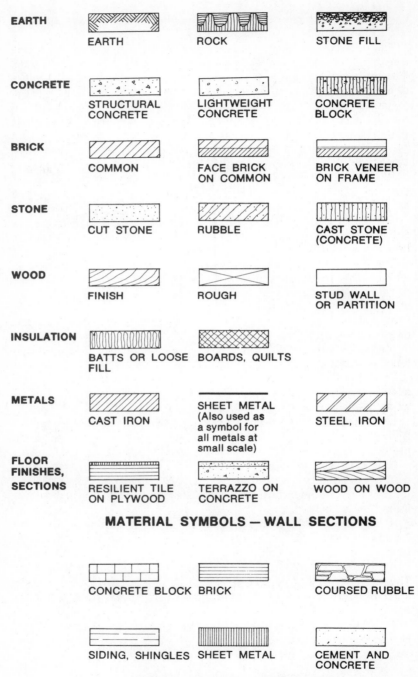

EARTH

EARTH   ROCK   STONE FILL

CONCRETE

STRUCTURAL CONCRETE   LIGHTWEIGHT CONCRETE   CONCRETE BLOCK

BRICK

COMMON   FACE BRICK ON COMMON   BRICK VENEER ON FRAME

STONE

CUT STONE   RUBBLE   CAST STONE (CONCRETE)

WOOD

FINISH   ROUGH   STUD WALL OR PARTITION

INSULATION

BATTS OR LOOSE FILL   BOARDS, QUILTS

METALS

CAST IRON   SHEET METAL (Also used as a symbol for all metals at small scale)   STEEL, IRON

FLOOR FINISHES, SECTIONS

RESILIENT TILE ON PLYWOOD   TERRAZZO ON CONCRETE   WOOD ON WOOD

MATERIAL SYMBOLS — WALL SECTIONS

CONCRETE BLOCK   BRICK   COURSED RUBBLE

SIDING, SHINGLES   SHEET METAL   CEMENT AND CONCRETE

Fig. 20   **MATERIAL SYMBOLS — ELEVATION**

(Arrow 11). Complete details about the electric service are set forth in the specifications.

The floor plans which illustrate the horizontal sections of the house are not sufficient to show all the construction details the builder needs. Many of the details are better shown in a wall section. This is a picture of the house sliced on a vertical plane to reveal the wall construction.

Special symbols are used to indicate the many different materials that are used. Many of the symbols little resemble the actual materials. (Fig. 20)

For example, the architects use unique traditional symbols to indicate earth, rock and stone fill.

Concrete, brick and stone all have special symbols. The symbol for common brick is widely spaced diagonal lines. The symbol for face brick is narrower spaced diagonal lines.

The construction of a window sill (Arrow 12) is almost impossible to show on a horizontal section of an elevation. It is quite easy to visualize how the window sill is made in this wall section. (Fig. 21)

The finished wood symbol shows the wood grain. Rough wood is indicated by crossed diagonal lines (Arrow 13).

The actual studs in a stud wall are not shown on the plan. Just the stud wall symbol is used.

Insulation is indicated by two different symbols. One is used for batts or loose fill (Arrow 14), another for boards and quilts.

Sheet metal is indicated by a solid line. This symbol is used for all types of metal when the scale is small.

The best place in the plans to see the floor finishes is the wall section. Resilient tile on plywood, terrazzo on concrete and wood on wood each have different symbols.

Look carefully at the top of the wall section (Arrow 15); find the two 2″ x 6″ headers and the two 2″ x 4″ plates. A header is a framing member which goes across the top of an opening and carries the load.

Details of how the house frame is connected to the foundation are clearly shown in the wall section (Arrow 16). The plate is bolted to the top of the foundation. The bolt is shown as two parallel broken lines imbedded in the concrete.

At the bottom of the foundation (Arrow 17) you can see how the foundation wall sets into the footing and how the basement floor is built.

A complete set of house plans includes elevations of all four sides. This is necessary because often the right and left or front and back sides of a house are quite different. This is especially true when there is a porch or garage on one side or in the rear of the house.

At first glance elevations appear to be sketches of the house. However, the materials are indicated by symbol rather than as they appear in real life. (Figs. 22 and 23)

For example the symbol for brick is narrow spaced horizontal lines. This is different from the way bricks appear in real life. If you did not

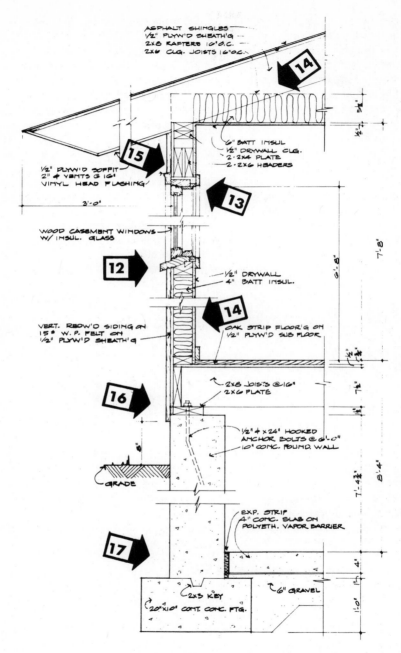

ASPHALT SHINGLES
1/2" PLYW'D SHEATH'G
2X8 RAFTERS 16" O.C.
2X6 CLG. JOISTS 16"O.C.

14

6" BATT INSUL
1/2" DRYWALL CLG.
2·2X4 PLATE
2·2X6 HEADERS

15

1/2" PLYW'D SOFFIT
2" ⌀ VENTS @ 16'
VINYL HEAD FLASHING

3'-0"

13

WOOD CASEMENT WINDOWS
W/ INSUL. GLASS

12

1/2" DRYWALL
4" BATT INSUL.

14

VERT. RBDW'D SIDING ON
15# W. P. FELT ON
1/2" PLYW'D SHEATH'G

OAK STRIP FLOOR'G ON
1/2" PLYW'D SUB FLOOR

16

2X8 JOISTS @16"
2X6 PLATE

1/2" ⌀ X 24" HOOKED
ANCHOR BOLTS @ 6'-0"
10" CONC. FOUND. WALL

GRADE

EXP. STRIP
4" CONC. SLAB ON
POLYETH. VAPOR BARRIER

17

6" GRAVEL

2X3 KEY
20"X10" CONT. CONC. FTG.

Fig. 21   TYPICAL WALL SECTION
SCALE 1 1/2" = 1'-0"

372

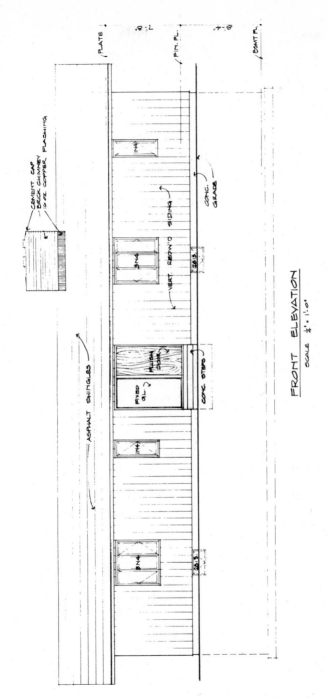

FRONT ELEVATION
SCALE ¼" = 1'-0"

Fig. 22

373

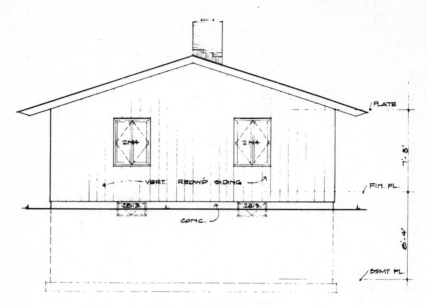

LEFT SIDE ELEVATION
SCALE ¼" = 1'-0"

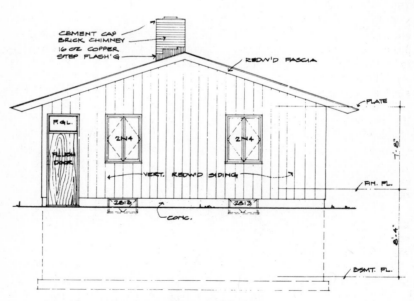

Fig. 23   <u>RIGHT SIDE ELEVATION</u>
SCALE ¼" = 1'-0"

THESE PLANS ARE FOR EDUCATIONAL PURPOSES ONLY

know the symbol you might think a brick house had clapboard siding. The symbol for brick on an elevation is also different from the symbol for brick in a wall section.

Other materials that have special symbols when shown in an elevation are concrete block, coursed rubble, siding, shingles, sheet metal, cement and concrete.

The symbols for doors and windows in the elevations are similar to how they appear in a real house. The arrows on a sliding window indicate the direction the window slides to open. How glass doors open is also indicated the same way. The dotted lines on casement windows and awning windows show on which side the hinges are attached. (Fig. 24)

The two-pane glass door (Arrow 18) shows that this door has one fixed pane and one movable pane that opens by sliding to the right. The three-pane glass door (Arrow 19) shows a door with three panes. One is fixed and the other two open by sliding to the left. (Fig. 25)

The importance of elevations is highlighted by the rear elevation of our house. It gives a different overall impression of the house. It clearly shows the placement of the large glass doors and the wooden steps leading to the rear yard. (Fig. 25)

Doors look in elevation drawings just about the same as they look in real life. The six-panel entrance door with window lights and entrance door frame shown here are used on colonial style houses. (Fig. 26)

A plot plan indicates the position of the house on the lot. It also shows the natural contours of the land before grading which are indicated by broken lines (Arrow 20) and the proposed grading which is indicated by solid lines (Arrow 21). (Fig. 27)

Plot plans are known by several other names around the country, such as a plotted plate plan or a plate.

The plot plan also reveals the location of the underground pipes to the street (Arrow 22).

Other site improvements shown are the gravel driveway and gravel walk. Typical other site improvements shown on plot plans are sidewalks, curbs, patios, fences, etc.

Trees that are to remain and trees to be removed are also shown respectively by their solid and unfilled circle symbols.

Last, but of course most important, are the lot lines and dimensions. The compass direction of each line should also be shown. In states where the country is divided into sections, these references may be shown, too.

The last two parts of a complete set of house plans are the "Door and Window Schedules" and "The Specifications."

As previously pointed out, the plans do not show the size of the windows. On each indicated window opening there is a reference number that refers to the door and window schedule. It is here that the size, type, manufacturer and other details about the window are found. (Fig. 28)

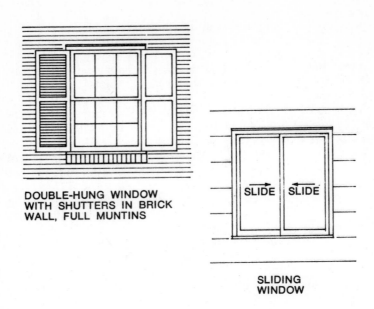

DOUBLE-HUNG WINDOW
WITH SHUTTERS IN BRICK
WALL, FULL MUNTINS

SLIDE | SLIDE

SLIDING
WINDOW

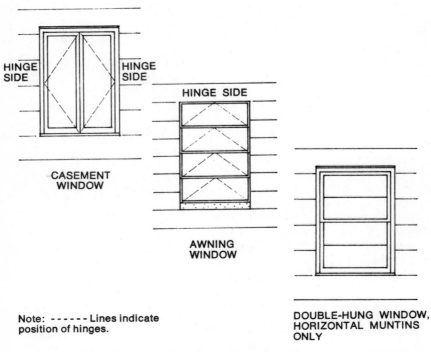

HINGE
SIDE

HINGE
SIDE

CASEMENT
WINDOW

HINGE SIDE

AWNING
WINDOW

DOUBLE-HUNG WINDOW,
HORIZONTAL MUNTINS
ONLY

Note: - - - - - - Lines indicate
position of hinges.

Fig. 24 **WINDOW INDICATIONS AS SHOWN IN ELEVATION**

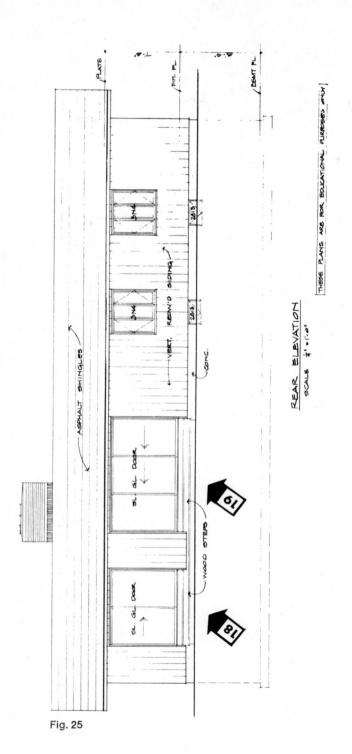

REAR ELEVATION

SCALE ¼" = 1'-0"

THESE PLANS ARE FOR EDUCATIONAL PURPOSES ONLY

Fig. 25

377

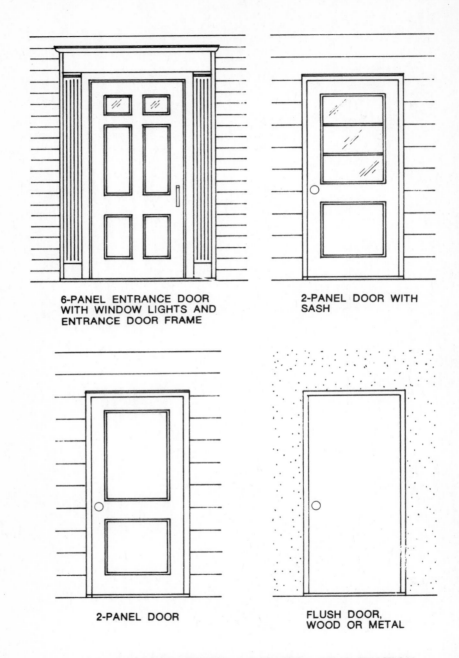

6-PANEL ENTRANCE DOOR
WITH WINDOW LIGHTS AND
ENTRANCE DOOR FRAME

2-PANEL DOOR WITH
SASH

2-PANEL DOOR

FLUSH DOOR,
WOOD OR METAL

Fig. 26   **DOOR INDICATIONS AS SHOWN IN ELEVATION**

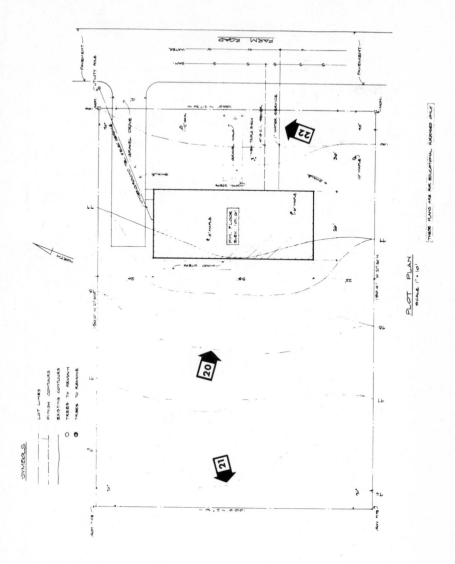

PLOT PLAN
SCALE 1" = 10'

SYMBOLS

LOT LINES
FINISH CONTOURS
EXISTING CONTOURS
TREES TO REMAIN
TREES TO REMOVE

Fig. 27

379

DOOR AND WINDOW SCHEDULE

NOS.	LOCATION(S)	TYPE	NOMINAL SIZE	MANUFACTURER & NO.	REMARKS
1 4	East & West Walls of Living Room	3 Panel Sliding Glass Door	12'-4" x 6'-11"		Complete Unit with 5/8" Insulating Glass & Screen, Operating Hardware
2	South Wall of Living Room	2 Panel Sliding Glass Door	6'-2" x 6'-11"		do
3	do	2 Units Fixed (Flr. to Clg.)	6'-4" x 6'-11"		Fixed Units Matching Sliding Glass Door; 5/8" Insulating Glass
5 18 22	South Wall of Dining, 2 Bedrooms & Unfin. Space (future 3rd. Bedroom)	Fixed "Picture" Window	8'-2" x 5'-6"		1" Insulating Glass
6 19	East Wall of 3.R. =2 & Unfinished Space (future 3rd Bedroom)	Fixed Center Horiz Slider Each Side	7'-7" x 4'-2"		Complete Unit with Storm Panels, Screens (Side Units Only), Operating Hardware
7 8	Sidelights of Front Entry Door	Fixed Semi-Obscure	1'-6" x 6'-8"		1" Insul. Glass Unit of L.O.F. 9/32" Rough Plate For Outside Sheet, 7/32" Crystal Inside Sheet
9	Master Bath	Aluminum Horiz. Slider	4'-0" x 5'-0"		Complete Unit with 5/8" Insulating Glass, Screen, & Operating Hardware
10 21	West Wall of Master Bedroom, Dining & Rec. Rm.	2 Panel Sliding Glass Door	3'-2" x 6'-11"		See Doors No. 1 & 4 Above
12	East & West Walls of L. R. (High)	Fixed Trapezoidal Unit	See Remarks		5/8" Insulating Glass (Fixed), Approximate Size 4'-0" x 4'-9" x 3'-7" Contractor to Field Check B/4 Ordering
14 15	North Wall of L.R. Balcony (Clearestory)	Fixed "Picture" Window	5'-1" x 3'-6"		5/8" Welded Insulating Glass
16	South Wall of Rec Room	Fixed	6'-10" x 4'-8"		Fixed Unit with Removable Double Glazing

Fig. 28

The "Specifications" provide all the other essential facts needed by the builder such as kinds of materials to be used and methods of construction.

The plans in this section were drawn especially for this book by Ronald Noe, a New Haven, Connecticut, architect. They are for educational purposes only and are not intended to be used to construct an actual house. Any such use is a violation of the copyright laws.

COMMON PROBLEMS OF HOUSES

Salespeople, appraisers, potential buyers, government employees, students and others are often asked to inspect a house and give an opinion about its condition. The best way to do this is to look for specific problems rather than just checking the house in an unorganized, random manner.

In this appendix are listed the 33 most common problems found in houses and how to identify them. There are also some suggested solutions. Many problems can be corrected easily at low cost while others are so costly or difficult to correct that the occupant must be prepared to accept the unpleasantness of living with them.

1. Termites

The termite has been around longer than man himself. They attack wood structures above the ground. Termites do not like light, so they prefer wood that touches the ground which can be reached directly from their subterranean nests. Termites will also travel to above-ground wood through cracks in the masonry foundation, or as a last resort, through shelter tubes the insects build on the exterior surface.

Termites look substantially different from flying ants with which they are often confused. Termites have wings similar in shape and size, patterned with many small veins, and without stigmae. The middle of the body is wide, and the antennae are not "elbowed." Ants have unpatterned varied size wings with stigmae. The mid-sections of their bodies are very narrow, and their antennae are "elbowed."

The best way to check for termites is to hire a professional. The FHA, VA and other lending institutions require professional termite inspections in many areas of the country. Many reliable exterminators guarantee their work for five years.

Termites work fairly quickly. If they are caught in the early stages of infestation, they may be stopped for a few hundred dollars.

2. Inadequate Wiring

The problem starts with inadequate voltage and amperage in the house; it should be a minimum 220 to 240 volts and 100 amperes or more if the house is large, or if it has major electrical appliances such as ranges and clothes dryers, or electric heat and air conditioning.

Lack of sufficient branch circuits to the various rooms of the house and appliances can be corrected by installing a bigger distribution panel and additional wiring.

The use of fuses with higher ratings than necessary is a sign that the wiring is under the needed capacity.

Insufficient wall outlets in the rooms leads to the dangerous use of extension cords and monkey plugs.

Lack of outside outlets is an inconvenience.

Knob and tube wiring must be viewed suspiciously. It is often old and its insulation has a tendency to crack, leaving exposed wires which are very dangerous. Most houses with old knob and tube wiring are ready for rewiring.

Switches should be placed to permit walking into any area of the house, lighting a path and extinguishing the lights without retracing steps.

3. Poor Heating

The major causes of poor heating are insufficient insulation and an inadequate or poorly functioning heating system. Insulation may often be added, as may storm windows and weather stripping.

The best way to see if a house is heating comfortably is to visit it on a cold day.

The condition of the furnace is often reflected in its appearance. An old furnace encased in asbestos probably is a potential trouble maker.

An adequate clean furnace without rust may require minor repairs but usually has lots of good life left.

A free or nominally priced inspection of the heating system often is available from fuel suppliers in the area. The performance of many furnaces can be improved with a good cleaning, minor adjustments and/or replacement of clogged filters.

4. Inadequate Hot Water

This is a common complaint when water is heated by the heating unit; sometimes an additional storage tank is helpful.

Undersize or low grade hot water heaters will produce insufficient hot water. A larger tank either purchased outright or rented from the utility company will correct the problem.

5. Defective Septic System

Your nose is often the best test, walk around the area of the tank and sniff; odors are a sure sign of trouble. Be suspicious of an overly green lawn in the area of the leeching field. Another sign is a toilet that flushes slowly.

It doesn't hurt to ask how often the system has to be pumped out. In many towns, the local health officer is very knowledgeable about many systems in his jurisdiction and of problems in a particular neighborhood.

Be sure to determine the location of the cleanout from the current owner. This saves a lot of digging and searching if it is buried.

Septic system problems may sometimes be corrected by simply pumping out the tank. Sometimes new leeching fields are required. Unfortunately, there are situations such as when the soil absorption rate is poor or the water table is close to the surface when nothing can be done to make the system function properly.

6. Defective Plumbing and Plumbing System Noises

Plumbing suffers from the two major problems of leaking and clogging with rust and mineral deposits. Leaking can be detected by visual inspection. Old style iron or steel pipes are much more likely to develop leaks than corrosion-resistant copper and bronze. You can detect iron and steel pipes with a small magnet. It will be attracted to an iron or steel pipe, but not to a copper or bronze pipe.

Insufficient water pressure can be caused by either clogged pipes, **383** too small a water main from the street, low water pressure in street main or problems in the well or plumbing system.

Test the water pressure by turning all faucets in the highest bathroom on full and then flushing the toilet. If a substantial reduction of flow from the faucets results, it is a sign of trouble and the system should be checked by a plumber to determine the cause and the cost to correct it.

Look for stains in the bathtub and lavatories; these are a sign of rusting pipes or unsoftened hard water. If you suspect hard water, have a sample professionally tested by a firm selling water softeners. They may also recommend the equipment necessary to correct the situation and give an estimate of the cost.

Look for leaks under sinks which may be caused by simple things like a loose washer or may indicate a cracked fixture.

A high-pitched whistling sound made when the toilet is flushed is caused by the valve in the toilet closing too slowly. A simple adjustment by a plumber should eliminate the noise.

A sucking sound that occurs when water runs out of a fixture often is made by a siphoning action in the trap caused by improper venting of the waste stock. If unclogging the vent doesn't eliminate the noise, a major change in the vent system is necessary.

A hammering noise in the water pipes when the water is turned off is caused by a sharp build-up of pressure in the pipe. This is a serious problem which, if uncorrected, will result in broken or leaking pipes.

In areas where there is high pressure in the water mains, capped pipe sections filled with air called water hammer arresters should be installed at the time of initial construction to provide a cushion of air in the system. A hammering noise is a sign that either the needed air chambers were not installed or they have become filled with water and no longer operate effectively. Draining the system and restoring the air in the air chambers may solve the problem; otherwise it may be necessary to have a plumber install air chambers or other pressure reduction devices to eliminate the hammering noise.

The sound of running water is caused by undersized pipes and pipes run in walls which are not sound insulated. Wrapping the pipe with a noise insulation material may help. If the noise is very objectionable the pipe may have to be replaced with a larger one.

7. A Wet or Damp Basement

Dampness in the basement reduces its usefulness, is unpleasant and affects the comfort of the rooms overhead. A basement that is wet or damp only part of the year can usually be detected any time by careful inspection. Check all walls for a powdery white mineral deposit a few inches off the floor. Only the most diligent cleaning will remove all these deposits after a basement has been flooded.

Look for stains along the lower edge of the walls and columns, on the furnace and hot water heater, etc. Be suspicious if nothing seems to be stored on the basement floor. Finally, use your nose; mildew odor is hard to eliminate.

The causes of wet and damp basements are numerous. Some are easily corrected and others are almost impossible to correct.

In areas where the soil drainage is poor or the water table is near the surface of the ground, it is necessary to have well-constructed footing and foundation drains to maintain a dry basement. They should be installed when the house is constructed since it is a major expense to do it afterwards. The same is true of a vapor barrier under the basement floor, which is easily installed during construction but impossible to add afterwards.

Cracks in the floor and walls can be patched with various compounds whose advertised claims should be taken with a grain of salt. A more drastic step is to dig down and repair the wall from the outside.

What appears at first to be a major water problem might be traced to a leak in a window or the hatch door. A simple caulking job will stop the leaks. Water will leak in at the bottom of a window well which does not drain properly in a heavy rain storm. Extending the drain line or deepening the dry well will eliminate the problem.

The ground around the house should slope away from the foundation wall so ground water will not collect along the edge of the foundation.

If there is an edge of roof line without a gutter, water may be running off and collecting near the foundation wall. The water collected by the gutter flows into the leaders and must be diverted away from the foundation wall. The leader should run into a sewer drain, dry well or splash pan in that order of preference.

Dampness and mildew may also be caused by moisture condensing on the walls, ceiling and pipes. Proper ventilation eliminates this problem.

8. Leaks Above the Basement

Water may leak through the roof for a variety of reasons. Asphalt shingle roofs may leak in a high wind if light grade shingles were used. As these shingles get older they curl, tear and become pierced with holes.

Wood shingles curl, split, loosen, break and fall off the roof. Asbestos shingles crack, break and become lost. Metal roofs become rusted, bent and pierced with holes. Roll and built-up roofs become loose, torn, patched and worn through.

The condition of the roof is usuallly apparent upon close examination. If the house has an uninsulated attic, light may shine in through holes in the roof. Also, look for water stains on the ceilings and be suspicious of any ceiling that has been recently repainted.

The valleys and flashing will leak when they become loosened, rusted, worn or pitted with holes.

Leaking gutters allow water to run down and through the exterior walls.

Windows and doors are another source of leaks. Often a simple recaulking will solve the problem.

Water may penetrate through a masonry wall when the joints become **385** soft or cracked. Some masonry walls will leak in a driving rain storm and must be treated with a waterproofing material.

9. Defective House Framing

The best thing one can do about a house with a defective frame is not to buy it. If, unfortunately, you encounter one, seek the advice of a competent contractor as to what may be done to arrest the problem.

After a house is a few years old, you can often detect visual signs of defective framing. One sign is bulging exterior walls, which are best seen by standing at each corner of the house and looking along the wall. If you don't trust your eyes, make a plumb line out of a key and string and hold it against the wall.

Stand back from the house and look at the ridge line. If it sags in the middle, trouble is developing.

Window sills that are not level are a sign of settling, defective framing or original sloppy carpentry. A careful house inspection should include the opening and closing of *every* window. Sticking windows may be a sign of settling or defective framing.

Check all doors; look at the bottoms. Have they been resawn to allow free movement after sagging of the frame caused them to jam?

A sure sign of trouble is a large crack developing on the outside of the house between the chimney and the exterior wall. Another tip-off to defective framing is cracks running outward at an angle from the upper corners of window and door frames.

Sagging and sloping floors may be detected visually or by putting a marble on the floor and watching to see if it rolls away. This may be a sign of defective framing or may be caused by weak or defective floors.

Cracks in the walls other than those just discussed should be a cause of concern, but in themselves are not conclusive evidence of framing problems. All houses settle unless built upon solid rock; rare is the house that does not develop some wall and ceiling cracks. These should be of concern only when accompanied by some of the other signs of defective framing.

If you suspect the house has defective framing, get some professional advice to confirm your opinion.

10. Weak and Defective Floors

We have just covered the problem of sagging and sloping floors caused by defective framing. There are other reasons for floor troubles. Perhaps the floor joists are too small or lack support from adequate bridging thereby causing the sagging or sloping.

Floors that have been exposed to water may warp and bulge upwards.

Wide cracks between the floorboards are a sign of poor workmanship or shrinkage caused by wood that was improperly dried or not stored correctly at the time of installation.

Fortunately, floors that are rough, stained, discolored, blemished, burned and gouged usually can be cured by refinishing.

11. Defective Windows

In Problem No. 9 we emphasized the need to open and close every window in the house and noted that defective, hard-to-open windows may be a sign of defective framing. Dust streaks or water stains around the window trim are evidence of leakage. Of course, it is possible that the water came in because the window was left open in the rain.

In the process of opening and closing each window you may discover some with missing locks, window lifts or counter balance weights. You may also discover that it is difficult to reach over the kitchen sink to open that window if it is the double-hung type.

A window in the bathroom over the tub or toilet lets in uncomfortable drafts.

Windows in children's rooms should be high enough to be safe, yet low enough to allow escape in the event of a fire.

12. Loose or Defective Interior Plastering

In Problem No. 9 we also discussed cracked plaster caused by defective framing and settling. As long as the cracked plaster is tight to the wall, it may be sufficient to just patch and redecorate the crack.

Bulging plaster on the ceilings is dangerous and should be repaired. When suspected, it can often be detected by pressing a broom handle against the ceiling and feeling if there is any give in the plaster.

13. Broken or Missing Storm Windows, Screens and Shutters

The only way to be sure if there is a complete set of these and their condition is to count and inspect them.

14. Mortar Joints Between Bricks or Stones Cracked or Loose

Mortar that is not perfectly installed will, in time, become soft, crack and fall out. This will in turn weaken the wall and allow water to seep through.

The joints can be inspected visually and by poking them with a pointed instrument. Defective joints should be repointed.

15. Cracked Pavement in Sidewalks, Driveways, Terraces, Patios, Steps, etc.

Cracks can be easily detected by visual inspection. In cold climates water seeps into the crack, freezes and expands, and enlarges the crack if left unattended.

16. Peeling and Blistering Paint and Peeling Wallpaper

Most exterior paints become chalky with age but should not blister or peel. Peeling and blistering is a sign of trouble, which has been caused by incorrect initial application or moisture in the wood.

Paint blisters filled with dark, colored water are a sign of moisture in the wood underneath.

Peeling wallpaper likewise may be caused by poor initial application or moisture in the wall. To prevent recurrence of the problem, first the source of moisture must be located and corrected, the paint or paper may be reapplied.

17. Cracked, Loose or Leaking Bathroom Tiles

The principal area where the problem occurs is around the tub, especially when there is a shower splashing water on the tile wall. Defective grout will permit the water to seep behind the tile and loosen the glue. New waterproof adhesives help eliminate this.

Tiles set in plaster also are less likely to present problems. Here is an area where initial good workmanship will produce lasting results while shoddy work will have to be redone.

18. Inferior Grade Bathroom Fixtures

A good grade fixture produced by a major manufacturer, when properly installed, should give many years of trouble-free service. Look for the name of the national manufacturer stamped on each fixture: a fixture without a name stamped on it often is the lowest grade or "second."

In general, a cast iron tub is better than steel. It can be identified by knocking on the bottom with your knuckles. There is a distinct difference in the sound of a steel and iron tub. In general, the bigger the tub, shower stall, or lavatory sink is, the better it is.

Wash-down toilets are a sign of cost-cutting. Best materials for toilets are vitreous china and enameled cast iron. Enameled steel is a cheaper product.

19. Too Few Bathrooms

Every family's requirements are slightly different. If the bathroom is a social center, the family will need fewer than if Mom likes to put on her make-up in peace and quiet. Naturally, the number of children and their ages also makes a difference.

A three-bathroom house without at least one full bath and one lavatory is now substandard. A two-floor house should have a full bathroom on each floor. There should be at least one bathroom for every two people with an extra bathroom or lavatory for each bedroom over three.

20. Garage and Driveway Deficiencies

It is dangerous to have a driveway that enters the public street at a blind curve. A driveway that slopes upward to the street may be dangerous, especially in winter.

It is dangerous, if not unlawful, to back into a public street, so there must be room to turn the car around in the driveway. The driveway should be so designed that cars parked in it will not obstruct the walkway to the house.

Though once very common, we now know it is inconvenient to have a garage that is detached from the house with no connecting, sheltered accessway.

21. Poor Security and Privacy

Millions of tract houses have been built with large picture windows facing the street which, unless covered with closed drapes, result in a major loss of privacy.

To check if activities inside the house may be seen by passers-by on the street, walk slowly by the house trying to look inside the various areas of the house. Then make the same test for privacy from the neighbors' view by walking slowly around the lot line, trying to look into the house. Sometimes a substantial improvement can be made by planting a few shrubs or building a fence.

Bathrooms should not be visible from the living room and halls.

The danger of opening the door to a stranger is obvious. In order to safely determine who is at the door without opening it, there should be some glass in the door, a sidelight, peephole or an inter-com unit. An additional safety feature is a door chain that permits the door to be opened a crack to talk with a stranger before deciding to let him in.

Most professional thieves will get into a house no matter what kind of locks are installed or other precautions are taken. However, many robberies are not performed by professionals and good hardware with locks which are difficult to pick will often act as a satisfactory deterrent.

An automatic burglar alarm system connected to the police station or private security system provides maximum burglar protection. The kind that sounds an alarm on the premises may scare away a prowler before he has a chance to take anything.

Electric, gas and water meters should be outside the house to eliminate the need to allow strangers to enter the house to read them.

22. A Poor Floor Plan

Here is a list of some of the most common floor plan deficiencies:
1. Front door enters directly into living room.
2. No front hall closet.
3. No direct access from front door to kitchen, bathroom and bedrooms without passing through other rooms.
4. Rear door not convenient to kitchen and difficult to reach from street, driveway and garage.
5. No comfortable eating space for family in or near the kitchen.
6. The separate dining area or dining room is not convenient to kitchen.
7. Stairway located between levels of a room rather than in a hallway or foyer.
8. Bedrooms are located so as to be visible from living room or foyer.
9. Walls between bedrooms not soundproof; best way to accomplish this is to have them separated by a bathroom or closets.
10. Recreation room or family room poorly located.
11. No access to the basement from outside of the house.
12. Outdoor living areas not accessible from kitchen.
13. Walls cut up by doors and windows so it is difficult to place furniture around room.

23. Inadequate Kitchen

Many kitchens suffer from one or more of the following inadequacies, listed in order of most common occurrence:
1. Insufficient base cabinet storage space.
2. Insufficient wall cabinet storage.
3. Insufficient counter space.
4. No counter beside the refrigerator.
5. Not enough window area (at least 10 percent of floor area).
6. Poorly placed doors that waste wall space.
7. Traffic through work area.

8. Too little counter space on either side of sink.
9. No counter beside range.
10. Insufficient space in front of cabinets.
11. Distance between sink, range and refrigerator too great.
12. Range placed under a window; this is unsafe.

24. Insufficient and Poorly Planned Closets and Storage Space

A crazy closet is one with doors that are too narrow, shelves too high and clothes poles so low that coats and long dresses trail the floor.

If there is no coat closet near the main entrance, where will guests put their coats?

If the living room is going to be used, it needs storage for books, records, fireplace wood, etc.

A dining room without storage for linens, dishes and silverware is inconvenient.

Ironing boards and dangerous laundry and cleaning products require special safe storage space.

A bathroom without storage for towels, soap, extra toilet paper, cosmetics, medicines, dirty clothes, soon becomes a cluttered mess.

Nothing is more inconvenient than bedrooms without adequate closets.

Houses without attics or basements often lack dead storage areas for trunks, boxes, extra furniture, sleds, storm windows, etc.

Outdoor living is made unpleasant by lack of nearby storage for the barbeque, outdoor furniture, garden tools and lawnmower.

25. House Too Small for Growing Family's Needs

A growing family has three choices of ways to solve the need for additional space:

1. A house that can be expanded or with additional areas which may be finished into living space.
2. A house that is bigger than that currently needed with the extra space to be used in the future.
3. Moving to a bigger house.

26. Safety Hazards

The number of household accidents is staggering and unfortunately many of them result in death or permanent injury. Here are a few common sources of household accidents that can be eliminated:

1. Closets and cupboards that latch shut so they can only be opened from the outside and can trap children inside.
2. Doors that open out over stairs without a landing.
3. Steep, poorly lighted basement stairs without a handrail.
4. Bathroom light fixtures with pull strings or switches that can be reached from the tub or shower.
5. Swimming pools that are not fenced completely with at least a four-foot fence and a gate that locks (the neighbor's pool should be fenced too).
6. No adequate, convenient space to lock up dangerous cleaning products, medicines and other poisonous things so young chil-

dren cannot reach them.
7. Too little headroom on stairs.
8. Porches, patios and stairwells without strong handrails around them.
9. Changes in floor levels in the house which are only one or two steps.
10. Stair risers of unequal size.
11. Stairs without adequate lights which may be turned on and off from both top and bottom of the stairs.

27. Fire Hazards

The ten most common causes of residential fires are listed below. An examination of this list should suggest many ways to reduce these hazards:

1. Smoking.
2. Defective and overloading electric wiring.
3. Heating and cooking equipment.
4. Children playing with matches.
5. Open flames (fireplaces, candles, etc.)
6. Flammable liquids.
7. Arson.
8. Chimneys and flues.
9. Lighting.
10. Rubbish in cellar and attic (spontaneous combustion).

28. Declining Neighborhood

Some early warning signs of a neighborhood starting to decline are:
1. Increasing average age of population.
2. Unusual number of "For Sale" signs, where permitted.
3. Construction of new homes stopped before all vacant land used up.
4. Conversion of large homes into multiple dwellings or rooming houses.
5. Increasing ratio of rented houses.
6. Breakdown of enforcement of zoning regulations and deed restrictions.
7. Declining reputation.
8. Decrease in rental rates.

29. Nuisances and Adverse Influences

The following is a list of nuisances that, when too close to a house, will decrease its value. On the other hand, having them a few blocks away may be very convenient.
1. Fire stations.
2. Vacant houses.
3. Schools.
4. Funeral homes.
5. Stores.
6. Apartment houses.
7. Motels.

8. Restaurants and bars.
9. Utility wires and poles.
10. Noisy highways.
11. Hospitals.
12. Offices.
13. Commercial buildings.
14. Gas stations.
15. Industries.

30. Noise

Noisy plumbing is discussed in Problem No. 6.

Steps to decrease noise in sleeping areas are discussed in Problem No. 22.

Often it is possible to put some insulation material between the studs to soundproof an existing wall.

Noisy hot air heating ducts are caused by having direct connection to the furnace without a short piece of canvas between the furnace discharge and the beginning of the ducts.

31. House Poorly Located and Oriented on Lot

The poor location and orientation of a house on its lot substantially detracts from its livability. Unfortunately, the vast majority of existing houses are not located or oriented the best possible way.

One problem is that an individual house will be nonconforming with the neighborhood if it is oriented to the lot by taking into consideration topography, view, sun, trees, etc. when all the other houses on the street are lined up in a row facing the street.

32. Vermin

Rats, mice and a variety of other rodents and animals like the same type of food and shelter as humans. Often when such rodents move into a house, they will remain there happily until expelled. Vermin may transmit diseases, destroy property and are a source of annoyance. To control vermin the house should be made as rodent-proof as possible. Food and water sources should be made inaccessible. Many types of poisons are available for household use. For extreme problems an exterminator will kill off the existing population and help prevent future infestations.

In addition to termites (previously discussed in Problem No. 1) a partial list of insects which often invade our homes includes: flies, mosquitoes, cockroaches, fleas, ticks, moths, millipedes and centipedes, crickets, earwigs, ground beetles, ants, spiders, scorpions, mites, wasps, bees, hornets, yellow jackets, bedbugs, lice and silver fish. Besides being a nuisance, insects may transmit diseases and destroy property. The best long-run methods of control are to close up or screen any holes through which insects may enter and to eliminate sources of food and water. Households poisons will produce temporary relief but an exterminator may be needed.

33. Wood Rot and Poor Ventilation

Sooner or later, except in very dry climates, wood that directly touches the ground will begin to rot. Older homes often contain rotted wood. It can be easily detected by poking the wood with a sharp instrument like an ice pick.

Poor ventilation will cause wood below the floor, in the attic and behind the siding to rot. It will also cause paint to peel and will decrease the effectiveness of the insulation. Often, ventilation problems can be corrected by installing ventilators on the roof, in the attic, in the crawl areas and under the cornice eaves.

BIBLIOGRAPHY

Agan, Tessie and Luchsinger, Elaine *The House: Principles/Resources/Dynamics* Philadelphia, J. P. Lippincott Co., 1965.

American Institute of Real Estate Appraisers *Appraisal Terminology and Handbook* Chicago, 1967.

Anderson, L. O. *Wood Frame House Construction* Los Angeles, Craftsman Book Co. of America, 1971.

Andrews, Richard B. *Urban Land Policy* New York, Free Press, 1972.

Architects' Emergency Committee *Great Georgian Houses of America* New York, Dover Publications, 1970. Vol. 1 and 2.

Asher, Benjamin *The American Builders Companion* New York, Dover Publications, 1969. Reprint of 1827 edition.

Babb, Janice and Dordick, Beverly F. *Real Estate Information Sources* Detroit, Gale Research Co., 1963.

Badzinski, Stanley, Jr. *Stair Layout* Chicago, American Technical Society, 1971.

Ballinger, Richard M. and York, Herman (eds.) *The Illustrated Guide to the Houses of America: A Region-By-Region Survey of Contemporary and Traditional Residential Houses* New York, Hawthorn Books, 1971.

Barrows, Claire M. *Living Walls — How to Appreciate and Install Wallpaper and Wall Coverings* New York, Wallpapering Council, 1968.

Baumgart, Fritz *A History of Architectural Styles* New York, Praeger Publishers, 1969.

Bellis, Herbert F. and Schmidt, Walter A. *Blue Print Reading for the Construction Trade* New York, McGraw-Hill Book Co., 1968.

Bennett, George Fletcher *Early Architecture of Delaware* New York, Bonanza Books, 1932.

Boyce, Byrl N. *Real Estate Appraisal Terminology* Cambridge, Mass., Ballinger Publishing Company, 1975.

Brown, Elizabeth Mills *The Historic Houses of Wooster Square* New Haven, Conn., New Haven Preservation Trust, 1969.

Bunton, Alice Bice *Bethany's Old Houses and Community Buildings* Bethany, Conn., Bethany Library Association, 1972.

Burbank, Nelson L. and Phelps, Charles *House Carpentry Simplified* New York, Simmons-Boardman Publishing Co., 1958.

Burbank, Nelson L. and Shaftel, Oscar *House Construction Details* New York, Simmons-Boardman Publishing Corp., 1959.

Callender, John Hancock *Before You Buy a House* New York, Crown Publishers, 1953.

Camesasca, Ettore and Quigly, Isabel *History of the House* New York, G. P. Putnam's Sons, 1971.

Chamberlain, Samuel *Open House in New England* New York, Bonanza Books, 1958.

Collins, John J., et al. *Planned Unit Development with a Home Association* Washington, D.C., U.S. Federal Housing Administration, 1970.

Cook, Olive *The English House Through Seven Centuries* London, Thomas Nelson and Son, Ltd., 1968.

Cooper, James R. and Guntermann, Karl R. *Changing Neighborhoods: Social Evolution or Threat To Social Stability?* REAL ESTATE APPRAISER, November-December, 1972, p. 7-13.

Cutler, Laurence Stephen and Cutler, Sherrie Stephens *Handbook of Housing Systems for Designers and Developers* New York, Van Nostrand Reinhold Company, 1974.

Czolowski, Tel and Ramsey, Bruce *Under All is the Land* Chicago, NATIONAL ASSOCIATION OF REALTORS®, 1972.

Daniels, G. Emery *The Home Comfort Handbook* New York, Popular Library, 1966.

Davidson, Marshall B. *The American Heritage History of Notable American Houses* New York, American Heritage Publishing Co., 1970.

Davis, Deering; Dorsey, Stephen P. and Hall, Ralph Cole *Georgetown Houses of the Federal Period 1780-1830* New York, Bonanza Books, 1956.

Deering, David *Alexandria Houses 1750-1830* New York, Bonanza Books, 1956.

Deering, David *Annapolis Houses 1700-1775* New York, Bonanza Books, 1957.

Denton, John H. *Buying or Selling Your Home* New York, M. Barrows and Co., 1961.

Dietz, Albert G. *Dwelling House Construction* Cambridge, Mass., M.I.T. Press, 1971.

Downing, Antoinette F. and Scully, Vincent J., Jr. *The Architectural Heritage of Newport, Rhode Island* New York, Bramhall House, 1967.

Drury, John *The Heritage of Early American Houses* New York, Coward-McCann, 1969.

Duncan, Kenneth *Home Builders and Home Buyers — Blue Print for Happy Home Ownership* New York, Funk and Wagnalls Co., 1951.

Duncan, Kenneth *The Home Builders Handbook* New York, D. Van Nostrand Co., 1948.

Dunham, Clarence W., and Thalberg, Milton D. *Planning Your Home for Better Living* New York, Whittlesey House, 1945.

Everard, Kenneth Eugene *An Identification of Areas of Knowledge About Which Home Buyers Need Understanding* Doctoral dissertation, Indiana University, School of Education, 1962.

Ferguson, Emmie *Old Virginia Houses, The Mobjack Bay Country and Along the James* New York, Bonanza Books, 1955. Vol. 1 and 2.

Fleming, John *The Penguin Dictionary of Architecture* Middlesex, England, Penguin Books, 1966.

Fletcher, Banister *A History of Architecture on the Comparative Method for Students, Craftsmen and Amateurs* London, B. T. Batsford, Ltd., 1924.

Forest Products Laboratory and Central Mortgage and Housing Corporation *Canadian Wood Frame House Construction* Ottawa, Central Mortgage and Housing Corp., 1968.

Fowler, Glenn *How to Buy a Home — How to Sell a Home* New York, The Benjamin Co., 1969.

Frazer, Smith J. *White Pillars, The Architecture of the South* New York, Bramhall House, 1951.

Graham, Kennard C. *National Electric Code and Blue Print Reading* New York, American Technical Society, 1960.

Greiff, Constance M. *Lost America* Princeton, N.J., Pyne Press, 1971.

Griffin, Al *So You Want To Buy a House* Chicago, Henry Regnery Co., 1970.

Gutheim, Frederick *One Hundred Years of Architecture in America: 1857-1957* New York, Reinhold, 1957.

Hammond, Ralph *Ante-bellum Mansions of Alabama* New York, Bonanza Books, 1951.

Hoag, Edwin *American Houses: Colonial, Classic and Contemporary* Philadelphia, J. B. Lippincott Co., 1964.

Hoagland, Henry E. and Stone, Leo D. *Real Estate Finance* Homewood, Ill., Richard D. Irwin, 1973.

Huntoon, Maxwell C. *PUD — A Better Way for the Suburbs* Washington, D.C., Urban Land Institute, 1971.

Institute of Real Estate Management *Energy Conservation Recommendations* Chicago, 1975.

Isham, Norman M. and Brown, Albert F. *Early Connecticut Houses* New York, Dover Publications, 1965.

Ishimoto, Kiyoko and Ishimoto, Tatsuo *The Japanese House — Its Interior and Exterior* New York, Bonanza Books, 1963.

Johnstone, Kenneth B., et al. *Building or Buying a House — A Guide to Wise Investment* New York, Whittlesey House, 1945.

Jones, Rudard A. *Basic Construction and Materials Takeoff* Urbana, Ill., University of Illinois, Small Homes Council — Building Research Council, 1970.

Katz, Robert D. *Design of the Housing Site: A Critique of American Practice* Urbana, Ill., University of Illinois, Small Homes Council — Building Research Council, 1967.

Kelly, Frederick J. *Early Domestic Architecture of Connecticut* New York, Dover Publications, 1963.

Kent, Robert Warren *Choose Your House From These 104 Full Color Photos* Boston, Lee Institute, 1965.

Kent, Robert Warren *How to Choose Your House and Live Happily Ever After!* Boston, Lee Institute, 1965.

King, A. Rowden *Realtors' Guide to Architecture — How to Identify and Sell Every Kind of Home* Englewood Cliffs, N.J., Prentice-Hall, 1957.

Kirkpatrick, W. A. *The House of Your Dreams — How To Plan And Get It* New York, McGraw-Hill Book Co., 1958.

Knowles, Jerome *Single-Family Residential Appraisal Manual* Chicago, American Institute of Real Estate Appraisers, 1967.

Laas, William *Lawyers' Title Home Buying Guide* New York, Popular Library, 1969.

Lancaster, Clay *The Architecture of Historic Nantucket* New York, McGraw-Hill Book Co., 1972.

Lewis, David M. and Boulton, David R. *Houston — A City Without Zoning* REAL ESTATE APPRAISER, November-December, 1970, p. 39-43.

Lionel, Henry and Williams, Ottalie D. *Old American Houses 1700-1850, How to Restore, Remodel and Reproduce Them* New York, Bonanza Books, 1957.

Lockwood, Charles *Bricks and Brownstone: The New York Row House, 1783-1929 — An Architectural And Social History* New York, McGraw-Hill 1972.

Maass, John *The Gingerbread Age, A View of Victorian America* New York, Bramhall House, 1957.

Marshall, Robert A. *Before You Buy A House* Washington, D.C., Kiplinger Washington Editors, 1964.

Marshall and Swift Publication Co. *Residential Cost Handbook — Rapid Method of Computing Residential Costs* Los Angeles, 1971.

May, Arthur A. *The Valuation of Residential Real Estate* Englewood Cliffs, N.J., Prentice-Hall, 1966.

Mencher, Melvin, ed. *The Fannie Mae Guide to Buying, Financing & Selling Your Home* Garden City, N.Y., Doubleday, 1973.

Miller, Hope Ridings *Great Houses of Washington, D.C.* New York, Clarkson N. Potter, 1969.

Moger, Byron and Burke, Martin *How To Buy A House* New York, Lyle Stuart, 1969.

Mix, Floyd M. *House Wiring Simplified* South Holland, Ill., Goodheart-Willcox Co., 1973.

Murray, Robert W. *How to Buy the Right House at the Right Price* New York, Collier Books, 1965.

Nash, Joseph *The Mansions of England in the Olden Time* New York, Bounty Books, 1970.

National Institute of Real Estate Brokers *Real Estate Advertising Ideas* Chicago, 1973.

National Institute of Real Estate Brokers *Showing Property and Types of Homes* Chicago, 1946. Including the following articles:
  Lakeview Subdivision — Its Restrictions by Stewart B. Matthews
  Sale and Subdivision of Large Estates by Edmund D. Cook
  Showing Property with a Purpose by Lorin T. Driscoll
  Types of Homes by Victor D. Abel

National Institute of Real Estate Brokers *Today's Home — Know It, Show It, Sell It* Chicago, 1952.

Newcomb, Rexford *Old Kentucky Architecture: Colonial, Federal, Greek Revival, Gothic and other Types Erected Prior to the War Between the States* New York, Bonanza Books, 1970.

Parker, Alfred Browning *You and Architecture — A Practical Guide to the Best in Building* New York, Dial Press, 1968.

Perl, Lila *The House You Want, How to Find It, How to Buy It* New York, David McKay Co., 1965.

Peters, Frazier Forman *How to Buy A House — and Get Your Money's Worth* New York, Garden City Books, 1950.

Poor, Alfred Easton *Colonial Architecture of Cape Cod, Nantucket and Martha's Vineyard* New York, Dover Publications, 1970.

Porter, Sylvia *Money Book: How to Earn It, Spend It, Save It, Invest It, Borrow It—and Use It to Better Your Life* Garden City, N.Y., Doubleday, 1975.

Post, Emily *The Personality of a House (The Blue Book of Home Charm)* New York, Funk and Wagnalls Co., 1948.

Pothorn, Herbert *Architectural Styles* New York, Viking Press, 1971.

Pratt, Dorothy and Pratt, Richard *A Guide to Early American Homes, North and South* New York, Bonanza Books, 1956.

Ramsey, Charles G. and Sleeper, Harold R. *Architectural Graphic Standards* New York, John Wiley and Sons, 1970.

Randall, Anne *Newport — A Tour Guide* Newport, R.I., Catboat Press, 1970.

Reiff, Daniel D. *Washington Architecture 1791-1861: Problems in Development* Washington, D.C., U.S. Commission of Fine Arts, 1971.

Ring, Alfred A. *Real Estate Principles and Practices* Englewood Cliffs, N.J., Prentice Hall, 1972.

Rogers, Tyler S. *The Complete Guide to House Hunting* New York, Avon Books, 1963.

Schuler, Stanley *Home Renovation: Making Your Home More Attractive, Modern and Valuable* Reston, Va., Reston Publishing Co., 1974.

Schwartz, Robert and Cobb, Hubbard H. *The Complete Homeowner* New York, Macmillan Co., 1965.

Scott, John S. *A Dictionary of Building* Middlesex, England, Penguin Books, 1969.

Scott, Mary Wingfield *Houses of Old Richmond* New York, Bonanza Books, 1941.

Scully, Vincent *American Architecture and Urbanism* New York, Praeger Publishers, 1969.

Semenow, Robert W. *Questions and Answers on Real Estate* Englewood Cliffs, N.J., Prentice Hall, 1971.

Sleeper, Catharine and Sleeper, Harold *The House for You to Build, Buy or Rent* New York, John Wiley and Sons, 1948.

Small Homes Council *Circular Series* Urbana, Ill., University of Illinois. Including the following pamphlets:

B3.0	*Fundamentals of Land Design*	D7.0	*Selecting Lumber*
C1.1	*Hazard-Free Houses for All*	D7.2	*Plywood*
C1.5	*Living with the Energy Crisis*	F2.0	*Basements*
C2.5	*Split-Level Houses*	F2.5	*Termite Control*
C5.1	*Household Storage Units*	F3.0	*Wood Framing*
C5.32	*Kitchen Planning Standards*	F4.4	*Crawl-Space Houses*
C5.4	*Laundry Areas*	F4.6	*Flooring Materials*
C5.6	*Bedroom Planning Standards*	F6.0	*Insulation in the Home*
C5.9	*Garages and Carports*	F6.2	*Moisture Condensation*
C7.3	*Pressure Treated Wood*	F9.1	*Counter Surfaces*

F11.0	*Window Planning Principles*	G3.5	*Fuels and Burners*
F11.1	*Selecting Windows*	G4.2	*Electrical Wiring*
F11.2	*Insulating Windows and Screens*	G5.0	*Plumbing*
F12.3	*Roofing Materials*	G6.1	*Cooling Systems for the Home*
F17.2	*Brick and Concrete Masonry*	H1.0	*Interior Design*
G3.1	*Heating the Home*		

Society of Real Estate Appraisers *Real Estate Appraisal Principles and Terminology* Chicago, 1969.

Springer, John L. *The Home You've Always Wanted At a Price You Can Afford* Englewood Cliffs, N.J., Prentice-Hall, 1962.

Stephen, George *Remodelling Old Houses Without Destroying Their Character* New York, Knopf, 1974.

Sunset Books *Cabins and Vacation Houses* Menlo Park, Calif., Lane Books, 1971.

Taeuber, Conrad, et al. *Density: Five Perspectives* Washington, D.C., Urban Land Institute, 1972.

The Reader's Digest Association *Reader's Digest Complete Do-It-Yourself Manual* New York, 1973.

Thompson, Elizabeth Kendal, ed. *Apartments, Townhouses and Condominiums* McGraw-Hill, New York, 1958.

Underwriters' Laboratories, Inc. *Building Materials List* Chicago. (Revised regularly.)

United States Savings and Loan League *Construction Lending Guide* Chicago, 1971.

United States Savings and Loan League *Savings and Loan Fact Book* Chicago. (Issued annually.)

U. S. Department of Agriculture Yearbook of Agriculture: *Handbook for the Home* Washington, D. C., U. S. Government Printing Office, 1973.

U. S. Department of Agriculture Yearbook of Agriculture: *Shopper's Guide* Washington, D. C., U. S. Government Printing Office, 1974.

U. S. Department of Agriculture, Extension Service *Simple Home Repairs* Washington, D. C., U. S. Government Printing Office, 1972.

U. S. Department of Agriculture, Forest Service *New Life for Old Dwellings*, Washington, D. C., U. S. Government Printing Office, 1975.

U. S. Department of Housing and Urban Development *Protecting Your Housing Investment* Washington, D. C., U. S. Government Printing Office, 1974.

U. S. Department of Housing and Urban Development *Rent or Buy?* Washington, D. C., U. S. Government Printing Office, 1974.

U.S. Department of Housing and Urban Development *Wise Home Buying* Washington, D.C., U.S. Government Printing Office, 1972.

Valorie, Carl M. *A Dream House Without A Nightmare* New York, Vantage Press, 1970.

Vanderbilt, Cornelius, Jr. *The Living Past of America: A Pictorial Treasury of Our Historical Houses and Villages That Have Been Preserved and Restored* New York, Bonanza Books, 1955.

Venturi, Robert *Learning From Las Vegas* Cambridge, Mass., M.I.T. Press, 1972.

Watkins, A. M. *Building or Buying the High-Quality House at the Lowest Cost* New York, Doubleday and Co., 1962.

Watkins, A. M. *How to Avoid the Ten Biggest Home-Buying Traps* New York, Meredith Press, 1960.

Weiss, Joseph Douglas *Better Buildings for the Aged* New York, Hopkinson and Blake, 1969.

Whiffen, Marcus *American Architecture Since 1780 — A Guide to the Styles* Cambridge, Mass., M.I.T. Press, 1969.

White, Norval and Willewsky, Elliot (eds.) *A.I.A. Guide to New York City* New York, Macmillan Co., 1967.

Whiton, Sherrill *Elements of Interior Design and Decoration* Philadelphia, J. B. Lippincott Co., 1963.

Wilhelm, John L. *Solar Energy, the Ultimate Powerhouse* NATIONAL GEOGRAPHIC, March, 1976, p. 381-397.

Williams, Henry Lionel and Williams, Ottalie K. *America's Small Houses (and City Apartments): The Personal Homes of Designers and Collectors* New York, Bonanza Books, 1964.

Williams, Norman, Jr. *The Structure of Urban Zoning and Its Dynamics in Urban Planning and Development* New York, Buttenheim Publishing Co., 1966.

Wren, Jack *Home Buyers Guide* New York, Barnes and Nobel Books, 1970.

Wright, Frank Lloyd *The Natural House* New York, Bramhall House, 1956.

Information on and photographs of specific styles were supplied by the following:

Carpenter Gothic: The Greater Eureka (California) Chamber of Commerce
Classic Colonial: Virginia State Travel Service
Georgian Colonial: Philadelphia Convention and Tourist Bureau
Garrison Colonial: The Farmington Museum
High Victorian Italianate: Oregon Historical Society
Japanese: Ann Carmichael, Montgomery, Alabama
Log Cabin: Hubert L. Duncan, Fayetteville, Ark.
Mobile Homes: Mobile Homes Association
Prairie Style: John G. Hoppe, Oak Park, Ill.
Pennsylvania Dutch: Gilbert Harrison and Shelley Harrison, Rydal, Pa.
Adobe: Mike Fiddes and Tess Fiddes, Santa Fe, N.M.
Southern Colonial: Virginia Chamber of Commerce
Swiss Chalet: Paul Manchester, Westport, Conn.
Most of the Victorian Styles: Historical Buildings Survey, Library of Congress

The following organizations and companies supplied background material, technical information and replied to specific requests for information:

Air-Conditioning and Refrigeration Institute
American Institute of Architects
American Plywood Association
Aromatic Red Cedar Closet Lining Manufacturers Association
Asphalt Roofing Manufacturers Association
Better Floors Council
Better Heating-Cooling Council
Better Homes and Gardens
Better Light — Better Sight Bureau
Carpet and Rug Institute
Cast Iron Soil Pipe Institute
Copper Development Association
Crane Co.
Edison Electric Institute
Electric Energy Association
Eljer Plumbingware Division, Wallace-Murray Corp.
Fir and Hemlock Door Association
Gypsum Association
Maple Flooring Manufacturers Association
Master Plan Service
National Association of Building Manufacturers
National Association of Home Builders
National Cellulose Insulation Manufacturers Association
National Concrete Masonry Association
National Electrical Contractors Association
National Electrical Manufacturers Association
National Forest Products Association
National LP-Gas Association
National Mineral Wool Insulation Association
National Oak Flooring Manufacturers Association
National Oil Fuel Institute
National Paint, Varnish and Lacquer Association

National Swimming Pool Institute
National Well Water Association
National Woodwork Manufacturers Association
Owens-Corning Fiberglass Corp.
Plastics Pipe Institute
Ponderosa Pine Woodwork Association
Red Cedar Shingle and Handsplit Shake Bureau
Steel Door Institute
Steel Window Institute
Southern Pine Association
Tile Council of America
Underwriters' Laboratory
Western Wood Moulding and Millwork Producers
Western Wood Products Association
Wood and Synthetic Flooring Institute

GLOSSARY — INDEX

Material that is added to cement to lengthen its setting time or increase its air-entraining capacity.

A type of brick, originally made by American Indians of the West, of adobe soil and straw hardened in the sun.

A coarse material, usually sand, stone and gravel, used as part of the ingredients of concrete.

A pipe filled with air, installed next to plumbing fixture to stop pipes from hammering.

An additive mixed into wet concrete to make it contain minute bubbles in order to improve the concrete's quality.

A unit used to measure the amount of electricity that flows through a wire. Houses commonly have 30, 60, 100, 150, 200 or more ampere services.

A bolt set into the top of a concrete foundation that extends through the sill and holds the sill down.

A process of applying an electrolytic oxide protective coating to aluminum.

In Greek architecture, the lowest part of the entablature. In modern use, it is the moulding above and on both sides of a door.

A well drilled through impermeable strata deep enough to reach water that will rise to the surface by its own internal hydrostatic pressure. Named after Artois, France, where these wells were first drilled.

A pit at the base of a chimney for fireplace ashes.

A type of window in which each light opens outward on its own hinge.

Piling up dirt and stone around the edge of a foundation and basement wall high enough to make a slope for drainage of water away from the foundation.

A type of framing system where the studs extend unbroken from the sill to the roof.

One of a string of small poles used to support the handrail of a stairway.

A row of balusters surmounted by a rail or cornice.

A piece of interior trim laid around the walls of a room next to the floor. Often it consists of three pieces of moulding called the base shoe, baseboard and base moulding.

405

A narrow board used to cover the space between siding boards or sheathing. Also a special type of board used for flooring.

A wall style copied from fortress walls with notches along the top used by soldiers to shoot through.

A window projecting outward from the wall of a house supported by an extension of the founda-tion. (See Oriel Window.)

A type of masonry joint that has a bead cut into it, primarily for decorative purposes.

A large piece of wood, steel, stone or other material used to support a house. It usually runs from foundation wall to foundation wall and is supported with poles or pillars.

A wall that supports the ceiling, floor or roof above it.

A glass-enclosed small room on the roof of a house used as a lookout, often reached through a trap door.

A mound of earth or pavement used to control the flow of surface water.

A type of house siding which is thicker on one edge than the other when installed.

A water faucet on the outside of the house with a screw nose to which a hose can be connected.

A bathroom fixture used to wash the perineal area after using the toilet.

A type of house. It has two stories. The lower story is partially below ground. The level at the front entry is halfway between the two stories.

A type of insulation usually 2 or 3 inches thick and 15 to 23 inches wide that comes in long rolls. Often it is nailed between the studs of an exterior wall. They are often made of mineral wool, wood fibers, glass fiber, cotton, eel grass or cattle hair.

A piece of sawed lumber usually 1 to 2 inches thick and 8 to 12 inches wide.

A type of house siding that consists of wide boards butted together with a narrow board nailed over the joint.

The part of the furnace that holds the water being heated by the ignited fuel.

A covered passage between the house and garage.

An arrangement of bricks on a wall where some are laid parallel to the wall and some perpendicular in order to give the wall additional strength. Some standard arrangements are common bond, English bond, English cross, Flemish bond and Flemish cross.

A small piece of wood or strips of metal nailed between joists or studs to give them lateral rigidity.

A flat fired clay tile that comes in a variety of types such as glazed, unglazed, non-vitrified, vitreous, etc. When attached to walls, floors and counter tops, may form a decorative, hard, durable surface.

Part of a waste disposal system that functions similarly to a septic tank. A covered cistern of stone, brick or concrete block. The liquid seeps out through the walls directly into the surrounding earth.

A type of moulding installed around the room at chair-back height to protect the walls from damage.

An extension at the top of the chimney made of tile or sheet metal used to improve the draft.

A permanent safety device used as a substitute for a fuse. It automatically turns off an electric circuit when it is overloaded.

A long board used as siding, thicker on one side than the other. It is installed so that the thick side overlaps the thin side.

Used in the CTS System to denote the number of families the unit being described will accommodate and whether the units are detached or have a party wall.

A door at the bottom of the chimney near the ash pit used to clean out ashes.

A line of columns at regular intervals usually used to support an architrave.

An outer door frame having an inside removable section for year-round utility. A screen panel is inserted in warm weather and a glass panel in winter.

A system that handles both storm water and sanitary sewage together. (See sewers.)

Any brick commonly used in an area for general building purposes. It usually is not specially colored or textured and may include clinker and over-burnt bricks. (See Clay Brick.)

A form of ownership in which each owner owns the fee to the individual unit and a percentage of the fee to the common areas.

In electricity, a wire or path through which electric current flows.

A type of heat distribution system consisting of

pipes with many fins attached at short intervals. The hot water or steam blows through the pipes, heating the fins which in turn heat the surrounding air.

Convenience Outlets *(i.)* 321, 366, *(i.)* 367, *(i.)* 368, *(i.)* 369
Conversation Circle *(i.)* 55-57, 59
Cooperative

A form of ownership in which each owner owns shares in the entire property and has the exclusive right to occupy part of the property.

Corbel 165, 168

A piece of wood or masonry projecting from a wall used to support some part of the house above it.

Cored Brick 240
Cork Tile 247-248, 336
Corner Beads 194
Cornice *(i.)* 132, *(i.)* 159

A molded projection at the top of an interior or exterior wall. The enclosure at the roof eaves or at the rake of the roof.

Cotswold Cottage *(i.)* 122
Cove Lighting 58

Lighting that is behind a concave moulding near the ceiling.

Cove Moulding 196, *(i.)* 197, 198

A concave or quarter round piece of moulding.

Cracked or Loose Mortar Joints 242
"Crapper's Valveless Water Waste Preventer" 80
Crawl Space 175, 219, *(i.)* 221

In basementless houses the open space between the underside of the floor joists and the ground.

Creole Style *(i.)* 131
Cross Bridging See Bridging
Crown Moulding *(i.)* 258
CTS System 28; 344-349

Class, Type, Style. A uniform method for describing houses.

Cupola See Belvedere
Curbs 375, *(i.)* 379
Damper 205, *(i.)* 206-207, 216

A metal plate in a flue that can be adjusted to regulate the draft. Also a device to control vibrations.

Dampness 175-176
Damp-Proofing See Vapor Barrier
Dead Load

The permanent weight which the structural parts of a building are subject to, such as the weight of the roof and upper floor of the bearing walls. (See Load, Live Load, and Total Load.)

Decay Resistant Wood 222
Deciduous Trees 232
Decorating 222-229
Defective Chimneys 208
Defective Framing 181
Defective Joints 242
Defective Plumbing 308-309
Defective Roofing 203
Defective Windows 269
Den 42
Dentil *(i.)* 127

Alternate square blocks and blank spaces on a cornice that give the appearance of teeth.

Detached Garage 102

411

In an electric system, a metal box containing the fuses (fuse box) or circuit breakers. It is used to divide a supply line into branch circuits. In a septic system, an underground box which divides the waste flowing from the septic tank and distributes it to the laterals leading to the disposal field.

The framing member at the bottom of a door opening that serves as a threshold.

A device attached to the wall or floor to prevent a door from opening too far and damaging the wall.

A projection built out from the slope of a roof, used to house windows on the upper floor and to provide additional headroom. Common types are the gable dormer and shed dormer.

A window with two sashes that move up and down independently in tracks.

A pin of wood or metal used to hold or strengthen two pieces of wood where they join.

Downspout, Downpipe 176, 209, *(i.)* 210

A pipe used for carrying rain water from the roof of the house to the ground or into a dry well or sewer connection.

Drains 172

Dressed Lumber *(i.)* 234

Lumber that has been machined and surfaced at the lumber mill.

Dressing Room 42

Drip Cap *(i.)* 267

A projection from the vertical face of an exterior wall used to make rain water drip from its outer edge rather than run down the wall face.

Driveways 24, 103, 239, 375, *(i.)* 379

Drops See Carved Drops

Drop Siding *(i.)* 183, 185, 234

A type of tongue and grooved exterior board siding.

Dry Wall 253-254, *(i.)* 372

A masonry wall laid up without mortar. An interior wall finished with something other than plaster. Commonly used to refer to a gypsum board finished wall.

Dry Well 172, 176

A hole in the ground lined with stone used to disburse waste or rain water into the ground.

Duct 274, *(i.)* 275, 277

In house construction, pipes of metal, plastic or other material used for distributing or collecting air.

Duplex Outlet 366, *(i.)* 367, *(i.)* 368, *(i.)* 369

Duplex Receptacle 322

Dutch Colonial Style *(i.)* 114

Dutch Door *(i.)* 114

A door that is divided horizontally in the middle so the halves may be opened and closed independently or together.

Early Georgian Style *(i.)* 125

Earth Symbols *(i.)* 370, 371

Earthquakes 180

Easing *(i.)* 197

Eastern Townhouse *(i.)* 148

Eastlake Style *(i.)* 144

Eave *(i.)* 191, *(i.)* 210

The lowest projecting part of the roof.

Eave Flashing *(i.)*. 210

Eave Vent *(i.)* 191

Eavetrough, Gutter 209

A channel of wood or metal on or at the edge of the roof to carry off water from rain and melting snow.

Economic Base 2

Education 3

Eelgrass 189

Egyptian Revival Style *(i.)* 137

Electric Air Cleaner 274

Electric Heat 283, 290, 315, 320

Electric System 312-325

artificial lighting, 324; branch circuits, *(i.)* 314, 316, 322; circuit breakers, *(i.)* 314, 316, 319; conduit, *(i.)* 314; convenience outlets, *(i.)* 321; distribution box, 315, 316; distribution panel, *(i.)* 314, 316; dimmer switches, 323;

The floor of an apartment building or row house that is about half below ground level.

In Greek architecture, that part of the building which rests horizontally upon the columns. Its major parts are the cornice, frieze and architrave.

A process of forming metal into a shape by forcing it through a hole in a die.
A piece of horizontal trim running along the perimeter of a flat roof. In Greek architecture, the horizontal band in the architrave.

A device at the end of a pipe to control the flow of water.

Federal Style *(i.)* 107
FHA-MPS

Federal Housing Administration, Minimum Property Standards. (See Introduction.)

Fiberboard 194-195, 201

A type of building board used for insulation, made of reduced fibrous material such as wood, cane or other vegetable fibers.

Fiberglass Siding 186
Field Tile *(i.)* 173

A porous tile laid under and around a foundation or slab to drain away the underground water.

Finial

The decorative top of a cupola or gable which usually forms the base of a flag pole or weather vane.

Finish Coat See Stucco
Fire Clays 239-240
Fire Protection and Safety 215-217
Fire Stopping *(i.)* 178, *(i.)* 207, 215

Material used to block the air spaces in walls to prevent the spread of fire through these air passages.

Firebrick 215-216, 265

A brick of fire clay which is made to withstand the effects of high heat used in fireplaces and heating chambers of furnaces.

Fireplaces 54, *(i.)* 56, *(i.)* 57, 205, *(i.)* 206, 208, 215-216, 239, 322
Flashing 182, *(i.)* 184, *(i.)* 210, 215, *(i.)* 372-374

A piece of material, usually metal or composition, used to protect, cover or deflect water from places where two materials join or form angles such as roof valleys.

Flat Roof 165, *(i.)* 169

A roof having a slope just sufficient to provide water drainage and a pitch not over 1 to 20. (See Roof Pitch.)

Flemish Bond 188
Flexible Conduit *(i.)* 317, 320
Floating Foundation, Floating Slab, Mat Foundation, Raft Foundation, Rigid Foundation 175

Special foundations with few or no footings used where the land is swampy or the soil unstable.

Flood Lights 324
Floor Beams *(i.)* 179
Floor Construction *(i.)* 184
Floor Framing *(i.)* 173, *(i.)* 197, 211
Floor Furnace 273

A type of heating unit that is installed directly under the floor with its grilled upper surface flush with the finished floor.

Floor Insulation *(i.)* 191
Floor Joists *(i.)* 173, 176
Floor Outlet *(i.)* 367
Floor Plan Deficiencies 51
Floor Plans *(i.)* 352, *(i.)* 364-365
Flooring 211-215, *(i.)* 213
Flue *(i.)* 207-208, 215-216, 363, *(i.)* 364-365

An enclosed passage in a chimney or any duct or pipe through which smoke, hot air and gases pass upward.

Flue Gases 205
Flue Lining *(i.)* 207, 208
Fluorescent Tubes 324, *(i.)* 367
Flush Door 260, *(i.)* 261, *(i.)* 378

A door without panels that has two flat surfaces. Often it has a hollow core.

415

Minerals used in the steel making process that have an affinity to the impurities in iron ore and cause it to separate and form into slag.

A door with panels which bend back on each other when opened.

A concrete support under a foundation, chimney or column that usually rests on solid ground and is wider than the structure being supported.

A bearing wall below the floor joists or below the first floor or below the ground which supports the rest of the house.

A class of house with four separate units connected by party walls.

A part of the building process that consists of putting together the lumber skeleton parts of the house.

A horizontal trim immediately below the cornice soffit. In Greek architecture, the part above the architrave and below the cornice.

Thin strips of wood used to level up a wall and provide air space between the wall and the plaster.

An electrical safety device that protects the wiring from overheating and becoming a fire hazard. When too much electricity is being drawn through the wiring, the fuse melts, stopping the flow of electricity.

A triangular gable formed by the ends of a ridged roof.

A two pitched roof style which has its slope broken by an obtuse angle so that the lower slope is steeper than the upper slope.

A style of entry hood found on Pennsylvania Dutch houses that are unsupported by columns.

Terms used to describe excessive ornamentation on a house.

A large horizontal beam, often carrying other beams and joists, on which the first floor is laid.

A thin type of mortar used to fill the spaces between non-resilient tiles.

Wood from deciduous trees. Does not refer to the hardness of the wood. Oak, birch and maple are commonly used in house construction for floors, cabinets, interior paneling and furniture.

The top-most piece of framing of a door or window frame.

A framing member that goes across the top of an opening and carries the load above. In masonry a brick, block or stone laid with its end toward the face of the wall.

The top framing member of a door or window.
(See Jamb.)
The floor of a fireplace and the floor imme-
diately in front of a fireplace.
The wood at the center of a log.

A part of a framing member that rests on the
wall plate.

A roof that rises by inclined planes from all four
sides of the house.
A door or wall built with two facing panels with
a space between them filled with air, insulation
or other filler.

A masonry wall made of two layers of masonry
separated by a small air space.
A piece of material on the sides of an awning
type window that reduces the draft.

Large brackets under the cornices common on the Italian Villa house style.

A hot water heating system.

An early Roman heating system consisting of small masonry flue in the walls and floor that distributed hot air from room to room.

Any material used to reduce the transmission of heat and cold or to reduce fire hazard.

Windows and doors made of movable glass louvers which can be adjusted to slope upward to admit light and air yet exclude rain or snow.

The framing pieces that line doors and windows. (See Head Jamb and Side Jamb.)

419

An operation of finishing off the surface of the mortar between courses of brick or stone with a special tool called a jointer. Common joint types are flush joints, concave joints, tooled joints, weather joints, "V" tooled joints, rabbet joints, cove joints, raked joints, stripped joints and struck joints.

Heavy pieces of timber laid horizontally on their edge to support the floors and ceilings.

An earthenware container filled with hot embers, covered with a cloth and usually placed under a table to provide heat. It is still used today in India and Turkey.

A large oven-like chamber used for baking, drying and hardening various materials such as lumber, brick and lime.

Lumber that has dried in a kiln from two days to several weeks, depending upon its thickness and grade, to reduce its moisture content.

An early method of electric wiring still used in some areas today. The wire is attached to the house frame with porcelain knob insulators and porcelain tubes.

In Hawaii and the West Coast, a veranda, porch or covered patio.

Strips of wood or metal often ⅜ inch thick and 1½ inches wide, used as a base for plaster.

The process of installing lath. It is also synonymous for lath.

A field of trenches dug into the ground and lined with tile, broken stone, gravel or sand, used to disburse waste water from a septic tank into the ground.

A type of stone used as a building material. Very popular in Indiana.

A pane of glass.

A floor covering material made of linseed oil, powdered cork, resins and color pigments which are coated onto burlap and baked under pressure for curing.

A variable weight which a house is subject to such as snow on the roof and people on the floors. (See Load, Dead Load and Total Load.)

Weight supported by a structural part such as a load bearing wall. (See Dead Load, Live Load, Total Load.) As an electrical term, power delivered by a motor and the electric current carried through a circuit.

A measurement of light. A 100 watt bulb gives off about 1600 lumens of light.

A roof with two slopes on all four sides. The lower slope is very steep. The upper slope is usually not visible from the ground.

A house of considerable size or pretension.
An ornamental facing around a fireplace.
A protruding shelf over a fireplace.
A term applied to anything constructed of stone, brick, tiles, cement, concrete and similar materials. Also, the work done by a mason who works in stone, brick, cement, tiles or concrete.

A type of insulation made by sending steam through molten slag or rock. Common types are glass, wool, rock wool and slag wool.

Moulding 232, *(i.)* 234, 257-260, *(i.)* 258

A long, narrow strip of wood or metal, plain, curved or formed with regular channels and projections, used for covering joints and for decorative purposes.

Mount Pleasant *(i.)* 127
MPS (FHA) See FHA-MPS
Mud Room 51
Mullions *(i.)* 124

Vertical bars between multiple windows.

Multiple Regression Analysis 344
Municipal Sewer System 309
Muntin *(i.)* 267

Thin bars of wood or metal between lights of glass in a window.

Music Systems 320
Nails 242, 245
National Academy of Sciences 360
Neighborhoods 4-9
  boundaries, 5; definition, 4; early warning signs, 7; fair housing, 8; land prices, 7; life cycles, 6-9; physical characteristics, 5; streets, 5; quality of life, 8; urban, 5
New England Colonial Style *(i.)* 113
New England Farm House Style *(i.)* 108
New Orleans Style *(i.)* 131
Newel, Newel Post 196, *(i.)* 197

The upright post at the bottom of a stairway that supports the handrail. (See Stairs.)

Newel Cap *(i.)* 197
Nogging *(i.)* 123

An ornamental finish on the top of a newel post. Used bricks placed between the timbers of wall construction. Called covered nogging when covered with stucco and exposed nogging when visible.

Noises in the Plumbing System 309
Nonbearing Wall *(i.)* 200

A wall that supports only its own weight.

Non-Metallic Cable *(i.)* 317, 320
Nonzoning 17
Nosing (Nose) 196, *(i.)* 197, 198

The rounded outer face of a stair tread.

Nuclear Energy 290
Nuisance Laws 10
Oak Flooring 238
Observatory See Belvedere
Octagon House *(i.)* 139
Octagonal Window *(i.)* 126
Ogee 209

Ornamental moulding on a gutter.

On Center, O.C., Center to Center 180, 193-196, 199, 212

A measuring term meaning the distance from the center of one structural member to the center of a corresponding structural member, such as studs or joists.

One-Pipe Return Tee Fitting 280, *(i.)* 282
One-Pipe Series Connected Heating System 280, *(i.)* 281
Ore

A natural mineral containing enough metal, coal or other useful materials to make mining it profitable.

A window and bay projecting from an upper story that is supported by corbels or brackets rather than a foundation wall.

The part of a roof that extends beyond the exterior wall. (See Cornice.)

A complete factory-built chimney, usually made of metal and insulation. Used primarily with prefabricated fireplace units.

The part of the wall of a house that rises above the roof line.

A coating of cement applied to the surface of a masonry wall to waterproof it.

Hardwood flooring consisting of small pieces laid in a decorative pattern.

A common wall in a house that separates two units in separate ownership.

An open, paved area used for outdoor living. It might be all or partially surrounded by parts of the house.

A short, hood-like roof between the windows of the first and second floor.

A drawing in which the lines are drawn to appear as the eye sees them in real life.

See Perspective Drawing

An iron ingot.

A rectangular column attached to the exterior side of a wall, often of the same color and material.

A horizontal structural member placed across the top of the studs in a frame wall. On the top floor it also serves as the place where the attic joists and roof rafters rest and are fastened to.

A large duct or air chamber in which the hot air from the furnace builds up pressure which forces it out through the ducts to the registers.

A building material consisting of thin sheets of wood glued together under pressure.

A process of troweling mortar into the joints between bricks or blocks to waterproof it.

An open space covered with a roof that is supported by columns.

A square or nearly square cut stone with beveled edges, set in the corners of masonry and stone houses.

A groove cut in the edge of a piece of wood so that another piece cut the same way will fit tightly into it. (See Joints.)

A type of conduit that carries electric wires.

A heating system that consists of pipes embedded in the walls, floors and ceilings.

A way that heat is transferred from a hot surface to the surrounding objects.

An exposed fixture usually made out of cast iron that transfers heat from the heating system by means of convection and radiation.

A sloping structural member of the roof that extends from the ridge or hip to the eaves and used to support the roof deck, shingles or other roof covering.

A horizontal member used as a guard such as a porch rail. A part of a door that runs horizontally. Based on its position it is known as a bottom rail, middle rail or top rail. Synonymous for stile. A part of a window that runs horizontally.

The outer edge of a sloped roof. The angle or slope of a roof rafter, known as the "rake of the roof."

Shingles of different widths.

Materials that resist heat, used to line furnaces and vessels used in iron and steel making.

A fixture installed at the end of an air duct that directs and controls the flow of air into the room.

A mathemathical formula used with the aid of a computer to analyze the effect independent variables (size, number of rooms, location, etc.) have upon dependent variables (sales price). A new technique being used to predict the sales price of houses.

The top structural member to which the rafters are nailed.

The part of a stair that is the vertical board under the stair stop. (See Stairs.) In plumbing systems, the vertical pipe that supplies steam to the radiators in the upper rooms.

The top of a house. Most common house roof types are gable, gambrel, hip, mansard and flat.

Boards or sheathing that are nailed to the rafters to which are fastened the roof covering or shingles, tiles or other material. (See Sheathing.)

The slope of the roof which is expressed in a ratio of vertical drop to horizontal distance. A 6-inch pitch or 6 in 12 pitch or 6 to 12 pitch means the roof rises 6 inches for each 12 inches of horizontal distance.

Material used to cover the upper exterior portion of a house to make it watertight.

The pipe system that carries the domestic waste
to a sewerage treatment plant or other disposal
area. (See Combination Sewer, Storm Sewer,
Septic System.)
A framework into which window frames are set.
The cord or chain that attaches the counter balance weights to each double-hung window sash.

A trade name for gypsum wallboard.

Thin pieces of wood or other material that are
tapered and oblong in shape, used as wall and
roof coverings.

A type of close rabbeted joint used to join two
boards together. (See Joints.)

The bottom framing piece of a stud wall laid on top of the rough flooring in platform framing. A type of moulding trim used to cover the junction of the floor and wall.

Pieces of glass at the sides of the front entrance door.

A framing member placed on top of the foundation wall that serves as a level base for the wall studs and floor joists. (See Window Sill and Door Sill.)

A quality toilet in which the flushing action is assisted by a jet of water.

A type of basementless foundation made by pouring concrete directly on the prepared ground surface and over the top of the footings.

A molten mass of fluxes and impurities that floats to the surface and is removed during the making of steel.

A large space over the damper that prevents the smoke from backing up into the fireplace. A part of a fireplace that prevents the cold air from flowing down the inside of the chimney into the fireplace.

Wood from trees known as conifer or coniferous because they have cones and usually needlelike leaves which are Evergreen. Does not refer to the softness of the wood. Pines, cedars and firs are commonly used in house construction for structural and framing lumber, sheathing and exterior siding.

A shelf-like support base of wood at the bottom of a window.

A brick, stone or block laid in a wall so its side is facing outward and is the exposed surface.

An inclined structural member that is shaped to support the treads and risers of a staircase.

A cement or plaster wall covering that is installed wet and dries into a hard surface coating.

A vertical framing member in a wall or partition.

Used in the CTS System to describe the design of the house based on historical or contemporary fashion.

A floor laid over the floor joist. The finish flooring is attached to the subflooring.

A hole or depression to collect water, usually in the basement floor.

An automatic electric pump installed in a sump hole that activates as the water in the hole reaches an excessive level and pumps it out to a drainage area.

An insect that resembles, but can easily be distinguished from, an ant. It lives underground and migrates into houses, eating wood.

A protective shield, usually of metal, placed so as to interrupt the migration path of termites from the ground into the house.

Terne Plate

A roofing material made of an alloy of tin and lead.

Terra Cotta 241

Terrazo 214, 239, 245, *(i.)* 370, 371

A floor covering of Portland cement into which is poured marble or stone chips which are polished to a smooth surface after the mixture has hardened.

Thatched Roof

A roof made of bundles of straw or shingles shaped to appear as if the roof was made of thatch.

Thermostat 58, 280, 286, 306, *(i.)* 367, *(i.)* 368

An electric, temperature-sensitive instrument that controls a heating or air-conditioning unit by turning it on and off as the temperature varies from the preselected setting. Automatic ones also turn the units off or on at preset times.

Three Prong Duplex Grounding Outlets 222

Three Prong Plug 322

Three-Way Switches 323, 366, *(i.)* 367

Thrust Block *(i.)* 197

Tie

Any structural member, piece of metal or other material used to hold two separated members together such as two beams or two layers of a masonry wall.

Tile *(i.)* 205, *(i.)* 213, 252, 333

A wide variety of building materials made of fired clay, cement, plastic, vinyl, stone, glass or other material used for floors, walls, drains and a number of other purposes.

Tile Grout 239, 240

Tile Laying 187

Timber 232

A large piece of lumber. Trees that are still growing.

Toe-nailed

A method used to drive in nails so that their heads will not be visible when the installation is complete.

Toilet Paper Holder 85

Toilets 80, *(i.)* 297, 301, 303, *(i.)* 304, 305, 362

Tongue and Groove *(i.)* 202

A type of joint where one half has notches cut into it (grooves) and the other matching piece has tongues that fit into the grooves.

Tongue and Groove Siding *(i.)* 183

Tooled Joint 185, *(i.)* 188

Top Late *(i.)* 178, 180, *(i.)* 262

Top Rail *(i.)* 258, *(i.)* 267

Towel Bars 85

Townhouse *(i.)* 149

Tracery *(i.)* 136

Transom, Transom Window

A small window over a door or another window.

Transom Bar

A horizontal bar that divides a window or other opening into two parts.

Trap *(i.)* 297, 308

A bend in a waste pipe that remains filled with water, making a water seal that prevents sewer gases from backing up into the house.

Tread 196, *(i.)* 197, 198

The upper horizontal board on each step of a flight of stairs. (See stairs.)

Treated Wood 217, 222

Trim — Wood or metal interior finishing pieces such as door and window casings, mouldings and hardware.

Trimmer — A floor joist or beam that supports the end of a header.

Trimmer Arch 205

Trowel — A tool used to spread cement, plaster or adhesive.

Trowel Joints 187, *(i.)* 188

Truss 199, *(i.)* 200 — A triangular arrangement of framing members usually preassembled and lifted intact into place. Used to support loads over long spans.

Tub-Shower Combinations 302-303, *(i.)* 362

Tudor Style *(i.)* 124

Turret 124 — A small decorative tower.

Two Panel Door *(i.)* 378

Two Pipe Reverse Return Heating System 280, *(i.)* 281

Two Pipe Steam Cast Iron Radiator *(i.)* 285

Two-Story House See House Types

Two-Way Snap Switch 323

Type 28-42; 345 — Used in the CTS System to denote the structural nature of the house.

Types of Houses See House Types

"U" Kitchen *(i.)* 63, 71

Underground Conduit Pipe 315

Underpinnings — Supports added to increase the load bearing capacity of a foundation.

Undersized Pipes 309

Unglazed Tile 240

Unit Wood Block Flooring *(i.)* 213

Unseasoned Lumber 329

Unsupported Hood *(i.)* 116

Urban Neighborhoods 6

Utility Room See Laundry

"V" Shaped Joints 187, *(i.)* 188

Vacuum Line *(i.)* 362, 363

Valleys — A depressed angle formed when the bottom of two roof planes meet.

Vapor Barrier *(i.)* 173, 175, *(i.)* 213, 215, *(i.)* 372 — Material or paint applied to a wall to prevent the passage of moisture into the wall.

Variances 15

Veneer — A decorative layer of wood, brick or other material that covers a structural material to improve its appearance such as on wood panels, brick and block walls, etc.

Vent Pipe *(i.)* 297, 308, *(i.)* 362 — A small, vertical pipe connected to plumbing fixtures that carries sewer gases through the roof to the outside air.

Ventilation 186, *(i.)* 191, 193

Vargeboard See Frieze

Vermiculite 190

Vertical Grain See Edge Grain

Vertical Load 180

Ceiling or roof beams that stick out through the exterior walls in Adobe style architecture.

A measurement of the pressure of electricity. Most houses use 110 to 120 volts or 220 to 240 volts.

An interior wall lining, usually of wood, that covers only the lower part of the wall.

Horizontal pieces of timber placed on top of a masonry wall and under joists and other load-bearing members to distribute their weight evenly to the wall.

A heating system in which air is heated inside a furnace and distributed throughout the house either by the natural force of gravity or by fans.

An inexpensive type of toilet depending upon water flowing down the inside of the bowl to clean it.

A pipe that carries waste from a bathtub, shower, lavatory or any fixture or appliance except a toilet.

A measure of electricity. It is volts × amperes.

An open railed platform over a peaked roof. First used on New England seacoast houses.

Treads of a curved staircase which are cut with one end wider than the other so they fit into place to make a curved form.

A masonry wall that separates two flues in the same chimney stack.

A chemical (hydrous aluminum silicate) used in water softeners which exchanges the minerals in the water for sodium chloride (common salt).